Resource-Oriented Agro-sanitation Systems

Naoyuki Funamizu
Editor

Resource-Oriented Agro-sanitation Systems

Concept, Business Model, and Technology

Springer

Editor
Naoyuki Funamizu
Hokkaido University
Sapporo
Japan

ISBN 978-4-431-56833-9 ISBN 978-4-431-56835-3 (eBook)
https://doi.org/10.1007/978-4-431-56835-3

Library of Congress Control Number: 2018942159

Printed on acid-free paper

This Springer imprint is published by the registered company Springer Japan KK part of Springer Nature
The registered company address is: Shiroyama Trust Tower, 4-3-1 Toranomon, Minato-ku, Tokyo 105-6005, Japan

Preface

WHO fact sheet on sanitation (updated July 2017) is reporting that (1) 68% of the world's population (5.0 billion people) used at least a basic sanitation service in 2015; (2) 2.3 billion people still do not have basic sanitation facilities such as toilets or latrines; (3) 892 million still defecate in the open, for example, in street gutters, behind bushes, or into open bodies of water. It is also reported that the penetration rate of sanitation is greatly different between regions. South Asia and sub-Saharan Africa are pointed out as regions with low penetration rates. In addition, 90% of those who are forced to excrete outdoor are pointed out as rural residents. In the 2030 Agenda for Sustainable Development (Sustainable Development Goals, SDGs), the target on sanitation is set as Goal 6 ("Ensure access to water and sanitation and sustainable management for all"). The increasing importance and recognition of sanitation have led to the need for special instruction of engineering and science students in their undergraduate and graduate levels, as well as practicing engineers and scientists, and a technical reference for project managers and government officials.

The book is an endeavor by the authors to assemble several research results on resource-oriented sanitation. Why resource-oriented sanitation? The answer is that, in order to add more value to sanitation system, resources in human excreta should be reclaimed and reused. Feces contain resources such as organic matter and nutrients for agriculture. Urine is also a source of nutrients. Gray water is very rich in water. This book is organized into six parts, each dealing with the concept of resource-oriented sanitation and technologies for reclaiming resources from human excreta and how to use them as well as case studies.

It is important to understand the concept of resource-oriented sanitation. In Part I, sanitation system is categorized into three generations: primitive sanitation, modern sanitation, and postmodern sanitation, and discussions on the requirement of new sanitation policy are introduced. Then, technologies required to realize the new sanitation policy are summarized.

Technologies discussed in this book

Target	Technologies
Feces	Composting
Urine	Storage, volume reduction, recovery of nitrogen
Gray water	Slanted soil system, membrane bioreactor, sand filter with geo-textile
Agriculture	Use of urine, use of reclaimed gray water

Several aspects relating to composting process are presented in Part II. Feces contain organic matter, nitrogen, pathogen, pharmaceuticals, and water, and therefore, the fate of those components should be understood for producing safe and valuable compost and rational design of composting-type toilet. These discussions are introduced in this Part II. Part III is related to urine. Composition of urine is presented from engineering point of view. And then volume reduction technology and recovery method of nitrogen and phosphorous directly from urine are introduced in this part. Three gray water reclamation technologies are introduced in Part IV. Part V is a part for agricultural technologies. There is not enough discussion on how to use collected urine and reclaimed gray water. This part introduces recycled resource utilization technology. Part VI presents two case studies from Burkina Faso and Indonesia. Burkina Faso case is a pilot study in rural area, and Indonesia case presents a pilot study in urban area.

This book is based on the results of the following three projects:

- Core Research for Evolutional Science and Technology of the Japan Science and Technology Agency (JST-CREST) (Project title: Development of Sustainable Sanitation System and Its Implementation to Asian Countries)
- Science research S of the Grants-in-Aid for Scientific Research of Japan Society for the Promotion of Science (JSPS) (Project title: Development of Resources Oriented Sanitation System Based on the Concept "Don't Mix" and "Don't Collect" Wastewater)
- Science and Technology Research Partnership for Sustainable Development of the Japan Science and Technology Agency and Japan International Cooperation Agency (JST-JICA-SATREPS) (Project title: Improving Sustainable Water and Sanitation Systems in Sahel Region in Africa)

I wish to thank all the contributors to these three projects for their sincere effort. I am also grateful to Ms. Ai Goto for her help and support in editing this book.

Sapporo, Japan
August 2017

Naoyuki Funamizu

Contents

Contributors

Maïmouna Traore Bologo International Institute for Water and Environmental Engineering (2iE), Ouagadougou, Burkina Faso

Sebastian Ignacio Charchalac Ochoa Division of Engineering Sciences, Western Regional Center (CUNOC), University of San Carlos of Guatemala (USAC), Quetzaltenango, Guatemala

Yosuke Deguchi Nippon Koei Co., Ltd, Tokyo, Japan

Naoyuki Funamizu Graduate School of Global Food Resources, Hokkaido University, Sapporo, Japan; Research Institute of Humanity and Nature, Kyoto, Japan

Umi Hamidah Research Unit for Clean Technology, Indonesian Institute of Sciences, Bandung, Indonesia

Nowaki Hijikata Graduate School of Engineering, Hokkaido University, Sapporo City, Hokkaido, Japan

Shinya Hotta Research and Development Center, Asahi Kasei Corporation, Kawasaki-ku, Kawasakishi, Japan

Aileen Huelgas-Orbecido De La Salle University, Manila, Philippines

Mitsuteru Irie Alliance for Research on North Africa, University of Tsukuba, Tsukuba, Ibaraki, Japan

Tadaharu Ishikawa Department of Environment and Technology, Tokyo Institute of Technology, Yokohama, Japan

Ryusei Ito Graduate School of Engineering, Hokkaido University, Sapporo, Hokkaido, Japan

Takashi Kakimoto Center for Environmental Science in Saitama, Saitama, Japan

Shinobu Kazama Department of Urban Engineering, The University of Tokyo, Tokyo, Japan

Miguel Ángel López Zavala Tecnológico de Monterrey, Water Center for Latin America and the Caribbean, Monterrey, N.L., Mexico

Amadou Hama Maïga International Institute for Water and Environmental Engineering (2iE), Ouagadougou, Burkina Faso

Ynoussa Maïga International Institute for Water and Environmental Engineering (2iE), Ouagadougou, Burkina Faso

Takako Nabeshima Research Faculty of Media and Communication Studies, Hokkaido University, Sapporo, Japan

Shingo Ogawa Graduate School of Engineering, Hokkaido University, Sapporo City, Hokkaido, Japan

Yu Onoda Institute of Environmental Ecology, IDEA Consultants, Inc., Ooigawa-cho, Shida-gun, Shizuoka, Japan

Masahiro Otaki Department of Human Environmental Science, Ochanomizu University, Tokyo, Japan

Muhammad M. Pahore Graduate School of Engineering, Hokkaido University, Sapporo, Hokkaido, Japan

Moustapha Sene Graduate School of Engineering, Hokkaido University, Sapporo, Hokkaido, Japan

Hitoshi Shibuya Division of Laboratory and Transfusion Medicine, Hokkaido University Hospital, Sapporo, Japan

Neni Sintawardani Research Unit for Clean Technology, Indonesian Institute of Sciences, Bandung, Indonesia

Seyram Sossou International Institute for Water and Environmental Engineering (2iE), Ouagadougou, Burkina Faso

Mariam Sou/Dakoure International Institute for Water and Environmental Engineering (2iE), Ouagadougou, Burkina Faso

Haruki Suzuki Division of Laboratory and Transfusion Medicine, Hokkaido University Hospital, Sapporo, Japan

Jovita Triastuti Research Unit for Clean Technology, Indonesian Institute of Sciences, Bandung, Indonesia

Ken Ushijima Northern Regional Building Research Institute, Hokkaido Research Organization (HRO), Asahikawa, Memuro, Japan

Part I
Concept and Business Model

Chapter 1
The Concept of Resources Oriented Agro-Sanitation System and Its Business Model

Ken Ushijima, Naoyuki Funamizu, Takako Nabeshima, Nowaki Hijikata, Ryusei Ito, Mariam Sou/Dakoure, Amadou Hama Maïga and Neni Sintawardani

Abstract This chapter proposes concept of the sanitation business model, based on the discussions; what is necessary to solve current world's sanitation issue and what is the limitation of current sanitation concept. Proposed policy was on the basis of the Postmodern Sanitation concept; (1) discuss a sanitation value chain which create and add a value to human excreta and its products, and build a sanitation business model to drive this sanitation value chain, (2) design the sanitation business model focusing on incentive for individual toilet users, (3) analyze current

K. Ushijima (✉)
Northern Regional Building Research Institute, Hokkaido Research Organization (HRO), Midorigaoka Asahikawa 078-8801, Japan
e-mail: ushijima-ken@hro.or.jp

N. Funamizu
Graduate School of Global Food Resources, Hokkaido University, Kita-9, Nishi-9, Kita-ku, Sapporo 060-8589, Japan

N. Funamizu
Research Institute of Humanity and Nature, Motoyama Kamigamo, Kita-Ku, Kyoto 603-8047, Japan

T. Nabeshima
Research Faculty of Media and Communication Studies, Hokkaido University, Kita-17, Nishi-8, Kita-Ku, Sapporo 060-0817, Japan

N. Hijikata · R. Ito
Graduate School of Engineering, Hokkaido University, Kita13-Nishi 8, Kita-Ku, Sapporo City, Hokkaido 060-8628, Japan

M. Sou/Dakoure · A. H. Maïga
International Institute for Water and Environmental Engineering (2iE), Ouagadougou, Burkina Faso

N. Sintawardani
Research Unit for Clean Technology, Indonesian Institute of Sciences, Bandung, Indonesia

N. Funamizu (ed.), *Resource-Oriented Agro-sanitation Systems*,
https://doi.org/10.1007/978-4-431-56835-3_1

user's value chain to find available potential resources, (4) connect these potential sanitation resources to next value chain, (5) make a financial plan based on market analysis of sanitation product, (6) find and organize facilitating organization to support individual toilet users' business. This concept was applied for the case of rural area in Burkina Faso, and reasonable agro-sanitation business model was designed based on material flow analysis and value flow analysis.

Keywords Resource-oriented sanitation · Value chain · Sanitation business model Sahel Africa

1.1 Introduction

At the moment of 2008, coverage of safe drinking water reached 87% in world average; however, the coverage of basic sanitation has reached to 52% only. Especially in rural area of sub-Saharan Africa, the coverage of basic sanitation is only 24% in 2008 (UN 2010). This is the lowest percentage in eight regions of the world. Although governments and related organizations including NGOs have been tackling sanitation issue, UNDP (2006) estimated that sub-Saharan Africa will achieve sanitation goal in 2076. This implies that current concept and approach may not be reasonable for providing mass of sanitation supply to those developing countries.

As another aspect, effective resource recycling is inevitable issue for world's sustainable future. In both developed and developing countries, resource recycling from domestic wastewater including human excreta can be said also as an inevitable topic. Some resource recycling sanitation technologies have been proposed and tested for last decades (e.g., Lopez et al. 2001; Winblad and Simpson 2004). These technologies are promising also as a suitable technology for developing countries, those have poverty problem, poor infrastructure, low efficiency and instability of government, and severe environmental condition, because it does not require expensive infrastructures such as pipe system. In fact, many engineers and practitioners have been considered and tried to apply it to developing countries, including sub-Saharan Africa (e.g., Rüd and Münch 2008; Matsebe and Osman 2012).

Thus, world's situation desires a sanitation system which has higher sustainability and applicability to developing countries, and many studies and practices have been done for diffusing it; however, we still cannot find large-scaled success (Haq and Cambridge 2012). Authors are considering that above-mentioned situations imply that current sanitation concept or scheme may not be sufficient to solve those stagnations in sanitation matter. This chapter aims to clarify the limitation of current sanitation system concept and the features of next generation of sanitation concept which is desired for future, and to discuss the sanitation business model as suitable approach for the next generation of sanitation. This chapter also describes case model for rural area of Burkina Faso to make practical discussion for sanitation business model.

1.2 Nature of Conventional Sanitation Concept and Required New Policy

1.2.1 Three Sanitation Generations

This chapter categorizes sanitation systems into three generations: (1) the Primitive Sanitation, (2) the Modern Sanitation, and (3) the Postmodern Sanitation. The system which relies on only natural purification and natural material cycle system is categorized into the Primitive Sanitation, such as open defecation, toilet on the river, and open ditch. These sanitation systems are quite simple and they require no or low cost for installation and management, however control of pollutants and pathogens is not sufficient especially in highly populated area because those systems do not completely separate human excreta from human living environment.

Next generation, the Modern Sanitation, is defined as the sanitation purposing safe separation and disposal of human excreta from human living environment. Urbanization and modernization made people difficult to continue the Primitive Sanitation habit. The Primitive Sanitation in highly populated area causes unsanitary conditions and makes it easy to cause serous pandemic, such as by Cholera (Griffith et al. 2006). Sewer system, which separates human excreta by piping system, was one of the critical solutions. Although it is known that some of early historical cities, such as Rome, have primitive sewer systems, those technologies did not continue after Rome empire collapsed, and we had to wait for modern sewer system development until eighteenth century (Lofrano and Brown 2010).

In eighteenth century, industrial revolution occurred in UK, and industrialization wave spread to all over the world from nineteenth to twentieth centuries. Industrialization enhanced development of the Modern Sanitation technology such as water flush toilet and relating plumbing technologies (Lofrano and Brown 2010; Maeda 2008); on the other hand, industrialization also enhanced urbanization and serious environmental pollution (Lofrano and Brown 2010). After this experience, we found that disposal of human excreta requires treatment before discharge to natural environment. Now the same concept, safe separation, treatment, and disposal of human excreta, is applied for decentralized system such as community plant, Johkasou, and septic tank. This chapter regards these decentralized sanitation systems also as the Modern Sanitation.

As the World Health Organization (WHO) defined that "Sanitation generally refers to the provision of facilities and services for the safe disposal of human urine and feces" (WHO 2013a), conventional approach for sanitation project including activities for MDGs can be said that these are basically aiming to replace the Primitive Sanitation by the Modern Sanitation.

1.2.2 Limitation of the Modern Sanitation Concept

Basically, the Modern Sanitation systems are managed as public services, and budget recovery rate is generally not good. For example, the Japanese centralized sewerage systems are managed by sewerage agency belongs to municipality, and the tariff is collected together with water tariff. However, total annual budgets of sewerage systems are covered by sewerage tariff only 38% in national average (MLIT 2013) though few people reject to pay. Because Japanese sewerage pipe network also has a role of flood water removal, which is completely purposed for public, rest parts of budget were covered by municipal budget based on tax.

Decentralized Johkasou system, which is one of successful case of decentralized systems in Japan, is also depending on governmental support even though the responsibility to manage the Johkasou belongs to individual users. Up to 90% of initial cost for the Johkasou system was subsidized by central or local government (JECES 2009). Furthermore, the Johkasou requires periodical sludge removal and the removed sludge is treated in night soil treatment plant which is managed by local government, originally build for treating excreta from Japanese traditional pit latrine (JECES 2009). Thus, decentralized Johkasou system also cannot function without governmental support and public service.

Additional to above, Japanese government established many organizations those support construction and management of sewerage system and Johkasou system. Some of those organizations were established to contribute for developing and imploving sewerage technologies or Johkasou technologies under collaboration with private sectors.

Thus, Japanese major two successful cases are basically worked as/on public services, which aims public benefit on the basis of effectively designed social and administrative system; this system requires (1) user's positive participation by paying tariff and tax, (2) high level of governance to realize the policy, (3) sophisticated organizations which support realizing governmental policy, (4) financial supporting system, (5) accountability and strategic advertisement, (6) economical background of government to support these system and (7) technologies.

Contrary examples can be seen in urban slum of Southeast Asian countries. In such area, water flush toilet has been diffused without treatment facility (Ishikawa et al. 2002; Harada 2007; Ushijima et al. 2007). They just discharge their toilet wastewater directly to river. However due to the nature of river system, direct discharge affects living environment of downstream people in adverse way, such as insanitary conditions, bad odors, but not directly affect the person who discharged. Ushijima et al. (2013a) concluded it as a reasonable result of people's value judgment, by clearly showing that installing individual treatment facility will give users only obligation to pay for management cost but no clear benefit in individual scale.

According to the comparison between Japanese case and urban slum's case, the modern sanitation system seems difficult to give a sufficient incentive to invest for treatment facility in individual level because feedback of individual investment will not directly come back to person who invested. Therefore, it can be concluded that well designed public service aspect is necessary for the modern sanitation concept, at least in some part, otherwise unfavorable disposal might occur following to reasonable value judgment by individual people. If the situation can not allow well designed public service, some kind of clear and direct incentive for individual is required.

1.2.3 Resource-Oriented Sanitation

No matter how much income they obtained, human urinates and defecates as daily activity, and it contains rich nutrients such as nitrogen and phosphorus (Lopez et al. 2002; Maksimovic and Guibert 2003). It is clear that human excreta are massive potential resource for agriculture in the world. Total amount of nitrogen and phosphorus in the world's human excreta corresponds to 40% of nitrogen and 19% of phosphorus in the world's fertilizer consumption [calculation based on FAO (2012) and Maksimovic and Guibert (2003)]. Furthermore, important feature is that this resource is almost equally distributed according to population and originally it belongs to people, not by specific company, nor government. Therefore, if we can make appropriate value chain for this resource and its products namely agro-sanitation products such as vegetable grown by human excreta, people can start their business based on these resources. This idea brings both (1) solution to end poverty and (2) provision of appropriate sanitation to those people, with respecting peoples' independent development.

Concepts and technologies to recycle human excreta have been proposed and tested in this decade and summarized information is available (e.g., GTZ 2010). These are regarded as promising concept to provide sustainable way of sanitation, and to provide low-cost, low-energy-consuming sanitation. Economical evaluations have also been done by some researchers [summarized by Mara (2008)]. However, most of current study discussed from sanitation technology side, it is that how to sell the toilet. Only a few studies discuss a system to create and add value from human excreta (Fogelberg et al. 2010; Hijikata et al. 2012; Ushijima et al. 2013a). Resource recycling technology is just a tool which has a potential to increase the value of human excreta. What brings benefit to individuals is sanitation value chain; therefore, discussion should be started from value chain of sanitation products.

Why current discussion does not include value chain seems due to the nature of the Modern Sanitation concept. Basically, the modern sanitation concept does not require strict discussion on value chain because it is usually designed as public service based on tariff and tax. Therefore, main task is to treat harmful or unpleasant materials and dispose them. We have to say that the modern sanitation concept basically does not have a sense to create sanitation value chain which adds the value and connect it to market. In fact, difficulty of current resource recycling sanitation practice or trial is exit design of sanitation products (Ushijima et al. 2011; Fogelberg et al. 2010). If we manage human excreta or agro-sanitation products as resources, its product should be treated strictly on business manner, including quality control, marketing, distribution network; otherwise, sanitation product would not have sufficient value in the market, and sanitation products would lose way to go.

One of interesting and important example is old Japanese model, called as Edo period sanitation model (Watanabe 1983; Onjo 2004; Mitsumata 2008). In Japan, human excreta had been used in agriculture from around tenth century, and particularly in Edo period (1603–1869), human excreta were traded with monetary value and this system was managed under the market system not by government. This system had been functioning until early twentieth century. Edo model is not able to separate human excreta completely, in this sense this model should be categorized to the Primitive Sanitation; however, its value chain system and material balance give us many useful ideas, and interestingly these were automatically optimized under the manner of market (Ushijima et al. 2012b).

1.2.4 Basic Policy of the Postmodern Sanitation

According to discussion on current sanitation system described above, new sanitation concept seems necessary to achieve breakthrough against limitation of current modern system and concept. The new concept enhancing independent development by low-income people, those who live in slum or rural area with traditions, will be the key to solve both of poverty and sanitation issue. This chapter proposes the Postmodern Sanitation concept which has basic policies as follows.

(1) Put a sanitation value chain in the center of discussion, and prepare sanitation business model to drive this value chain.
(2) Design the sanitation business model focusing on incentive for individual toilet users.
(3) Analyze current user's value chain including food, sanitation, and agriculture, and find available potential resources in and around sanitation material flow.
(4) Find effective way to connect these potential sanitation resources to next value chain, in the most cases it would be agriculture.
(5) Make a financial plan based on market analysis of sanitation product, and use it to support individual users as sanitation resource provider.
(6) Find and organize facilitating organization to support individual toilet users' business as resource provider.

1.3 Sanitation Business Model: Case of Rural Area of Burkina Faso

1.3.1 Case Field

In order to make it easy to understand the concept and its advantage of the sanitation business model, this section describes case study of rural area in Burkina Faso. In this country, the low (Low of public sanitation, No. 022-2005/AN) states that house owners have a responsibility to construct wastewater treatment facilities in their land and they have to manage it appropriately, however the low mentioned neither practical strategy nor detailed supporting system and technologies. Real situation is that proportion of population who has access to basic sanitation facility is only 6% in rural area of Burkina Faso (UN 2010). Although urination is done in shower space, open defecation is still common (Ushijima et al. 2012a). The under-five mortality rate is 11th highest in the world, 169 per 1000 live birth, and 19% of these deaths are caused by diarrhea (WHO 2010). Water and sanitation are clearly urgent issues in Burkina Faso. However, application of the modern sanitation system, which requires high level of governance and economic background, seems difficult in short period. GDP of this country is only 1149 USD capita^{-1} yr^{-1} [in Purchasing Power Parity (PPP) 2005 by WHO (2013b)] and Human Development Indicator (HDI) is 5th worst in 186 countries (UNDP 2013).

Burkina Faso is an agricultural country. More than 90% of population lives on agricultural activity (FAO 2012), and 33.8% of GDP is produced by agriculture (World Bank 2013). However, environment for agriculture is severe. Burkina Faso is located in the Sahel region and its climate belongs to the semiarid type, means low precipitation, furthermore irrigation facilities are very rare in rural area of Burkina Faso. The main soil type is Lixisols, which requires frequent fertilization and careful erosion control (Driessen and Deckers 2001).

The case field of this study is two rural villages in Ziniaré commune, Oubritenga province in Burkina Faso (Fig. 1.1). Ushijima et al. (2012a) observed and described the daily life in light of material and value flow relating to water, sanitation, and agriculture in and around rural household in the target area.

1.3.2 Current Value Chain in and Around Rural Household

According to Ushijima et al. (2012a), feature of daily life in target village can be described as follows. Main activity is crop cultivation to make food for self-consumption. However, crop cultivation is possible only in short rainy season, which generally corresponds June to September. One of the major income sources is vegetable production. They have two types of vegetable garden different from crop field. One is a small garden (e.g., 10 m^2, Fig. 1.2) located near from house. Products from this garden are vegetables for self-consumption but extra production

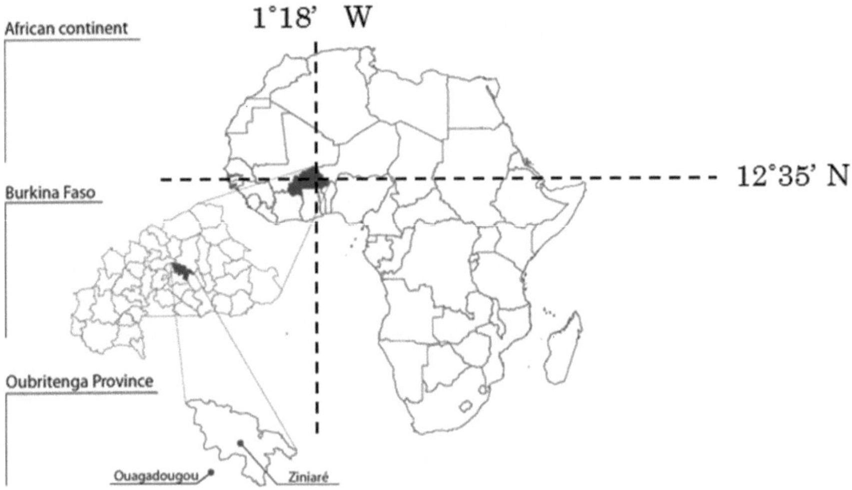

Fig. 1.1 Location of case field, Ziniaré, Oubritenga province, Burkina Faso [Reprinted from Ushijima et al. (2013b)]

Fig. 1.2 Small vegetable garden located near from house (Photograph by Ushijima)

Fig. 1.3 Large vegetable garden located along reservoir (Photograph by Ushijima)

is sold in the market. This small garden is generally managed by wives and available only in rainy season because there is no irrigation facility near the house. Other is comparatively large garden [e.g., 1000 m^2, Fig. 1.3)] located along reservoir or shallow well. This large garden is managed by men and women and available even in dry season. Products from this garden are vegetables for sale. Generally, people use most part of their workforce for crop production in rainy season, and for vegetable production in dry season. People use synthetic fertilizer only for vegetable gardens, but the available amount is limited by their individual economy. Generally, fertilizer use is low level (Crawford et al. 2006).

People's main water source is deep well, which exist 3–5 per hundreds of household. This water is used for drinking, cooking, bathing, and dishwashing. Some use this water for laundry also, but some uses water of reservoir or shallow well for laundry. All wastewater is disposed to ground, except for some part of dishwashing waste water which is given to livestock.

Figure 1.4 shows schematic picture of current material flow and value chain in and around rural household, based on the field data by Ushijima et al. (2012a). There find four important value flows: (1) people consume large part of workforce at crop field in order to obtain their food. Water for crop is critically important value but it depends on natural precipitation; therefore, it is not controllable. (2) Clear monetary income flow is vegetable production. However, this part requires cost for fertilizer and seeds. Location of the garden is also limiting factor. (3) People

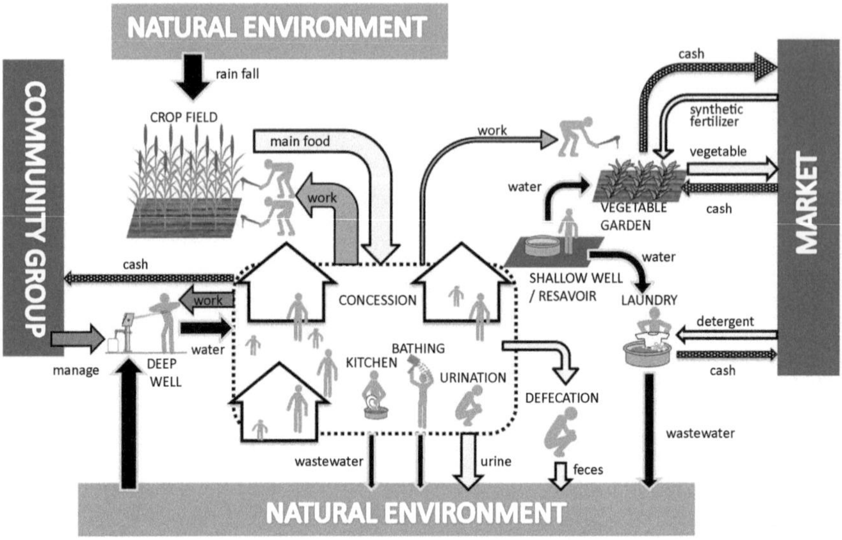

Fig. 1.4 Current material flow and value chain in and around rural household (Reprinted from Ushijima et al. (2015) with permission from the copyright holders, IWA Publishing)

consuming workforce in order to obtain daily water from well; however after using water, these are just disposed. (4) Human excreta are also just disposed, though it contains nutrients.

1.3.3 Framework of Sanitation Business Model for Rural Household in Burkina Faso

Straightway to connect sanitation value chain appears to be like Fig. 1.5. In order to create value in the market and obtain sufficient compensation, in this case, essential points are: (1) find good channel to the market, (2) producing high quality and quantity of agricultural product from currently available resources, (3) obtaining sufficient quality and quantity of fertilizer, (4) avoiding contamination of products and resource materials in each process. Additional to these, we can easily find that additional resources to this single value chain are necessary to each process; for example, farming needs a good quality of irrigation water, seeds, workforce, well-managed soil, and so on.

When back to the Fig. 1.4, we can find that (1) vegetable garden is currently existing channel to the market, (2) current vegetable garden lacks sufficient amount of fertilizer and irrigation water especially in dry season in small garden near the house, (3) human excreta are potential fertilizer source, (4) disinfection of sanitation products must be optimized for agricultural use, and (5) gray water is potential

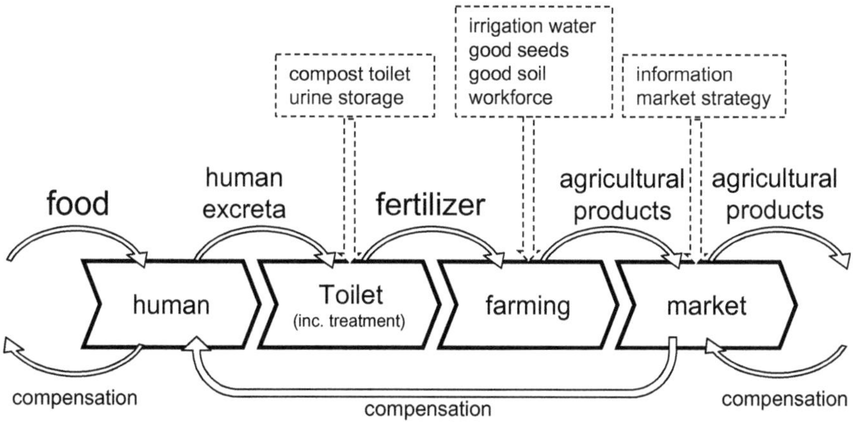

Fig. 1.5 Single sanitation value chain (Reprinted from Ushijima et al. (2015) with permission from the copyright holders, IWA Publishing)

water source for irrigation. As a result, promising new sanitation value chain network can be proposed as Fig. 1.6. In this flow, farmer will increase their income by vegetable cultivation. In this value chain, composting toilet and grey water treatment facilities function not as just sanitation facilities but also as some kind of asset to increase agricultural products, which can be called "agro-sanitation asset." In this model, farmers invest for agro-sanitation asset in order to increase their income from vegetable garden. If this value chain functions well, income increase becomes a great incentive for individual toilet users to invest and manage those agro-sanitation assets, which is also function as sanitation facilities. This incentivization would encourage overcoming the limitation of modern sanitation approach.

Ushijima et al. (2013b) quantitatively estimated available agricultural resources in this material flow and revealed that irrigation water is strong limiting factor (Fig. 1.7). In this target area, obtaining irrigation water source is a key factor for effective use of excreta resources. Hijikata et al. (2012) proposed one example of crop rotation for small vegetable garden based on this concept, and estimated optimum income increase as about 40–100 USD yr^{-1} $person^{-1}$ [converted in PPP 2005 (WHO 2013b)]. This increased income can be considerable for farmers when comparing with averaged GDP in Burkina Faso of 1149 USD $capita^{-1}$ yr^{-1}.

In order to realize this concept, more detailed business model should be proposed. Figure 1.8 shows picture of little bit more practical business model for this concept. In order to support this system, facilitating organization becomes key factor, who provides three key functions: (1) microfinance, (2) technical support for vegetable farming, and (3) marketing support, because (1) farmers generally do not have capitals to invest, and they needs to upgrade (2) their farming technique and (3) marketing ability to maximize their productivity and income. In this business model, farmers are provider of vegetable but also consumer of agro-sanitation asset.

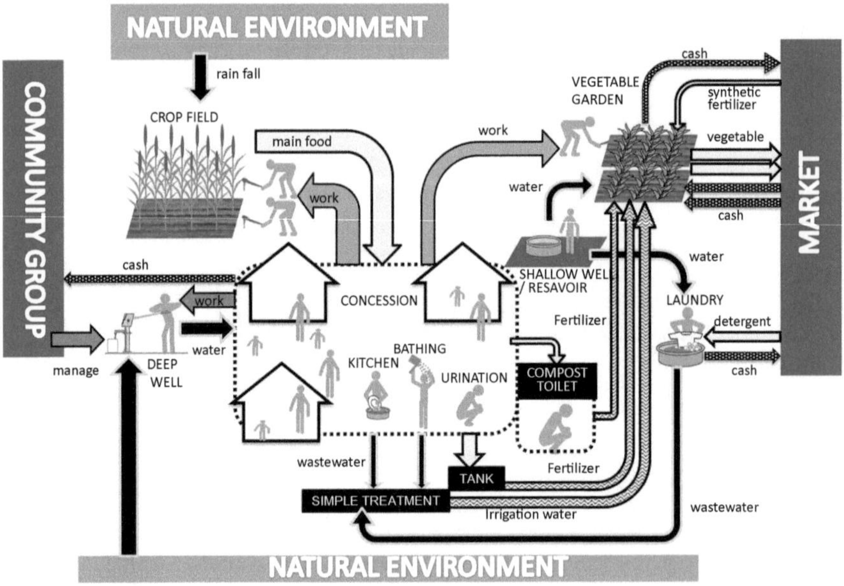

Fig. 1.6 Proposing sanitation value chain network in and around rural household (Reprinted from Ushijima et al. (2015) with permission from the copyright holders, IWA Publishing)

Fig. 1.7 Expected cultivatable area by using composting toilet, urine collection system, and grey water reclamation system (Reprinted from Ushijima et al. (2015) with permission from the copyright holders, IWA Publishing)

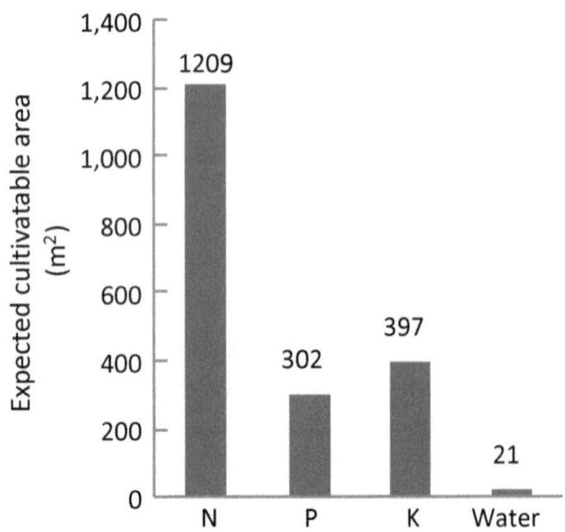

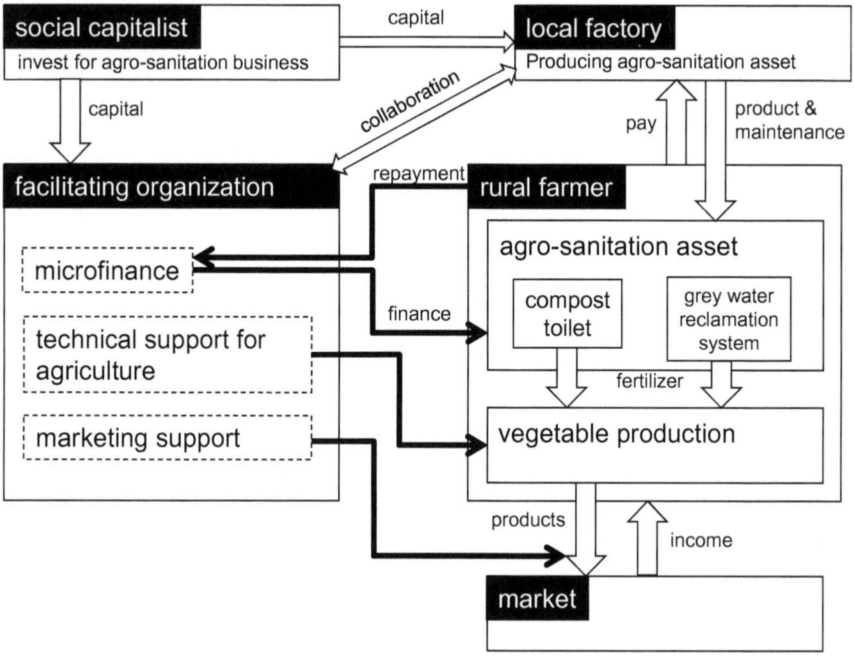

Fig. 1.8 Business model of agro-sanitation in rural area of Burkina Faso (Reprinted from Ushijima et al. (2015) with permission from the copyright holders, IWA Publishing)

1.4 Discussions

1.4.1 Difference of Benefit and Beneficial Actors

As described above, the Postmodern Sanitation concept proposes different viewpoint of actors and value from conventional concept. In the case of Burkina Faso, farmers were put as main actor of business model. Although some kind of support by government is necessary, the role and aim or benefit is completely different from the modern sanitation concept. Here, we would like to describe sanitation concept from four important actor's benefit:

- Public or government;
- Individual;
- Community;
- Commercial sector.

In the modern sanitation in Japan, each actor's benefit is:

- Government provided sanitation as public service, which aims public benefits such as public health, decrease of medical cost, environmental conservation.
- Individual: They obtained safety, comfortable toilet, which is supported by sewerage system or Johkasou system.

- Community obtained no clear benefit. Farmers had already obtained low-cost synthetic fertilizer, instead of human excreta which was used as important fertilizer from seventeenth century until early twentieth century.
- As private sector, water flush toilet bowl and related commodities have been sold by private companies, and sewerage construction-benefited construction-related companies. Relating consulting company obtained benefit from those.

On the other hand, current sanitation system in developing countries can be described as:

- Government does not provide sufficient sanitation service because they do not have sufficient governmental ability and/nor sufficient budget.
- Individuals in middle or middle–low-income class obtained safety, comfortable toilet; however, no or insufficient treatment facility is accompanied. Seriously low-income class, such as peasants in sub-Saharan Africa, could not obtain toilet itself.
- Community obtained no clear benefit. Farmers do not obtain any specific benefit from sanitation.
- Private companies sell toilet facilities for middle or middle–low class but seriously low-income class cannot be target.

In the business model for rural area of Burkina Faso, based on the Postmodern Sanitation concept, each actor would obtain benefit as follows:

- Government does not have to cover all sanitation service as public service.
- Individual obtains toilet by agro-sanitation business.
- Private companies might be able to sale toilet facility to seriously low-income class also.
- Community may obtain indirect benefit by community members' income increase.

1.4.2 Advantage of the Postmodern Sanitation Concept

It is clear that the Postmodern Sanitation system based on resource-oriented concept has higher sustainability in terms of resource management. But the most important feature of the Postmodern Sanitation based on value chain of agro-sanitation resources is that sanitation resources belong to people no matter how much income they obtain, as mentioned above. This is important advantage because people who do not have access to basic sanitation are usually live in low income and many of them are neglected from market economy. The Postmodern Sanitation is easy to involve those people into agro-sanitation value chain, which is productive and important channel to the market and income. Furthermore in this system, people are involved with being motivated by incentive of toilet ownership and expected income increase; therefore, it becomes also internal driving force to diffuse basic sanitation.

Cost aspect also seems better than installing the modern sanitation system. Whereas the modern sanitation system requires continuous input of budget to install and manage the system, the Postmodern Sanitation model, for example in the case study (Fig. 1.8), it just requires initial investments for facilitating organization and local factories, and these capitals are reusable also for toilet renovation unless the sanitation business is failed. This means that this system has high sustainability in economic aspect also.

In this model, central government does not have to manage whole project like as sewerage system and does not have to prepare huge budget. Governance issue becomes more local and market based, like as Edo period model (Ushijima et al. 2012b). Particularly in developing countries, local community based on ethnicity or traditional group sometimes has greater importance for people (e.g., Nakanishi 1991), and this gap sometimes makes conflicts (Hyden 1980, 2006; Nabeshima 2011). Local-based sanitation management is big advantage for this situation.

The Postmodern Sanitation concept with sanitation business model approach also has advantage in system design procedure. In the case study of Burkina Faso, it was easy to find that irrigation water was bottleneck of this sanitation value chain because discussion was started from value chain like Fig. 1.5. Sanitation business model based on value chain discussion gives us comprehensive and practical idea. Therefore, we could design sanitation system including grey water reclamation system and excessive fertilizer use in vegetable garden along reservoir from first picture. Requirements of each technology are also able to be identified directly through designing sanitation value chain and sanitation business model.

1.4.3 Effective Approach for the Postmodern Sanitation

On the basis of six policies written in 1.2.3, this study designed the sanitation business model for rural area of Burkina Faso in reasonable and feasible way. Particularly in the case study of Burkina Faso, material flow analysis (MFA) and value chain analysis (VCA) took important role for (3) analyzing current value chain and finding available potential resources and (4) finding effective way to connect to agricultural value chain. Furthermore, in sanitation value chain, estimation of exit capacity, that is market size for sanitation products, is critically important. If it was overestimated, sanitation products lose a way to go immediately. Accurate estimation of exit capacity is essential for the Postmodern Sanitation system. MFA and VCA give overviewing information about this. MFA and VCA were also useful to understand structure of value and benefit exchange in and around household that is necessary to (2) design the sanitation business model focusing on incentive for individual toilet users.

Building process of sanitation business model gives us image of (5) financial plans required for this system. There are many method and tools to build business model. For example, business model canvas, proposed by Osterwalder and Pigneur (2010), gives simple but comprehensive way to build business model. However, relationship between provider and consumer is not so simple in sanitation business model. We should carefully analyze and treat these multiple relationships of provider and consumer.

As an organizational aspect, some kind of facilitating organization seemed to be necessary. In the case study on Burkina Faso, we found that facilitating organization must have functions of microcredit for initial investment, marketing support to sale the agro-sanitation products in better conditions, and agricultural technical support for farmers to increase productivity and marketability. Promising candidates for this actor are NGO's, private company, farmers union, and so on. This part is very practical; therefore, reasonable way is to contact and corroborate with those practitioners.

1.4.4 Topics Require Further Accumulation of Knowledge and Know-how

The sanitation business model based on Postmodern Sanitation concept still has many topics to accumulate the knowledge and know-how. Major topics to be discussed more are (1) quality control, (2) risk management system, (3) marketing system, and (4) individual value judgment system.

Because agro-sanitation business model is the business, (1) quality control in all process is required in order to keep stable sales. At least the Postmodern Sanitation system contains motivating function for individual toilet users, but quality control needs to cover all sanitation value chain. The package of method and technology for quality control of sanitation products has to be developed.

The sanitation business model based on Postmodern Sanitation concept has also the risk of fail in business. If it was general industrial product, it means just fail of one company; however, fail of sanitation business means also collapse of sanitation system and social damage is much larger. In fact, Edo model, which was managed as market, collapsed due to emerge of low-priced synthetic fertilizer, and Japanese government had to take a role of human excreta management from that period (Onjo 2004; Mitsumata 2008). Business risk management has to be prepared well, of course, and furthermore, some kind of additional risk management system seems required.

Another uncertainty points are (3) actual market capacity and acceptability for agro-sanitation products and (4) value judgment system of individual toilet users. There are some studies on psychological resistance or acceptability of sanitation system (Rosenquist 2005; Wilke 2003); however, it is still difficult to definitely predict people's acceptability under giving incentive of income increase.

Additional to above, as sociological aspect, we should note that involving farmers to market system sometimes causes economical differentiation, and in worst case, it generates serious conflicts. Increasing income becomes great incentive for individual; however, it should be modified according to traditional community culture or norms, and some part of benefit may have to be shared by community.

1.5 Conclusions

This study challenged to describe the approach of sanitation business model based on the concept of Postmodern Sanitation, which is desired as the next generation of sanitation concept. Obtained conclusions are as follows.

- Human excreta are massive potential resource for agriculture in the world, and this resource is originally distributed for people equally. Therefore, this resource can be important business resource especially for low-income people, if appropriate value chain for sanitation resource and products was provided. This can be solution for both poverty issue and sanitation issue. However, current sanitation idea, the modern sanitation concept, seems difficult to involve the concept of value chain for sanitation resource.
- The Postmodern Sanitation concept was proposed under following policies;

(1) Put a sanitation value chain in the center of discussion, and prepare sanitation business model to drive this value chain.
(2) Design the sanitation business model focusing on incentive for individual toilet users.
(3) Analyze current user's value chain including food, sanitation, and agriculture, and find available potential resources in and around sanitation material flow
(4) Find effective way to connect these potential sanitation resources to next value chain, in the most cases it would be agriculture.
(5) Make a financial plan based on market analysis of sanitation product, and use it to support individual users as sanitation resource provider.
(6) Find and organize facilitating organization to support individual toilet users' business as resource provider.

- Sanitation business model based on MFA and VCA gives us comprehensive idea for the favorable Postmodern Sanitation system and also enables us to assess the feasibility of the system appropriately.
- Accumulation of knowledge and know-how is desired in the topic of, (1) method and technology of quality control for sanitation products, (2) development of some kind of risk management system to stabilize sanitation business, (3) agricultural marketing based on agro-sanitation products, (4) value judgment of individual toilet users, effective incentive, and motivating scheme.

Acknowledgements This concept was inspired through former research activities and discussions with Prof. Tadaharu ISHIKAWA, Prof. Mitsuteru IRIE, Mitsumasa YOKOTA, Shigeo OKAYA, Mayuko TOTSUKA, and Rui KOTANI. This study was supported by JST-JICA, JST-CREST, and JSPS.

References

Crawford EW, Jayne TS, Kelly VA (2006) Alternative approaches for promoting fertilizer use in Africa, agriculture and rural development. Discussion Paper 22. World Bank, Washington, DC

Driessen P, Deckers J (2001) Lecture notes on the major soils of the world. FAO (Food and Agriculture Organization in United Nations), Rome

FAO (Food and agriculture Organization) (2012) FAOSTAT. http://faostat.fao.org/. Accessed on Sept 2012

Fogelberg K, Montes J, Soto B (2010) From excrement to pines to mushrooms to money in Bolivia. Sustain Sanitation Pract 5:4–9, EcoSan club, Vienna

Griffith DC, Miller MA, Kelly-Hope LA (2006) Review of reported cholera outbreaks worldwide, 1995–2005. Am J Trop Med Hyg 75(5):973–977

GTZ (Deutsche Gesellschaft für Technische Zusammenarbeit) (2010) Basic overview of composting toilets (with or without urine diversion). Technology Review "Composting Toilets", Draft version (Last updated: 14 September 2010), GTZ

Haq G, Cambridge H (2012) Exploiting the co-benefits of ecological sanitation. Curr Option Environ Sustain 4:431–435. https://doi.org/10.1016/j.cosust.2012.09.002

Harada H (2007) A proposal of advanced sanitation system and attempts to improve Vietnamese sanitation. Dissertation for the Doctoral Degree. Kyoto University, Kyoto

Hijikata N, Ushijima K, Ito R, Sawadogo B, Funamizu N (2012) Design of vegetable garden for income estimation toward to agro-sanitation business model, a case study in Burkina Faso. In: Proceedings of dry toilet conference 2012, Tampere, 23–25 Aug 2012

Hyden G (1980) Beyond Ujamaa in Tanzania: underdevelopment and incaptured peasantry. University of California Press, Berkeley and Los Angeles, California

Hyden G (2006) African politics in comparative perspective. Cambridge University Press, New York

Ishikawa T, Nakanishi T, Ohta H, Qian X, Shinozawa K, Shimizu T, Ushijima K, Watanabe M, Yoshida M (2002) Chapter 6. Environmental improvement by recycling organic matter between urban and rural areas. In: Ohmachi T, Roman ER (eds) Metro Manila: In search of a sustainable future, University of the Philippines Press, Manila, pp 239–297

JSCES (Japan Education Center of Environmental Sanitation) (2009) Johkasou systems for domestic wastewater treatment, 4th edn. JACES, Japan

Lofrano G, Brown J (2010) Wastewater management through the ages: a history of mankind. Sci Total Environ 408(22):5254–5264. https://doi.org/10.1016/j.scitotenv.2010.07.062

Lopez Zavala MA, Funamizu N, Takakuwa T (2001) Onsite wastewater differentiable treatment system: modeling approach. Water Sci Technol 46(6–7):317–324

Lopez Zavala MA, Funamizu N, Takakuwa T (2002) Characterization of feces for describing the aerobic biodegradation of feces. J Environ Syst Eng 720:99–105. https://doi.org/10.2208/jscej.2002.720_99

Maeda H (2008) Suisen toile no sangyou shi (Industrial history of water flush toilet). The University of Nagoya Press, Japan (in Japanese)

Maksimovic C, Tejada-Guibert JA (ed) (2003) Frontiers in urban water management deadlock or hope (trans: Matsui S et al.). Gihodo shuppann, p 231

Mara DD (2008) Sanitation now: what is good practice and what is poor practice? In: Proceedings of the IWA international conference 'Sanitation Challenge: new sanitation and models of governance', Wageningen, Netherlands, 19–21 May 2008

Matsebe G, Osman A (2012) Ecological sanitation in urban South Africa: socio-cultural, design and operational challenge of Urine Diversion Dry (UDD) toilets and the impact on users' perceptions. In: Proceedings of dry toilet conference 2012, Tampere, 23–25 Aug 2012

Mitsumata N (2008) Fostering the material cycle between urban and suburban areas: the monetary transactions of night-soil in Kyoto during the Edo period. Doshisha Econ Assoc 60(2):259–281 (in Japanese)

MLIT (Ministry of Land, Infrastructure, Transportation and Tourism, Japan) (2013) Management of sewerage system. http://www.mlit.go.jp/crd/city/sewerage/data/keiei.html. Accessed on 13 April 2013 (in Japanese)

Nabeshima T (2011) Criticism of regional integration theory: through the analysis of society and nation-state in Africa. Media Commun Stud 60:13–34 (in Japanese)

Nakanishi T (1991) Suramu no keizaigaku (Economics in urban informal sector). University of Tokyo Press, Tokyo (in Japanese)

Onjo A (2004) On urbanization and the changes of urban-rural relations: nature, power and sewage disposal problems in Fukuoka, Japan: 1900–1930. Shien 141:1–28 (in Japanese)

Osterwalder A, Pigneur Y (2010) Business model generation: a handbook for visionaries, game changers, and challengers. Wiley, USA

Rosenquist LED (2005) A psychosocial analysis of the human-sanitation nexus. J Environ Psychol 25(3):335–346. https://doi.org/10.1016/j.jenvp.2005.07.003

Rüd S, Münck EV (2008) Ecological sanitation: selected example project from Sub-Saharan Africa, Asia and Europe. In: Proceedings of international conference 'Pathways towards Sustainable Sanitation in Africa', Ouagadougou, 24–27 Sept 2008

UN (United Nations) (2010) Millennium development goals report 2010. United Nations, Department of Economic and Social Affairs (DESA), New York

UNDP (United Nations Development Programme) (2006) Human development report 2006. United Nations Development Programme, New York

UNDP (United Nations Development Programme) (2013) Human development report 2013. United Nations Development Programme, New York

Ushijima K (2007) Study on water environment rehabilitation by introducing new toilet system in urban area of Southeast Asia. Dissertation for the Doctoral Degree, Tokyo Institute of Technology, Tokyo (in Japanese)

Ushijima K, Irie M, Sintawardani N, Triastuti J, Ishikawa T, Funamizu N (2011) Sanitation model for urban slum in Southeast Asia. In: Proceedings of 2nd IWA development congress & exhibition, Kuala Lumpur, 21–24 Nov 2011 (CD-ROM)

Ushijima K, Hijikata N, Ito R, Funamizu N (2012a) Effect estimation of dry-toilet application for rural farmer family in Burkina Faso. J Arid Land Study 22(1):99–102

Ushijima K, Funamizu N, Yokota M (2012b) Value network design approach for sanitation system in developing countries—lessons from two Japanese models. In: Proceedings of dry toilet conference 2012, Tampere, 23–25 Aug 2012

Ushijima K, Irie M, Sintawardani N, Triastuti J, Ishikawa T, Funamizu N (2013a) Sustainable design of sanitation system based on material and value flow analysis for urban slum in Indonesia. Front Environ Sci Eng 7(1):120–126. https://doi.org/10.1007/s11783-012-0460-5

Ushijima K, Sato R, Leray L, Hijikata N, Ito R, Funamizu N (2013b) Resource recycling system for human excreta based on the analysis of material flow related to agricultural production in rural household of Burkina Faso. J Water Policy Integr River Basin Manage 21:33–41 (in Japanese)

Ushijima K, Funamizu N, Nabeshima T, Hijikata N, Ito R, Sou M, Maiga AH Sintawardani N (2015) The postmodern sanitation—agro-sanitation business model as a new policy. Water Policy 17(2):283–298. https://doi.org/10.2166/wp.2014.093

Watanabe Z (1983) Toshi to noson no hazama (Interaction between urban and rural). Ronsosha, Tokyo (in Japanese)

WHO (World Health Organization) (2010) World health statistics 2010. World Health Organization, p 177

WHO (World Health Organization) (2013a) Health topics sanitation. http://www.who.int/topics/sanitation/en/. Sited on 25 Jan 2013

WHO (World Health Organization) (2013b) Choosing interventions that are cost effective. http://www.who.int/choice/costs/ppp/en/. Sited on 8 May 2013

Wilke I (2003) Psychology and sanitation: a personal perspective. In: Proceedings of the 2nd international symposium, Lübeck, 7–11 Apr 2003

Winblad U, Simpson-Hebert M (2004) Ecological sanitation—revised and enlarged edition. SEI, Stockholm

World Bank (2013) Burkina Faso at a glance. http://devdata.worldbank.org/AAG/bfa_aag.pdf. Accessed on April 2013

Chapter 2
Technologies for Resources Oriented Agro-Sanitation System—Overview

Naoyuki Funamizu

Abstract This chapter is an overview of the required technologies for resources oriented agro-sanitation system. The discussion includes the concept for building up the agro-sanitation system by selecting element technologies followed by the models for rural and urban area. The key points of the resources oriented sanitation are: (1) the feces, urine, and gray water are the important resources and properties of people; (2) technologies are used for improving their values; (3) recovered resources are used as agricultural resources. In the agro-sanitation system, separation of feces, urine, and gray water at the source is essential for resource recovery. And separated feces, urine and gray water are processed, respectively. The brief introduction of technologies for these resources' recovery is given in this chapter.

Keywords Resource-oriented sanitation · Technologies for urine
Feces and gray water

2.1 Introduction

As discussed in Chap. 1, the new postmodern sanitation, agro-sanitation, is required to achieve breakthrough against the current "Modern Sanitation" system. In the agro-sanitation, sanitation value chain is put in the center of the discussion, and human waste and wastewater are recognized as a resource not a waste. It should be noted that technologies are used for improving the value of human waste and wastewater, and for connecting sanitation resources to next value chain, agriculture. Therefore, the design of the technical system in the agro-sanitation includes the

---

N. Funamizu (✉)
Graduate School of Global Food Resources, Hokkaido University,
Kita-9, Nishi-9, Kita-Ku, Sapporo 060-8589, Japan
e-mail: funamizu@eng.hokudai.ac.jp

N. Funamizu
Research Institute of Humanity and Nature, Motoyama Kamigamo,
Kita-Ku, Kyoto 603-8047, Japan

N. Funamizu (ed.), *Resource-Oriented Agro-sanitation Systems*,
https://doi.org/10.1007/978-4-431-56835-3_2

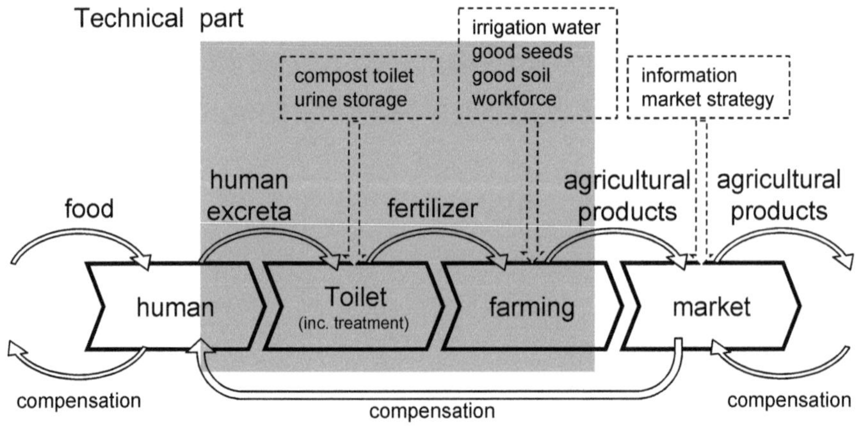

Fig. 2.1 Sanitation value chain and its technical part (made by author based on Ushijima et al. (2015) with permission from the copyright holders, IWA Publishing)

evaluation of human waste and wastewater as agricultural resources; selection of unit operations; and building up the technical system.

This chapter discusses the concept for building up the technical system in the agro-sanitation based on the analysis of characteristics of human feces, urine, and gray water. Then, the system configurations for rural area and urban slum area are introduced followed by overview of element technologies adequate to the system. Figure 2.1 is the modified version of Fig. 1.3 shown in Chap. 1. The technical system in the sanitation value chain works for improving the value of human excreta and producing sanitary products such as fertilizer and irrigation water. The performance of the technical system is evaluated as an agricultural resource. The technical system also works in farming part by introducing controlling measures for preventing salt accumulation in soil.

2.2 The Characteristic of Feces, Urine, and Gray Water

The human waste and wastewater effluent from a household are made up of contributions from various appliances, such as the toilet, kitchen sink, washbasin, bath, shower, and washing machine. The design of technical system for improving their value as an agricultural resource requires the following steps: (1) evaluation of their potential as resources by analyzing their characteristics; (2) setting pathway for resources production by selection of suitable waste among several wastes and wastewater from household; and (3) allocating adequate element technologies to each resource producing pathways.

2.2.1 Characteristics of Human Feces

The average daily feces production rates per capita per day found in the literature is approximately 130 g (wet basis) (Almeida et al. 1999; Del Porto and Steinfeld 1998). As to the constituents of feces, Lopez Zavala et al. (2002a, b) reported the average values for some bulk parameters of feces from the analysis of 26 feces samples as shown in Table 2.1. It should be noted from Table 2.1 that moisture value is 81.8% and this means that the main component of feces is water. TS shows the dry solids' content of feces and VS% is the percentage of organic matter in the dry solids. Human produces only 24 g of dry solids daily through feces. The data shows that the organic matter occupies 84.4% or 20 g per capita per day and this amount of organic matter should be stabilized for reuse of feces as a soil conditioner. COD is the other parameter showing organic matters in terms of oxygen requirement for their aerobic degradation.

Feces contains nitrogen and phosphorous which will be used for fertilizer. Total nitrogen content of feces is 60.1 mg/g (dry solid of feces) and 1.44 g daily and the content of inorganic type of nitrogen (NH_3–N and NO_3–N in Table 2.1) is low. Therefore, it can be said that its main form is organic nitrogen. Phosphorous production is about 108 mg daily through feces.

If we consider the feces as an agricultural resource, it should be also noted that feces contains pathogens which cause health risk to workers by direct contact to compost and to customers of agricultural products. Generally, pathogen content in feces strongly depends on the situation that person is infected by some disease or not. Some of pharmaceuticals are discharged directly feces, and the control of pharmaceutical contamination of soils is the other issue when we reuse feces as an agricultural resource (Kakimoto and Funamizu 2007; Kakimoto et al. 2007).

Table 2.1 Characteristics of feces (made by author based on the data from Lopez Zavala et al. 2002b)

Parameter	Average content	Production per capita per day
Moisture	81.8%	106 g
TS	18.2%	24 g
VS	84.4%	20 g
pH	7.5	–
COD	1.45 mgO_2/mg	35 gO_2
NH_3–N	3.4 mg/g	82 mg
T–N	60.1 mg/g	1.44 g
NO_3–N	0.03 mg/g	0.7 mg
Cl	4.2 mg/g	100 mg
PO_4–P	4.5 mg/g	108 mg

Units in mass/mass are expressed on dry basis

2.2.2 Characteristics of Human Urine

The average daily urine production rates per capita per day found in the literature is approximately 1.2 L (Almeida et al. 1999; Del Porto and Steinfeld 1998). The detailed constituents of urine from 189 volunteers in Hokkaido University within 4 h are reported in Chap. 9. Table 2.2 is the brief summary of the data described in Chap. 9, showing that

- Urine contains organic carbons and 30% of them are the carbon in Urea. 57% of organic carbon in urine did not identify.
- Second important constituent is nitrogen. In the fresh urine, most of them have urea form. But, during storage, urea is hydrolyzed and ammonification occurs. The ammonia nitrogen increases and pH also increases from pH 6 (fresh urine) to pH 9 (Hotta and Funamizu 2008). 7.6 g of nitrogen is produced daily as urea.
- Phosphorous production is 1.3 g per capita daily.
- It should be noted that there are lots of inorganic ions in urine. Main cation and anion are Na^+ and Cl^-, respectively. And these ions might accumulate in soil if we use urine as a fertilizer continuously.

Table 2.2 Characteristics of urine (made by author based on the data presented in Chap. 9)

		Average content		Production per capita per day
		Fresh urine	After 90 days storage	
pH	(–)	6.5	9.0	–
Total organic carbon	g-C/L	7.3 g-C/L	4.6 g-C/L	8.8 g-C
Carbon in urea	g-C/L	2.3 g-C/L	0.15 g-C/L	
Carbon in amino and organic acid	g-C/L	0.8 g-C/L	0.35 g-C/L	
Carbon in sugar	g-C/L	0.0 g-C/L	0.0 g-C/L	
Unknown	g-C/L	4.2 g/C-L	4.1 g-C/L	
Total nitrogen	g-N/L	6.3 g-N/L	6.3 g-N/L	7.6 g-N
Nitrogen in urea	g-N/L	5.4 g-N/L	0.3 g-N/L	
Nitrogen in ammonia	g-N/L	0.3 g-N/L	5.9 g-N/L	
Nitrogen in creatinine and uric acid	g-N/L	0.5 g-N/L	0.0 g-N/L	
Total phosphorous (phosphate)	g-P/L	1.1 g-P/L	0.6 g-P/L	1.3 g-P
Na^+	g/L	2.94	–	3.5 g
Cl^-	g/L	5.18	–	6.2 g
Ca^{2+}	mg/L	120	0.0	0.14 g
Mg^{2+}	mg/L	60	0.0	0.07 g
K^+	g/L	1.76	–	2.1 g

As in the case of feces, urine also contains estrogens, pharmaceuticals, and their metabolite. Therefore, when we use urine as an agricultural resource, their fate in soil and plants should be monitored.

2.2.3 *Characteristics of Gray Water*

Traditionally, wastewater effluent from a household has been divided into two fractions, blackwater (toilet wastes), and gray water (wastewater from kitchen sink, wash basin, bath, shower, and washing machine). Therefore, we also use the word "gray water" in this book. Almeida et al. (1999) summarized the contribution of each appliance to the daily discharge volumes and pollution loads as shown in Table 2.3. Gray water is classified into two groups, higher-loaded gray water (kitchen sink and washing machine), and lower-load gray water (shower, bath and wash basin). Higher-load gray water consists of a high proportion of the pollution load in terms of COD. Furthermore, higher-loaded gray water contains linear alkylbenzene sulfonates (LAS) which are a surfactant mostly used in laundry detergent formulations (Huelgas and Funamizu 2010).

Maiga et al. (2014) reported the characteristics of gray water generated in two rural areas ("Barkoundba" and "Kologoudiesse") located in the Sahel region in Burkina Faso. Their results showed that gray water is generated from three main sources (shower, laundry, and dishwashing) with average daily productions of 8–13 L/capita/day. Table 2.4 summarizes the result by Maiga et al. (2014), showing that wastewater from shower exhibited high contribution to all main parameters.

Table 2.3 Contribution of each appliance to the daily total discharge volumes and pollutants loads (% of total volume or mass per capita)

Appliance	Type of water	Volume	COD	NH_3–N	NO_3–N	PO_4–P	TS
Water closet	Blackwater	31	44	97	3.8	80	77
Kitchen sink	Higher-load gray water	13	23	0.3	38.	9.4	10
Washing machine		16	22	1.2	7.6	4.3	4.0
Wash basin	Lower-load gray water	13	1.7	0.1	11	1.3	2.1
Bath		16	2.5	0.6	15	1.1	1.3
Shower		12	6.4	0.7	25	4.1	5.1
Total		100 L	110 g	2.4 g	0.37 g	6.7 g	56 g

Made by authors based on the data from Almeida et al. (1999)

Table 2.4 Characteristics of gray water from main three in Kologoudiesse (% of total volume or mass per capita)

Appliance	Volume	COD	NH_3–N	NO_3–N	Total-P	TS
Shower	70	40	59	55	55	55
Dishwashing	18	28	17	15	18	20
Laundry	11	32	24	30	27	25
Total	10 L	22.0 g	0.27 g	0.17 g	0.11 g	13.8 g

Made by authors based on the data from Maiga et al. (2014)

2.3 The Concept for Building up the Technical Part in Sanitation Value Chain

In order to find effective way to connect the feces, urine, and gray water, which are potential sanitation resources, to next agricultural value chain, we examined the characteristics of feces, urine, and gray water and found that

- Most of nitrogen, potassium, and phosphorous which are important elements of fertilizer are existing in feces and urine.
- Feces are a potential agricultural resource for fertilizer or soil conditioner if the organic matter in feces can be stabilized and the pathogens can be inactivated.
- Urine contains fertilizer elements, but also pharmaceuticals and other micro-pollutants.
- Gray water is a source of water, because they, especially lower-load gray water, are not highly polluted and easy to reuse.

It is very important characteristics that the volume of feces and urine is very small if we compare the volume of gray water. The mass of feces is 130 g per capita per day and urine volume is 1.2 L per capita per day; on the other hand, the volume of gray water is 100 L as shown in Table 2.3. In the current centralized water-borne sewage systems, all the wastewater from household is mixed and transported to the centralized treatment plant. This means that nitrogen and phosphorous in feces and urine are diluted by gray water and water for flushing in the current system. Therefore, in the postmodern sanitation where feces and urine are recognized as agricultural resource, separation of feces, urine and gray water at source and dry toilet system is essential.

Lopez Zavala et al. (2002a) have proposed the onsite wastewater differential treatment system (OWDTS) based on differential management and treatment of household wastewater effluents, and this is a new approach for improving the traditional onsite wastewater treatment systems (OWTSs). Traditional OWTS has experienced two types of failure, operational and functional; the first, when the system does not remove wastewater from the home, and the last one, when the system continues to remove wastewater but does not properly treat the water prior to discharge into the environment. The inappropriate use and disposal of wastes from OWTS can have a number of adverse impacts that would include: spread of diseases, contamination of ground and surface water, degradation of soil and vegetation, decrease in amenity due to odors and insects, and potential litigation.

Figure 2.2 shows a hypothetical model for an onsite wastewater differential treatment system proposed by Lopez Zavala et al. (2002a). In this system, the fractioning of household wastewater into three types is essential. Thus, reduced-volume blackwater (feces and urine), higher-load and lower-load gray water are new concepts that are introduced in this model. Here, the treatment of blackwater conceives a change in the traditional way of using the water closet; in other words, the use of water in the proposed system is just to clean the toilet bowl, not to transport feces and urine; this is a very important change that is possible by using a composting toilet and urine storage or treatment. The essential and new concept in the OWDTS is to treat separately the three fractions of household wastewater and recover the resources. The importance of conceiving differential treatment for each fraction is indicated by the fact that:

1. *Blackwater* is practically eliminated from the household effluent by using the composting toilet and urine storage. It means freshwater use will be saved and organic matter, nutrients (Nitrogen and Phosphorous), and total solids load will be eliminated from the household effluent. Composting toilet is a non-water carriage toilet that uses natural biological decomposition to transform wastes (feces and toilet paper) into a relatively dry, nutrient-rich humus material called compost.

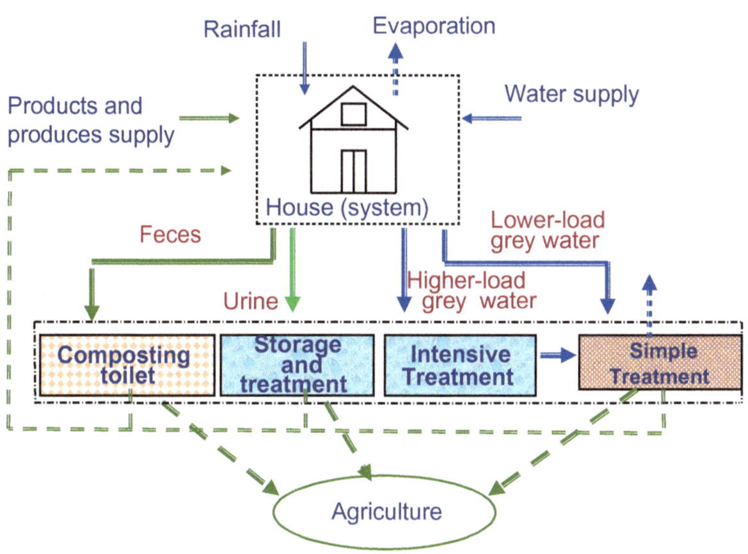

Fig. 2.2 Hypothetical model for an onsite wastewater differential treatment system (made by author based on Lopez Zavala et al. (2002a) with permission from the copyright holders, IWA Publishing)

2. *Lower-load gray water* could be treated by utilizing the natural capacity of soil microorganisms and/or reused by simple treatment units for gardening and agricultural irrigation.
3. *Higher-load gray water* with a pollution concentration of 1000–2000 mg-COD/l represents 30% of household effluent that needs any conventional treatment process for reaching acceptable quality (Almeida et al. 1999). Candidate technologies for higher-load gray water treatment must be robust to contamination from all likely sources.

The benefits of using an onsite wastewater differential treatment system (OWDTS) may be analyzed from the viewpoints of water and soil contamination prevention, conservation of resources, reduction of health risk for population, and economics. Such benefits may be the following:

1. The sources of pathogens, mainly toilet wastes, are eliminated from the wastewater stream; therefore, the groundwater contamination risk is reduced considerably.
2. The feces are converted into a stable organic matter source, probably free of pathogens, that may be recycled into the household when a garden area is available or be disposed in external systems such as urban green areas or agricultural land.
3. The stored urine may be used as liquid fertilizer directly with or without treatment. And in the urban area application, reduced volume of urine may be transported to suburban or rural area and be used as a fertilizer.
4. The reduction of blackwater volume is positive due to the saving and conservation of freshwater (probably drinking water) conventionally used in toilet flushing, and the reduction of wastewater flow rate to be treated.
5. Utilization of the natural capacity of soil organisms to biodegrade a fraction of the gray water organic load.
6. Reduced volume of gray water to be treated by conventional treatment processes, therefore, small size of treatment facilities, easier operation, and lower operation costs.
7. Safe and healthy disposal of treated wastewater into the soil and probable reuse of for gardening and/or agricultural irrigation.
8. "Zero wastewater discharge" into the centralized sewer system.

2.4 Models for Rural and Urban Area and Technologies for These Models

As discussed in the introduction, in the agro-sanitation, technologies are used for improving the value of human waste and wastewater, and for connecting sanitation resources to next value chain, agriculture. In the rural area, there is a farmland near houses and it is easy to connect sanitation resources to agriculture. On the other hand, in the urban area, the recovered resources should be transported to rural area. Therefore, there are two models for agro-sanitation, rural model and urban model.

2.4.1 Rural Model

Figure 2.3 is the rural model which is proposed for a rural area in Burkina Faso. And Fig. 2.4 is the model for a rural area in Japan. In these models, the concept is the same that the separation of feces, urine, and gray water is essential for recovering resources. Namely, the compost is produced from feces by composting type toilet; the gray water is reused as irrigation water after treatment. The separated urine is also used as a fertilizer with or without treatment.

2.4.2 Urban Model

Figure 2.5 is the model for urban slum area proposed by author. In the rural model, recovered nutrients and water are reused in the farmland easily; on the other hand, in the urban slum area, it is not easy to find the space for agricultural activities near houses. Therefore, the collection and transportation of recovered resources to the rural area are inevitable. To reduce the cost of the collection and transportation, volume reduction of feces and urine is important technical process in the urban model.

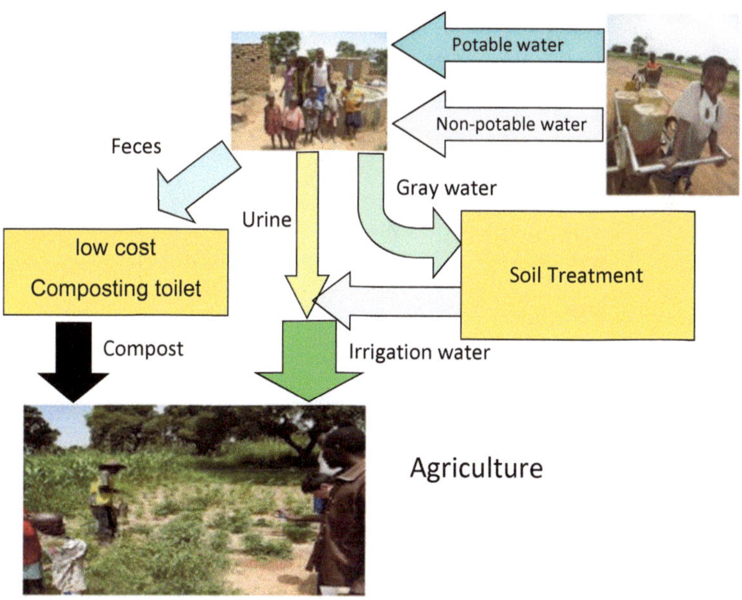

Fig. 2.3 Rural model proposed by author for Burkina Faso

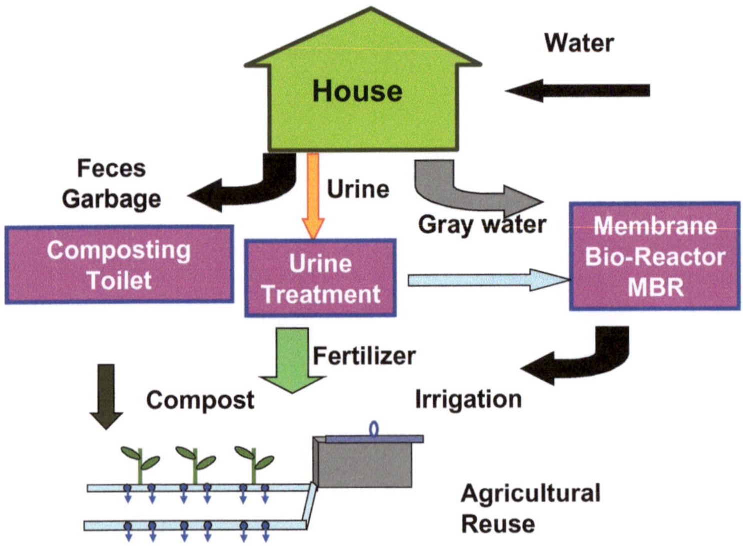

Fig. 2.4 Rural model proposed by author for Japan

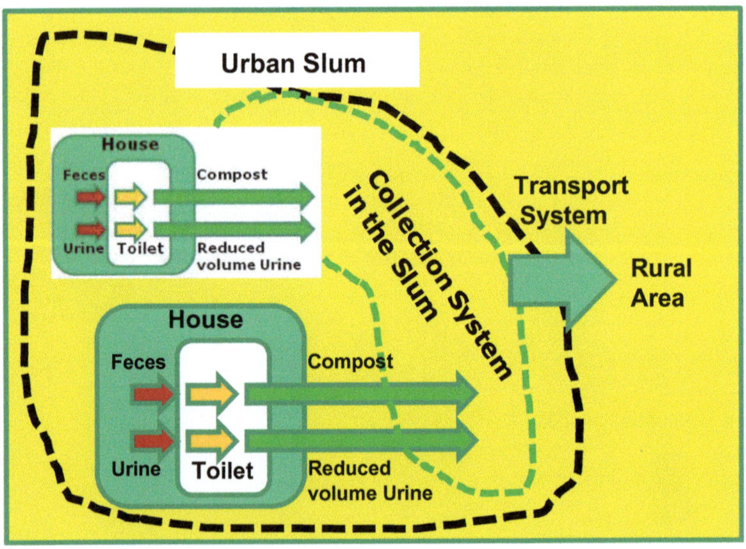

Fig. 2.5 Urban model proposed by author

2.4.3 Technologies for Feces

Solids and biosolids including feces are stabilized to (1) eliminate odors, (2) reduce pathogens, and (3) inhibit, reduce, or eliminate the potential for putrefaction, and the principal methods used for stabilizing feces are (1) composting, (2) anaerobic fermentation, (3) aerobic fermentation, and (4) alkaline stabilization (Metcalf and Eddy 2003). In our resources oriented agro-sanitation, composting and alkaline stabilization are adopted for both rural and urban models because of their simplicity. The feces are treated at toilet in the house; therefore, it is important that the structure is simple and the operation and maintenance of toilet should is easy.

As discussed in Sect. 2.2.1, main constituents in feces are water and organic matter. And the pathogenic microorganisms should be reduced for health risk management, and the micro-pollutants should be controlled in order to avoid the pollution of soil and crop by trace contaminants in compost. Therefore, the design and operation of composting type toilet should be based on the following scientific knowledge: (1) evaporation of water from compost; (2) biodegradation process of organic matter in feces which includes fate of carbon and nitrogen; (3) fate of pathogens; (4) fate of such micro-pollutants as pharmaceuticals and hormones. And these topics are covered from Chaps. 3 to 8 in this book.

2.4.4 Technologies for Urine

Separated urine is stored at onsite first, and then, it is used as fertilizer near house in the rural model or it is transported from urban area to rural area in the urban model. A topic of urine storage was found in several reports for phosphate precipitation in which main discussion was developed on crystallization and ammonification (Lind et al. 2000; Udert et al. 2003a, b; Harada et al. 2006). In this book, storage of urine is discussed in Chap. 9.

In the urban model, huge quantity of urine may be transported to farmland from urban area. Therefore, application of urine as fertilizer is problematic with regard to transportation (Lind et al. 2001; Maurer et al. 2006). To resolve the transportation issue, various steps have been suggested such as long-term local storage followed by truck transport in Sweden (Hanaeus et al. 1997). Likewise, some volume reduction techniques including reverse osmosis, electrodialysis, freezing and thawing, and thermal drying system have been tried (Maurer et al. 2006). However, all these techniques require high energy and operation and maintenances' cost; therefore, a new simple onsite volume reduction system (OVRS) by using atmospheric energy for evaporating water is introduced in Chap. 10.

Recovery of nutrient such as nitrogen and phosphorous from urine is also very important topic. With combination of struvite crystallization, Ganrot et al. (2007) showed efficiency of some minerals (zeolite, etc.) on nitrogen recovery from urine. In Chap. 11, new technology for recovering nitrogen as a slow nitrogen-releasing fertilizer is introduced.

2.4.5 Technologies for Gray Water

Treatment and reuse of gray water are widely practiced in many places. Reclamation of domestic gray water is of particular interest for water reuse in domestic gardens or small-scale irrigation systems since it generally contains lower concentration of organic matters and pathogens compared to mixed wastewater. Several treatment technologies for reclaiming gray water have been developed and selection of the adequate technology depends on the local situation such as electricity availability and economical background. Therefore, three typical technologies are discussed in this book: slanted soil system; sand filter with geo-textile; and membrane bioreactor. The slanted soil system discussed in Chap. 12 and sand filter is one of the simplest treatment technologies which will be operated without power supply.

The membrane bioreactor in Chap. 13 has become a popular treatment unit for domestic and industrial wastewaters especially in developed countries.

2.4.6 Technologies for Agricultural Use of Urine and Reclaimed Gray Water

The sustainability of use of urine and reclaimed gray water schemes using depends on the management of potential impacts of the urine and available water on soils, crops, and the environment. Knowledge of agronomics is used for making urine application and irrigation plan. Relevant topics in agronomics and water quality are (1) soil characteristics; (2) suspended solids; (3) salinity, sodicity, and specific ion toxicity; (4) trace elements and nutrients; and (5) consideration for crop selection (Metcalf and Eddy 2007). Since urine contains sodium ion, sodicity impact should be considered in urine application. Sodicity is a condition where the sodium is the dominant cation in the soil solution. Under sodic conditions, soil particles disperse and clay swell, with a result that the soil particles plug large pore spaces in the soil matrix and reduce the rate water and air enter the soil (Metcalf and Eddy 2007). Urine application is discussed in Chap. 15.

In the planning of irrigation by reclaimed gray water, making a balance between a potential amount of graywater and irrigation requirement in the climate is essential. The irrigation requirement is calculated by the simple mass balance of evapotranspiration, precipitation, and leaching. Gray water reuse for irrigation is discussed in Chap. 16.

References

Almeida MC, Butler D, Friedler E (1999) At-source domestic wastewater quality. Wat Sci Tech Urban Water 1:49–55. https://doi.org/10.1016/s1462-0758(99)00008-4

Del Porto D, Steinfeld C (1998) The composting toilet system book. Center for Ecological Pollution Prevention (CEPP), Concord

Ganrot Z, Dave G, Nilsson E (2007) Recovery of N and P from human urine by freezing, struvite precipitation and adsorption to zeolite and active carbon. Bioresour Technol 98(16):3112–3121. https://doi.org/10.1016/j.biortech.2006.10.038

Hanaeus J, Hellstrom D, Johansson E (1997) A study of urine separation in an ecological village in northern Sweden. Water Sci Technol 35:153–160. https://doi.org/10.1016/s0273-1223(97)00193-5

Harada H, Shimizu Y, Sato D, Miyagoshi Y, Matsui S, Matsuda T, Nagasaki T (2006) Development of a struvite-recovery plant from urine using a urine-diversion toilet. J Jpn Sci Water Environ 28(3):191–196 (in Japanese)

Hotta S, Funamizu N (2008) Inhibition factor of ammonification in stored urine with fecal contamination. Water Sci Technol 58(6):1187–1192. https://doi.org/10.2166/wst.2008.349

Huelgas A, Funamizu N (2010) Flat-plate submerged membrane bioreactor for the treatment of higher-load graywater. Desalination 250:162–166. https://doi.org/10.1016/j.desal.2009.05.007

Kakimoto T, Funamizu N (2007) Factors affecting the degradation of amoxicillin in composting toilet. Chemosphere 66:2214–2219. https://doi.org/10.1016/j.chemosphere.2006.09.060

Kakimoto T, Osawa T, Funamizu N (2007) Antibiotic effect of amoxicillin on the feces composting process and reactivation of bacteria by intermittent feeding of feces. Biores Technol 98:3555–3560. https://doi.org/10.1016/j.biortech.2006.11.029

Lind B, Ban Z, Byden S (2000) Nutrient recovery from human urine by struvite crystallization with ammonia adsorption on zeolite and wollastonite. Biores Technol 73(2):169–174. https://doi.org/10.1016/s0960-8524(99)90157-8

Lind BB, Ban Z, Byden S (2001) Volume reduction and concentration of nutrients in human urine. Ecol Eng 16:561–566. https://doi.org/10.1016/s0925-8574(00)00107-5

Lopez Zavala MA, Funamizu N, Takakuwa T (2002a) Onsite wastewater differentiable treatment system: modeling approach. Water Sci Technol 46(6–7):317–324

Lopez Zavala MA, Funamizu N, Takakuwa T (2002b) Characterization of feces for describing the aerobic biodegradation of feces. J Environ Syst Eng, JSCE, No. 720/VII-25:99–105

Maiga Y, Moyenga D, Ushijima K, Sou M, Maiga AH (2014) Greywater characteristics in rural areas of the Sahelian region for reuse purposes: the case of Burkina Faso. Revue des Sciences de l'Eau (J Water Sci) 27(1):39–54

Maurer M, Pronk W, Larsen TA (2006) Treatment processes for source separated urine. Water Res 40:3151–3166. https://doi.org/10.1016/j.watres.2006.07.012

Metcalf & Eddy (2003) Wastewater Engineering, 4th edn. McGraw- Hill, pp 1499–1501

Metcalf & Eddy (2007) Water reuse. 995 and 959, McGraw-Hill

Udert KM, Fux C, Munster M, Larsen TA, Siegrist H, Gujer W (2003a) Nitrification and autotrophic denitrification of source-separated urine. Water Sci Technol 48(1):119–130

Udert KM, Larsen TA, Biebow M, Gujer W (2003b) Urea hydrolysis and precipitation dynamics in a urine-collecting system. Water Res 37(11):2571–2582. https://doi.org/10.1016/s0043-1354(03)00065-4

Ushijima K, Funamizu N, Nabeshima T, Hijikata N, Ito R, Sou M, Maiga AH, Sintawardani N (2015) The postmodern sanitation—agro-sanitation business model as a new policy. Water Policy 17(2):283–298. https://doi.org/10.2166/wp.2014.093

Part II
Technologies for Feces: Composting Type Toilet

Chapter 3
Fate of Carbon in Composting Process

Miguel Ángel López Zavala

Abstract In this chapter, the activity of microorganisms in the composting toilet during the degradation of feces is presented through the quantification of reductions in total solids (TSs), volatile solids (VSs), and chemical oxygen demand (COD) during batch tests in laboratory-scale composting reactors. TS, VS, and COD reductions are on the order of 56, 70, and 75%, respectively, irrespective of the organic loading regarded. Characterization of organic matter attending biodegradability is a common practice when a bio-kinetic model is used to describe the reaction rates at which organic matter of wastewater is biodegraded. This chapter deals with the transformation of organic matter contained in feces as an indicator of the fate of carbon in composting process; two approaches for evaluating the transformation of organic matter of feces are discussed based on physico-chemical determinations and biological response during aerobic degradation. Application of those procedures led to the description of feces organic matter transformation in terms of chemical oxygen demand (COD) and mass units.

Keywords Aerobic biodegradation of feces · Bio-toilet · Characterization of feces Composting process · Organic loading · TS · VS · COD reductions

3.1 Composting Toilet

Bio-toilet is the name of composting toilet using sawdust as an artificial soil matrix for bioconversion of human excreta into compost which can be used either as organic fertilizer rich in N, P, and K, or as a soil conditioner (Kitsui and Terazawa 1999; Del Porto and Steinfeld 1998). The bio-toilet consists mainly of a toilet bowl, composting reactor, ventilation and mixing systems. The composting reactor is

M. Á. López Zavala (✉)
Tecnológico de Monterrey, Water Center for Latin America and the Caribbean,
Av. Eugenio Garza Sada Sur No. 2501, Col. Tecnológico,
64849 Monterrey, N.L., Mexico
e-mail: miganloza@itesm.mx

N. Funamizu (ed.), *Resource-Oriented Agro-sanitation Systems*,
https://doi.org/10.1007/978-4-431-56835-3_3

filled with sawdust, and a button activates the mixer after using the toilet. The human excreta, including toilet paper, are incorporated quietly and odorless into the sawdust matrix where the aerobic biodegradation occurs. Kitsui and Terazawa (1999) reported that sawdust creates an environment for bacteria thriving since high porosity, high void volume ratio, high water and air retention, and high drainage characterize the matrix. The high bacterial tolerance of the woody materials allows the longtime use of sawdust. The low apparent density of sawdust mass saves energy when mixing is performed. However, the key characteristic is the essential biodegradability of sawdust that makes it to be an excellent soil conditioner (Kitsui and Terazawa 1999). Figure 3.1 shows a schematic representation of the bio-toilet system.

The bio-toilet system differs from conventional composting systems in

(a) Human excreta are treated.
(b) The composting reactor of the bio-toilet system is provided with heating and mixing systems that ensure a continuous thermophilic-aerobic biodegradation process and almost uniform temperature distribution.

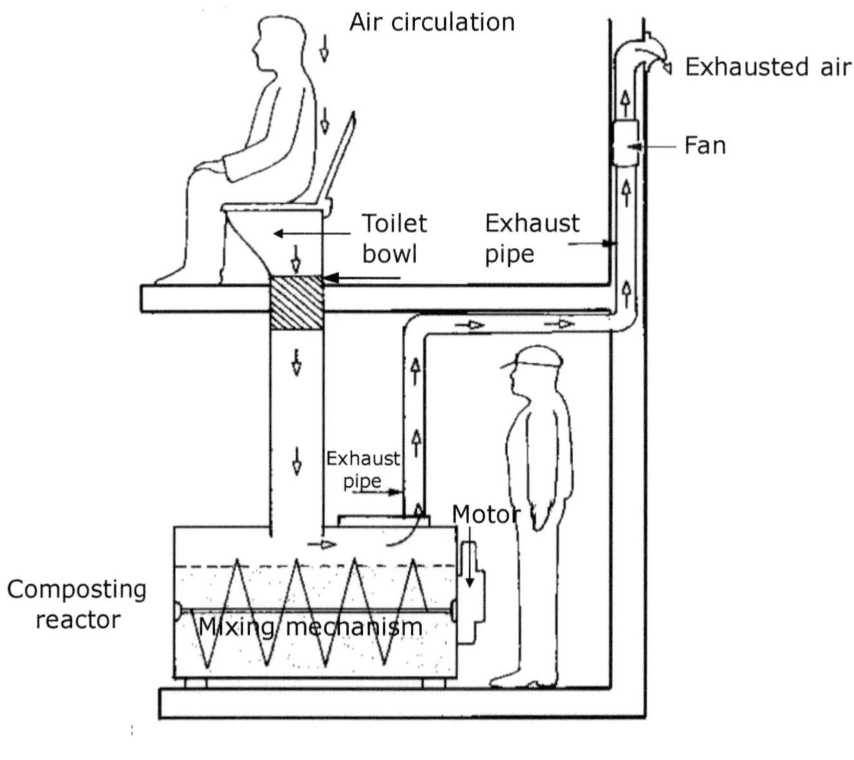

Fig. 3.1 Bio-toilet system and its components (reprinted from Lopez Zavala et al. (2005) with permission from Elsevier)

(c) The moisture content in the composting reactor is kept in the range 50–60% by heating, mixing, and ventilation.
(d) The system is managed with the aim of accelerating decomposition, optimizing efficiency, and minimizing any potential environmental or nuisance problems (odor).
(e) Traditional composting systems have batch configurations where drying is an important process for the proper management and operation, whereas the bio-toilet is a continuous feeding system where both drying and biodegradation rate of organic matter are important because urine and feces are daily added into the composting reactor.

Affordability of the bio-toilet system depends on several factors; among them, the capability of microorganisms for reducing and stabilizing the organic matter contained mainly in feces is of great significance for optimizing efficiency, for minimizing any potential environmental or nuisance problems (odor), and for design and operation purposes. The capability of microorganisms to degrade the organic matter contained in feces depends, among other factors discussed below, on the characteristics of feces itself.

3.2 Characteristics of Feces

The performance of the bio-toilet system, and more specifically biodegradation rates, is affected by several factors, including the substrate bioavailability that is mainly associated with the biodegradability of organic matter contained in feces. Organic matter is mainly constituted by carbon. Thus, the transformation of organic matter into the reactor of the bio-toilet system could be an indicator of the fate of carbon in the composting process.

There is a wide range of variability in the content of feces from person to person and place to place. Factors such as nutrition, climate, health, age, and lifestyle are responsible for that variability. In this respect, from the analysis of 26 feces samples (Lopez Zavala et al. 2002), Table 3.1 shows average values for some bulk parameters that indicate in a certain way the organic load and nutrient availability of feces. Data of moisture content, TS, and COD are in accordance with those reported by other researchers (Almeida et al. 1999). As seen, approximately 82% of water and 18% of solids constitute feces. 84% of solids are volatile solids (organic material), and 16% are fixed solids (inorganic material). Fixed solids and non-biodegradable VS remain after biodegradation without any transformation.

Characterization of feces has been traditionally carried out regarding the concentration of different constituents or values of several parameters that indicate in certain way the organic load and nutrient availability, as shown in Table 3.1. However, this way of characterizing feces does not allow judging the biodegradability of organic matter contained in feces. Thus, the biodegradability of feces should be assessed attending its biodegradable character, non-biodegradable and

Table 3.1 Pollution load of excreta

Parameter	Unit	Average	Almeida et al. (1999)
Moisture	%	81.8	79.2
TS	%	18.2	20.8
VS	%	84.4	–
pH	–	7.5	–
EC	μS/cm	115.9	–
COD	mg/mg	1.45	1.38
NH_3-N	mg/g	3.4	7.2
T-N	mg/g	60.1	–
NO_3-N	mg/g	0.03	0.14
Cl	mg/g	4.2	–
SO_4	mg/g	1.1	–
PO_4-P	mg/g	4.5	27.9

Units in mass/mass are expressed on a dry basis

biodegradable organic matter. This fractioning can be determined through the performance of simple batch tests; measuring the chemical oxygen demand (COD); monitoring the oxygen utilization rate (OUR); and simulating OUR profiles. Additionally, biodegradability may be assessed in terms of mass units by measuring simple physical parameters such as total solids (TSs) and volatile solids (VSs).

3.3 Biological Activity in the Composting Reactor of the Bio-Toilet System

The biological activity in the bio-toilet system was assessed by Lopez Zavala et al. (2005) through the quantification of reductions in total solids (TSs), volatile solids (VSs), and chemical oxygen demand (COD) during batch tests in laboratory-scale composting reactors conducted under controlled environmental conditions and for several organic loading conditions expressed in terms of feces–sawdust ratio (*F*/*S*).

3.3.1 Respiration Rate Profiles

Figure 3.2 shows the experimental OUR profiles obtained for each organic loading evaluated. Three stages were observed in all experimental OUR profiles: (a) rapid increase stage from the origin up to the peak (from point A to B), (b) rapid decrease stage from the peak down to the beginning of stabilization of the profile (from point B to C), and (c) steady stage, between points C and D. The OUR peak was reached in all trials between 6 and 12 h after starting the batch tests. In general, higher OUR at the peak occurred at higher organic loading; however, it is observed that OUR at

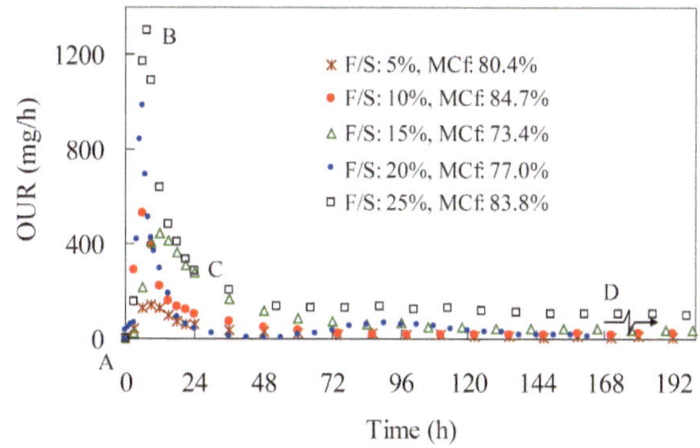

Fig. 3.2 Experimental OUR profiles for five different organic loadings (reprinted from Lopez Zavala et al. (2005) with permission from Elsevier)

the peak corresponding to *F*/*S* = 10% was higher than that obtained at *F*/*S* = 15%. As seen, feces degraded at *F*/*S* = 15% had lower moisture content; so that, this factor seems to be the responsible for such trend. Rapid decrease stage ended approximately 48 h (2 days) after the beginning of the batch tests. Steady stage was extended until the end of the tests, which was determined based on monitoring of the oxygen concentration in the output of the experimental reactors; when oxygen concentration recovered the level registered at the beginning of the batch tests, the practical "end" of the biodegradation process was set. For the *F*/*S* = 5%, it occurred approximately 1 week after starting the tests, whereas for the *F*/*S* = 25%, it was registered 2 weeks after the beginning.

3.3.2 TS, VS, and COD Reductions

Figure 3.3 shows the TS reductions obtained for the five organic loads evaluated. TS reduction is correlated with the TS of feces used in each organic loading, for instance, 6.5 g TS feces for *F*/*S* = 5%. As seen, there is a clear linear correlation between TS reduction and organic loading; furthermore, the TS reduction was

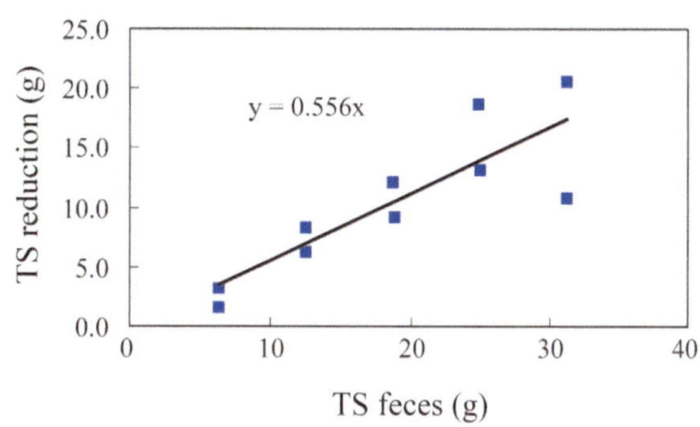

Fig. 3.3 Total solids (TSs) reduction during the aerobic biodegradation of feces (reprinted from Lopez Zavala et al. (2005) with permission from Elsevier)

approximately 56%, irrespective of the organic loading. It means that about 44% TS of feces remained in the composting reactor after biodegradation.

Similarly, the VS reductions measured in each trial were correlated with the organic loading expressed in terms of VS of feces, as depicted in Fig. 3.4. Like TS, a clear linear correlation between VS reduction and organic loading was observed; however, the VS reduction was nearly 70%, irrespective of the organic loading regarded. Thus, 30% VS remained in the composting reactor after biodegradation.

COD is the measure of the concentration of organic material because it alone provides a link between electron equivalents in the organic substrate, the biomass, and oxygen utilized. Furthermore, mass balance can be made in terms of COD (Henze et al. 1987). Results of mass balance made in terms of COD for each organic load are plotted in Fig. 3.5; like TS and VS, COD reduction showed a linear correlation to feces COD; however, the COD reduction totalized 75%, irrespective of the organic loading. It means that 25% of biological inner material remained in the composting reactor.

It is important to mention that biodegradation took place only on feces because experimental tests conducted only with sawdust, under the same conditions, and with addition of essential nutrients for microorganisms such N, P, and K, decomposition (COD reductions) of sawdust was not observed (Fig. 3.6). This is in agreement with results reported by other researchers (Kaiser 1996; Kitsui and

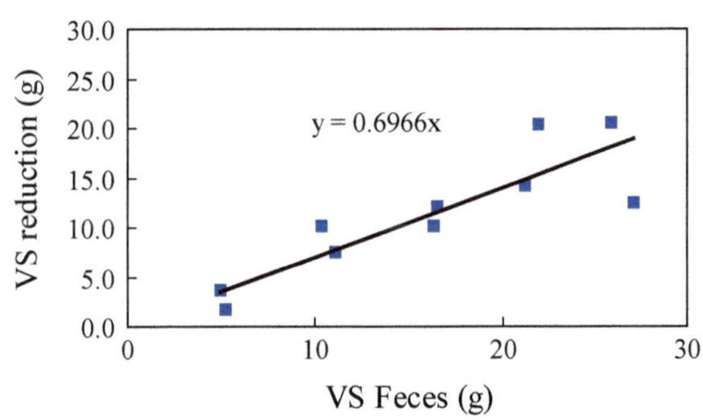

Fig. 3.4 Volatile solids (VSs) reductions during the aerobic biodegradation of feces (reprinted from Lopez Zavala et al. (2005) with permission from Elsevier)

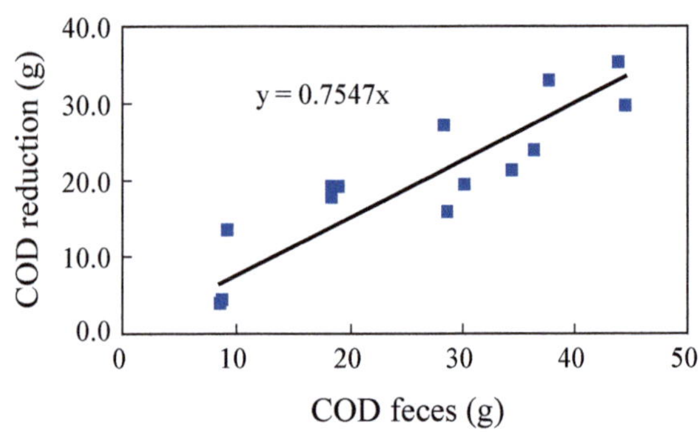

Fig. 3.5 Chemical oxygen demand (COD) reduction during the aerobic biodegradation of feces (reprinted from Lopez Zavala et al. (2005) with permission from Elsevier)

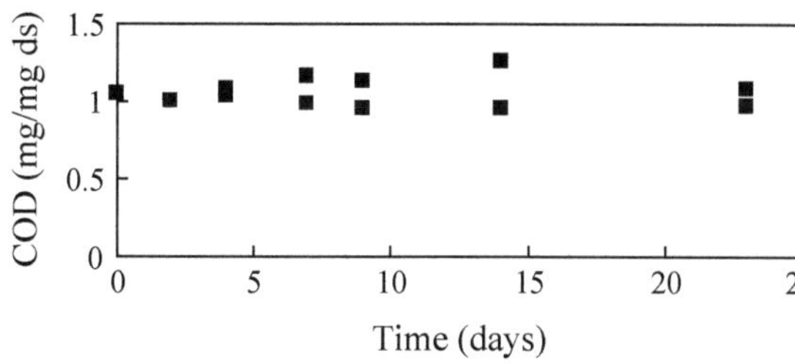

Fig. 3.6 Tolerance of sawdust to bacterial decomposition (reprinted from Lopez Zavala et al. (2005) with permission from Elsevier)

Terazawa 1999). Microorganisms responsible of degrading sawdust such as actinomycetes and fungi stop growing at thermophilic temperatures (Kaiser 1996). Sawdust has a high bacterial tolerance, and bacteria are the microorganisms responsible for biodegradation in the composting reactor of the bio-toilet system (Kitsui and Terazawa 1999).

Above results show that organic loading is not a limiting factor for aerobic biodegradation of feces in the bio-toilet system. Thus, limitations to increase the organic loading beyond the $F/S = 25\%$ will be imposed by constraints related to high moisture content levels rather than the organic loading itself because increases in the quantity of feces to be degraded imply necessarily high levels of moisture content in the sawdust matrix.

On the other hand, the practical implication of above results can be visualized as follows. The feces production rate per adult per day is approximately 130 g (wet basis) (Almeida et al. 1999), and feces are constituted by 82% of water (Lopez Zavala et al. 2002); therefore, approximately 23.5 g TS is excreted per adult per day. As shown above, 44% TS remains in the composting reactor after biodegradation, so that it is expected that every day 10.3 g TS is becoming part of the bulky material of the composting reactor. Regarding that sawdust can be used as a matrix per long time, 6 months or 1 year (Terazawa 1997), and the biotoilet is used by 4 household members, the amount of TS accumulated in the composting reactor of the biotoilet yields approximately 7.6 and 15 kg for 6 months and 1 year, respectively; this mass occupies a volume in the composting reactor that must be considered during the process of design and operation of the bio-toilet.

3.4 Transformation of Organic Matter Based on COD Measurements

COD is undoubtedly the superior measure of the concentration of organic material because it alone provides a link between electron equivalents in the organic substrate, the biomass, and oxygen utilized. Furthermore, mass balance can be made in terms of COD (Henze et al. 1987). When characterization is thought for describing the composting process through the use of a bio-kinetic model, it is important to note that there is a close relationship between characterization of feces and the model and constants used. The quality of characterization may be evaluated based on biological response (oxygen consumption) during biodegradation, so that the level of fractioning will be given by modeling and simulating the oxygen utilization rate. Thus, the most proper characterization will be the one that allows better model predictions.

The total organic matter contained in feces can be measured as COD, which can be divided depending on the complexity and the use of the model. Lopez Zavala et al. (2002) found that feces are constituted by slowly biodegradable particulate organic matter (X_S) and non-biodegradable (inert) particulate matter (X_I). Thus, the total COD of feces includes the following components:

$$\mathrm{TCOD} = X_I + X_S \tag{3.1}$$

During biodegradation, X_S is transformed into readily biodegradable organic matter (S_S) by hydrolysis; growth of heterotrophic biomass (X_H) occurs at the expense of S_S associated with oxygen utilization; under the death-regeneration approach as shown in Fig. 3.7, decaying biomass is split into two fractions: inert matter (X_{IP}) and slowly biodegradable matter (X_S); the latter is subsequently hydrolyzed into readily biodegradable substrate. The death-regeneration model implies that, even when all the substrate originating from the feces has been oxidized, there remains an oxygen consumption associated with the growth on substrate released from decay. The amount of new biomass formed from released substrate is always less than the amount of biomass lost (Henze et al. 1987). If the oxygen consumption rate at this stage is quite low, approximating toward to the initial level, it might be able to assume that the fraction associated with biomass is negligible and only two fractions of inert material remain at the end of the biological reaction, X_I and X_{IP}.

The total biological inert material remaining at the end of the biological reaction (X_{IT}) can be expressed as follows:

$$X_{IT} = X_I + X_{IP} = X_I + Y_H(X_S)f_{XI} \tag{3.2}$$

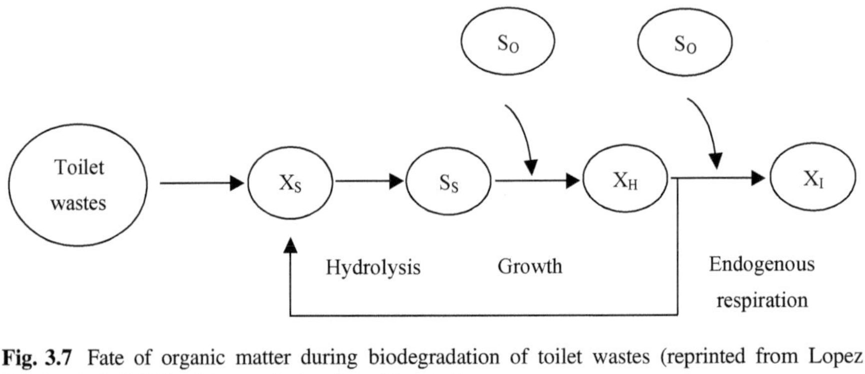

Fig. 3.7 Fate of organic matter during biodegradation of toilet wastes (reprinted from Lopez Zavala et al. (2004) with permission from Elsevier)

where

X_{IT}	Total biologically inert material that remains in the experimental reactor, g COD.
Y_H	Yield coefficient for heterotrophic biomass, g COD (g COD)$^{-1}$.
f_{XI}	Inert COD production coefficient in endogenous respiration, g COD (g COD)$^{-1}$.
$Y_H(X_S)$	Quantity of biodegradable organic matter that is converted into biomass, g COD.

Substituting Eq. (3.1) in Eq. (3.2) and rearranging terms yields

$$X_S = \frac{\mathrm{TCOD} - X_{IT}}{1 - Y_H f_{XI}} \tag{3.3}$$

Estimation of X_S requires the prior determination of stoichiometric parameters such as biomass yield coefficient Y_H and the inert COD production coefficient in endogenous respiration f_{XI}. Estimation of them constitutes a real challenge due to the difficulty for measuring directly the biomass growth in the sawdust matrix; so that, conservative values of these parameters are adopted from those reported in the literature of activated sludge systems, mainly determined at 20 °C. Regarding the uncertainty associated with the impact of temperature on yield coefficient, most engineers assume it to be independent of temperature (Grady et al. 1999). Thus, Y_H and f_{XI} were set as 0.63 g COD (g COD)$^{-1}$ and 0.10 g COD (g COD)$^{-1}$, respectively (Vanrolleghem et al. 1999; Spanjers and Vanrolleghem 1995; Kappeler and Gujer 1992).

Based on COD reductions observed during batch test performance for every *F*/*S* ratio, the fraction of total biologically inert material (X_{IT}) remaining in the experimental reactor is estimated. Thus, substituting this known term and values of Y_H and f_{XI} in Eq. (3.3), fractions X_S, X_I, and X_{IP} are determined from Eqs. (3.3), (3.1), and (3.2), respectively.

For easier comparisons, all values of fractions shown below are expressed in decimal fractions (relative values respect to TCOD). Lopez Zavala et al. (2002) evaluated the effect of organic loading on the aerobic biodegradation of feces using sawdust as a matrix, and they found that the fraction of total biologically inert material (X_{IT}) remaining in the experimental reactor after biodegradation was 25%. Substituting this value and those adopted for stoichiometric parameters in Eq. (3.3) yields

$$\frac{X_S}{\mathrm{TCOD}} = \frac{1 - X_{IT}/\mathrm{TCOD}}{1 - Y_H f_{XI}} = \frac{1 - 0.25}{1 - 0.63 * 0.10} = 0.80$$

Therefore, from Eqs. (3.1) and (3.2), the values of following fractions were obtained:

$$\frac{X_{\mathrm{I}}}{\mathrm{TCOD}} = 0.20 \quad \text{and} \quad \frac{X_{\mathrm{IP}}}{\mathrm{TCOD}} = 0.05$$

3.5 Transformation of Organic Matter in Terms of Mass Units

Knowledge of the amount and character of the compost produced during the aerobic biodegradation of feces allows proper design and operation of the bio-toilet system, since quantity of remaining residue is increasing with the time and the composition of such residue determines the quality of the compost generated. Characterization of feces based on mass units may be an affordable way for evaluating the quantity and composition of the residue accumulated in the composting reactor. Hypothetical composition of feces regarding mass units (g) assumes that feces are composed by fixed solids (X_{FS}) that are biologically inert, non-biodegradable volatile solids (X_{NB}) that are also biologically inert, and biodegradable volatile solids (X_{B}) which are used by heterotrophic biomass to growth; fixed solids (X_{FS}) and non-biodegradable volatile solids (X_{NB}) remain in the composting reactor without any transformation. During biodegradation, fixed solids production (X_{FSP}) occurs since some chemical species (S, P) contained in feces change to their oxidized form trapping oxygen in their molecular structure (Lopez Zavala et al. 2002). Biologically inert volatile solids (X_{BIP}) are produced in endogenous respiration and, besides X_{FSP}, remain in the composting chamber. Thus, the total mass of feces solids (TS) includes the following components

$$\mathrm{TS} = X_{\mathrm{FS}} + (X_{\mathrm{NB}} + X_{\mathrm{B}}) \tag{3.4}$$

And the total mass of solids remaining in the experimental reactor encompasses

$$\mathrm{TS}_{\mathrm{R}} = (X_{\mathrm{FS}} + X_{\mathrm{FSP}}) + (X_{\mathrm{NB}} + X_{\mathrm{BIP}}) \tag{3.5}$$

X_{FS} is estimated gravimetrically on the base of total solids and volatile solids measurements as follows:

$$X_{\mathrm{FS}} = \mathrm{TS} - \mathrm{VS} \tag{3.6}$$

where TS and VS are expressed in mass units (g).

On the other hand, the total mass of remaining volatile solids (VS_{R}) may be expressed as follows:

$$\mathrm{VS}_{\mathrm{R}} = X_{\mathrm{NB}} + X_{\mathrm{BIP}} = X_{\mathrm{NB}} + Y'_{\mathrm{H}}(X_{\mathrm{B}})f'_{\mathrm{XI}} \tag{3.7}$$

where

Y'_{H} Y_{H} expressed in g VS (g VS)$^{-1}$.
f'_{XI} f_{XI} expressed in g VS (g VS)$^{-1}$.
$Y'_{\mathrm{H}}(X_{\mathrm{B}})$ Quantity of biodegradable volatile solids that is converted into biomass, g VS.

Substituting VS = $X_{\mathrm{NB}} + X_{\mathrm{B}}$ in Eq. (3.7) yields

$$X_{\mathrm{B}} = \frac{\mathrm{VS} - \mathrm{VS_R}}{1 - Y'_{\mathrm{H}} f'_{\mathrm{XI}}} \tag{3.8}$$

Then, X_{NB} and X_{BIP} are computed from Eq. (3.7).

Estimation of X_{B} requires the prior determination of the stoichiometric parameters Y'_{H} and f'_{XI}. Those values adopted in COD characterization may be used but expressing them in mass units, g VS (g VS)$^{-1}$. Conversion factors are used for that purpose.

As seen, it is not required to know X_{FSP} for estimating the fractions that constitute feces; however, it can be determined as follows:

$$X_{\mathrm{FSP}} = \mathrm{TS_R} - \mathrm{VS_R} - X_{\mathrm{FS}} \tag{3.9}$$

For easier comparisons, all values of fractions shown below are expressed in decimal fractions (relative values respect to TS). As shown in Table 3.1, volatile solids constitute approximately 84% of feces; therefore, from Eq. (3.6) the fraction (X_{FS}) was computed as 16%.

The conversion factor VS to biomass was estimated as 0.76 g VS (g COD)$^{-1}$ from activated sludge literature (Gujer et al. 1999). Based on VS and COD measurements, the conversion factor VS to substrate was approximately 0.58 g VS (g COD)$^{-1}$, and adopting same ratio for VS to X_{BIP}, the stoichiometric parameters expressed in mass units yield Y'_{H} = 0.48 g VS (g VS)$^{-1}$ and f'_{XI} = 0.13 g VS (g VS)$^{-1}$.

On the other hand, Lopez Zavala et al. (2002) found that approximately 70% of feces VS was depleted during batch test performance. Substituting this value and those for stoichiometric parameters in Eq. (3.8) yields

$$\frac{X_{\mathrm{B}}}{TS} = \frac{(\mathrm{VS} - \mathrm{VS_R})/TS}{1 - Y'_{\mathrm{H}} f'_{\mathrm{XI}}} = \frac{0.84 * 0.70}{1 - 0.48 * 0.13} = 0.63$$

Hence, from Eq. (3.7), the values of following fractions were computed:

$$\frac{X_{\mathrm{BIP}}}{\mathrm{TS}} = 0.48 * 0.63 * 0.13 = 0.04 \quad \text{and} \quad \frac{X_{\mathrm{NB}}}{\mathrm{TS}} = 0.84 * 0.30 - 0.04 = 0.21.$$

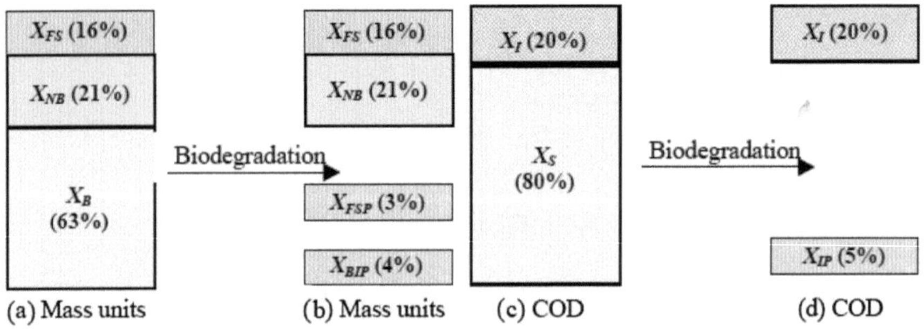

Fig. 3.8 Theoretical composition of feces and its transformation during biodegradation [reprinted from Lopez Zavala et al. (2002)]

Lopez Zavala et al. (2002) also found that total solids remaining in the experimental bioreactor (TS_R) counted approximately 44%; thus, substituting these values in Eq. (3.9) yields

$$\frac{X_{FSP}}{TS} = 0.44 - 0.84 * 0.30 - 0.16 = 0.03.$$

Figure 3.8 summaries schematically the results obtained by applying the two methods of characterizing feces. Characterization based on mass units may be useful for assessing the quantity and character of residue that accumulates in the composting reactor of bio-toilet systems, whereas characterization of feces based on COD measurements is convenient for modeling purposes, a technique that allows the establishment of criteria for the proper design and operation of bio-toilet systems.

References

Almeida MC, Butler D, Friedler E (1999) At-source domestic wastewater quality. Urban Water 1 (1):49–55. https://doi.org/10.1016/S1462-0758(99)00008-4

Del Porto D, Steinfeld C (1998) The composting toilet system book. Center for Ecological Pollution Prevention (CEPP), Concord Massachusetts

Grady CPL, Daigger GT, Lim HC (1999) Biological wastewater treatment, 2nd edn. Marcel Dekker, Inc, USA

Gujer W, Henze M, Mino T, Van Loosdrecht M (1999) Activated sludge model no. 3. Water Sci Tech 39(1):183–193. https://doi.org/10.1016/s0273-1223(98)00785-9

Henze M, Grady CPL Jr, Gujer W, Marais GvR, Matsuo T (1987) Activated sludge model no. 1. IAWPRC scientific and technical reports, no. 1. IAWQ, London

Kaiser J (1996) Modelling composting as a microbial ecosystem: a simulation approach. Ecol Model 91:25–37. https://doi.org/10.1016/0304-3800(95)00157-3

Kappeler J, Gujer W (1992) Estimation of kinetic parameters of heterotrophic biomass under aerobic conditions and characterization of wastewater for activated sludge modeling. Water Sci Tech 25(6):125–139

Kitsui T, Terazawa M (1999) Environmentally-friendly toilets for the 21st century, bio-toilet. In: Proceedings of the 10th ISWPC, Yokohama, III, pp 120–121

Lopez Zavala MA, Funamizu N, Takakuwa T (2002) Characterization of feces for describing the aerobic biodegradation of feces. J Environ Syst Eng JSCE No. 720/VII-25:99–105

Lopez Zavala MA, Funamizu N, Takakuwa T (2004) Modeling of aerobic biodegradation of feces using sawdust as a matrix. Water Res 38(5):1327–1339. https://doi.org/10.1016/j.watres.2003.10.028

Lopez Zavala MA, Funamizu N, Takakuwa T (2005) Biological activity in the composting reactor of the bio-toilet system. Biores Technol 96(7):805–812. https://doi.org/10.1016/j.biortech.2004.07.009

Spanjers H, Vanrolleghem PA (1995) Respirometry as a tool for rapid characterization of wastewater and activated sludge. Water Sci Tech 31(2):105–114. https://doi.org/10.1016/0273-1223(95)00184-o

Terazawa M (1997) Biological cleanup and recovery of organic wastes using sawdust as an artificial soil matrix. Division of Environmental Resources, Graduate School of Agriculture, Hokkaido University, Japan

Vanrolleghem PA, Spanjers H, Petersen B, Ginestet P, Takacs I (1999) Estimating (combination of) activated sludge model no. 1 parameters and components by respirometry. Water Sci Tech 39(1):195–214. https://doi.org/10.1016/s0273-1223(98)00786-0

Chapter 4
Fate of Nitrogen in Composting Process

Shinya Hotta and Naoyuki Funamizu

Abstract Fate of nitrogen in the composting process of aerobic biodegradation of feces was evaluated. A biodegradability of the fecal nitrogen was investigated; the nitrogen was characterized into three fractions: a type originally inert in biological activity (N_{XI}), an inert type reproduced by endogenous respiration of heterotrophic microorganism (N_{XIB}), and a slowly biodegradable type (N_{XS}). In fresh feces, 75% of fecal nitrogen was contained in the N_{XS}, whereas the other 25% was in N_{XI}. In the aerobic biodegradation of N_{XS}, 66% of fecal nitrogen mineralized into ammonia while the other 9% was transformed into biologically inert type (N_{XIB}); totally, 34% of fecal nitrogen (N_{XI} and N_{XIB}) remained in the composted material. Contribution of inorganic nitrogen of nitrates was ignored in mass balance evaluation of the Composting-Type Toilet.

Keywords Fecal nitrogen · Biodegradability · Aerobic biodegradation

4.1 Introduction

In the Composting-Type Toilet, organic matter of feces transforms to either an inorganic matter or more stable type of organic matter. Lopez Zavala et al. (2002) reported that approximately 75% of fecal organic carbon was mineralized to CO_2 gas in the composting process, whereas the other 25% remained in the composted

S. Hotta (✉)
Research and Development Center, Asahi Kasei Corporation,
Kawasaki-ku, Kawasakishi 210-0863, Japan
e-mail: nihaozhenye@hotmail.com

N. Funamizu
Graduate School of Global Food Resources, Hokkaido University,
Kita-9, Nishi-9, Kita-ku, Sapporo 060-8589, Japan

N. Funamizu
Research Institute of Humanity and Nature, Motoyama Kamigamo,
Kita-ku, Kyoto 603-8047, Japan

N. Funamizu (ed.), *Resource-Oriented Agro-sanitation Systems*,
https://doi.org/10.1007/978-4-431-56835-3_4

material. In addition to the carbon behavior, nitrogen transformation was also an important process in the Composting-Type Toilet. Hotta and Funamizu (2007, 2009) discussed the fate of nitrogen compounds during composting process. The aim of this chapter is investigation of biodegradability of fecal nitrogen and summarization of the fate of the nitrogen in the composting process of human feces.

4.2 Materials and Methods

4.2.1 Test Materials

In order to secure natural composting condition, sawdust matrix used in batch tests had been conditioned in advance because virgin sawdust was not suitable material for the composting process. In the conditioning, we used the mechanical composting device in which sawdust matrix was intermittently turned and mixed with daily input of feces. The conditioning had been continued for approximately half a year at room temperature. After the conditioning, the sawdust matrix in the mechanical composting device was packed into a couple of small plastic bags and stored in a freezer at −30 °C in order to secure identical condition of the sawdust matrix on every batch test described below. Fresh feces collected from normal individual were mixed in a plastic vessel and stored in a refrigerator at 5 °C for a couple of days until it was used. Each test condition of feces sawdust (FS) ratio was summarized in Table 4.1. Initial moisture content of composting material (mixture of feces and sawdust) was adjusted to 60% with addition of distilled water.

Table 4.1 Materials condition in the experiments

	FS ratio (%)	Feces (g-dry)	Sawdust (g-dry)	Moisture content of feces (%)	Moisture content of sawdust (%)
RUN1	20	18.0	90	87.9	34.4
RUN2	5	5.1	103	87.9	34.4
RUN3	15	14.1	94	84.3	34.4
RUN4	20	17.3	87	86.6	34.4
RUN5	20	17.3	87	86.3	34.4
RUN6	15	14.1	94	80.2	34.4
RUN7	10	9.8	92	80.2	34.4
RUN8	2	2.3	102	90.0	34.4
RUN9	15	13.3	87	90.0	34.4

4.2.2 Condition of the Butch Test

Each butch test was conducted until biodegradable organic matter of feces completely decomposed to inorganic or stable matter. Figure 4.1 illustrated test device. The device enabled to control several important conditions of the composting process: the initial moisture content was maintained around 60% (55–65%) throughout every butch test; the reaction temperature was fixed at 50 °C with the air supply of 100 ml/min. The test conditions were summarized in Table 4.2. In order to quantify volatilized ammonium, a bottle of sulfuric acid (1.2N) was placed downstream of the reactor and the bottle was periodically exchanged. The system was equipped with sensors of O_2 and CO_2 in order to measure the oxygen utilization rate (OUR) and CO_2 production rate (CO_2 PR). Content of moisture, carbon, and nitrogen was measured at the beginning (for composting material) and the end (for composted material) of each test.

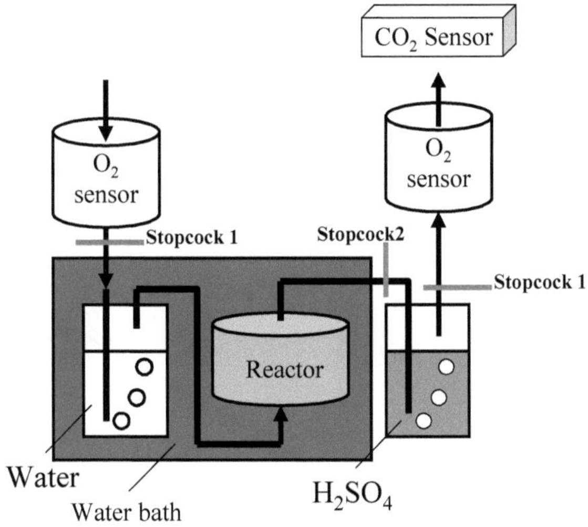

Fig. 4.1 Device for the batch test (reprinted from Hotta et al. (2007) with permission from the copyright holders, IWA Publishing)

Table 4.2 Test conditions

Temperature (°C)	Moisture content during the tests (%)	Airflow (ml/min)	Pressure in the reactor (atm)
50	60–73	100	1.25

4.3 Summary of Test Results

4.3.1 Component of Fecal Nitrogen in Human Feces

Figure 4.2a showed the components of carbon and nitrogen in human feces. As seen, component of carbon and nitrogen was constant with a high correlation in more than 50 samples; feces of 1 g-dry contained approximately 500 mg of carbon and 70 mg of nitrogen. Figure 4.2b showed that the correlation of carbon and nitrogen in human feces was 0.14 g-N/g-C. It was therefore concluded that the feces contained a constant component of both carbon (0.5 g-carbon/g-feces) and nitrogen (0.07 g-nitrogen/g-feces) and that the correlation of fecal carbon and nitrogen could be approximated to 0.14 g-N/g-C.

4.3.2 Oxygen Consumption in the Composting Process

Figure 4.3 showed time courses of CO_2 PR (Fig. 4.3a) and OUR (Fig. 4.3b) in RUN1 at FS 20%. The OUR completely corresponded with CO_2 PR. Therefore, it

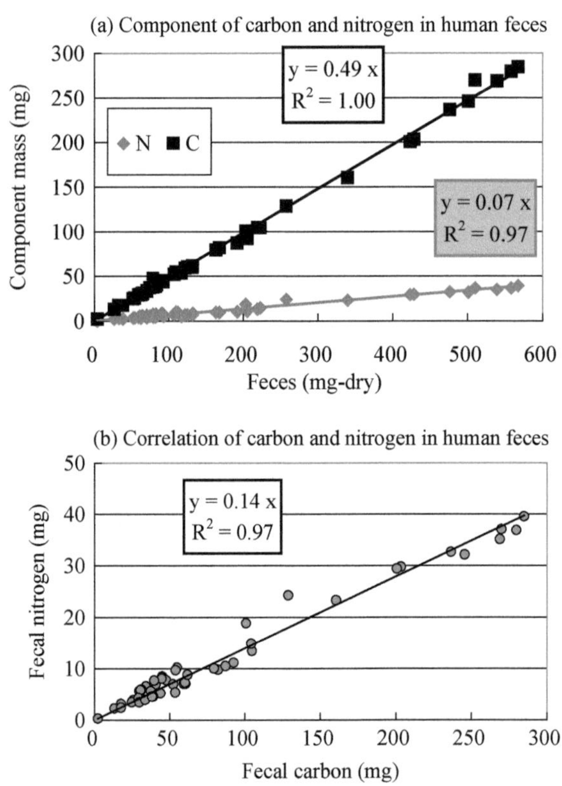

Fig. 4.2 Content of carbon and nitrogen in human feces

Fig. 4.3 Time course of the OUR and CO_2 PR in RUN1

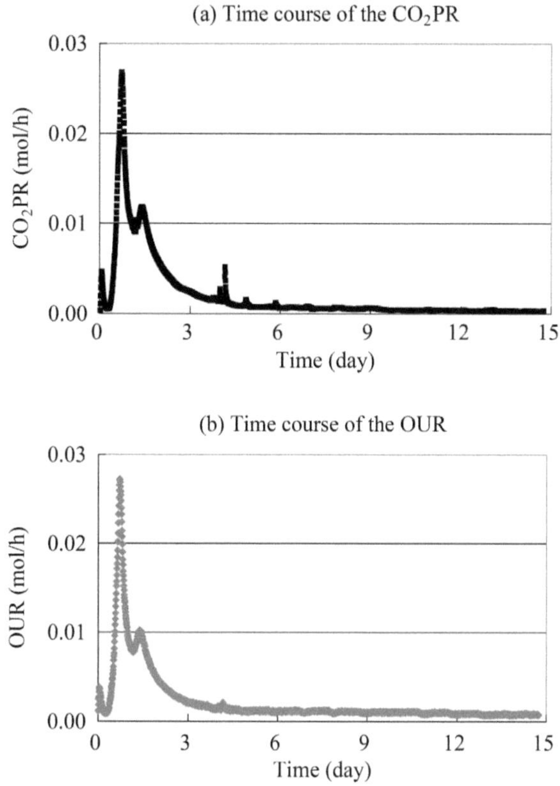

was concluded that no further oxygen was consumed for nitrification except for CO_2 production by heterotrophic microorganisms. No nitrification during the aerobic biodegradation of feces agreed with results reported by Lopez Zavala et al. (2005). Figure 4.4 showed the time course of the OUR in RUN1 to RUN9. As seen, each OUR was on the quite low level at the end of all tests. Therefore, we concluded that the all biodegradable organic matter had been mineralized or

Fig. 4.4 Time course of the OUR in RUN1 to RUN9 [reprinted from Hotta and Funamizu (2007)]

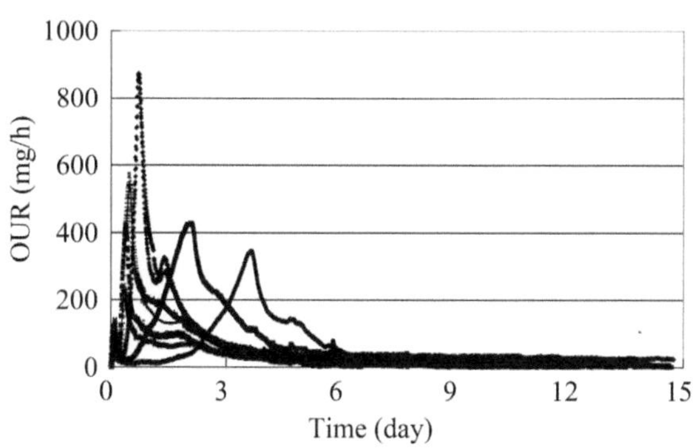

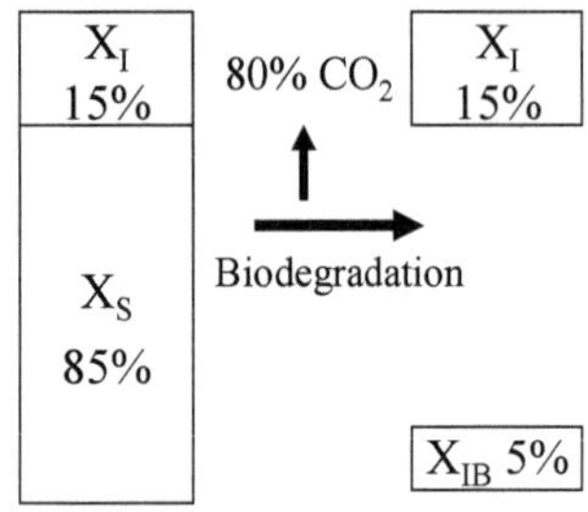

Fig. 4.5 Fate of fecal carbon during composting process of feces [reprinted from Hotta and Funamizu (2009)]

decomposed to biologically inert type throughout the butch tests. Figure 4.4 secured that no more biodegradable organic matter remained at the end of the test. Fate of carbon matter in this study was summarized in simple illustration of Fig. 4.5 which was used for characterization of fecal nitrogen later.

4.3.3 *Fate of Fecal Nitrogen in the Composting Process*

In this study, three types of nitrogen were prepared for evaluation: (a) nitrogen with carbon matter of X_S, X_I, and X_{IB}; (b) ammonia nitrogen in the liquid phase of the composting material; and (c) ammonia nitrogen volatilized into gas phase. Figure 4.6a showed the production of ammonia nitrogen against the input nitrogen in human feces. The production was obtained by sum of (b) and (c). As seen, approximately 66% of fecal nitrogen decomposed to ammonia in the composting process. Figure 4.6b secured that further 34% remained as a biologically inert type of organic matter.

Nitrogen contents in each of X_S, X_I, and X_{IB} were estimated by calculations. In order to simplify the calculation procedure, it is assumed that there was no difference of character between two inert components of X_I and X_{IB}. In the calculation, two unknown factors were prepared as nitrogen contents in X_I (=X_{IB}) and X_S whose unit was "g-C": N_{XI} (=N_{XIB}) and N_{XS} whose unit was "g-N/g-C".

Calculation was conducted on 100 mg of fecal carbon. From Fig. 4.2b, there was 14 mg-N in 100 mg-C in feces. The 14 mg-N was substituted into X_S and X_I (=X_{IB}) before biodegradation in Fig. 4.5. Then, Eq. (4.1) was obtained:

$$15N_{XI} + 85N_{XS} = 14. \tag{4.1}$$

Since 34% of fecal nitrogen remained as shown in Fig. 4.6b, then 14 mg-N of fecal nitrogen transformed into the 4.9 mg-N of an inert type (N_{XI} and N_{XIB}) and 9.1 mg-N of ammonia. The nitrogen of 4.9 mg-N was substituted into X_I (=X_{IB}) after biodegradation in Fig. 4.5. Then, Eq. (4.2) was obtained:

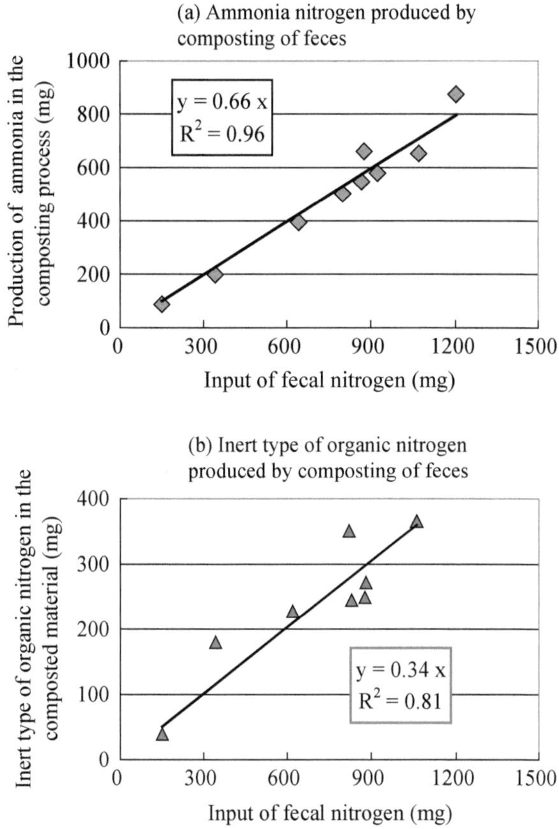

Fig. 4.6 Transformation of fecal nitrogen during the composting process of feces

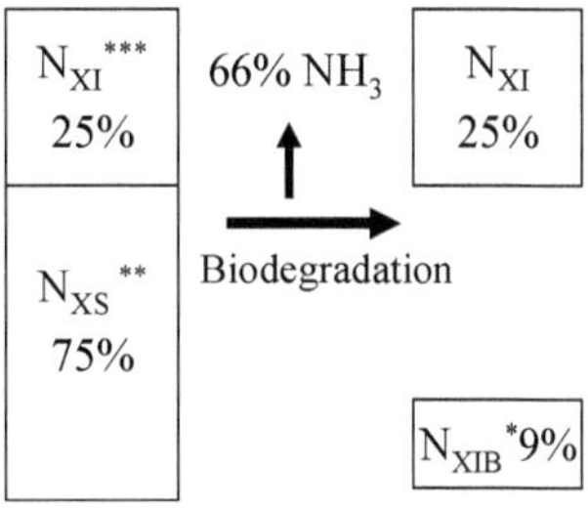

Fig. 4.7 Fate of fecal nitrogen during composting process of feces [reprinted from Hotta and Funamizu (2009)]

$$15N_{XI} + 5N_{XIB} = 4.9. \tag{4.2}$$

Since it was assumed that the N_{XI} was equal to N_{XIB}, Eq. (4.2) can be converted to $15N_{XI} + 5N_{XI} = 4.9$. Thus, the value of N_{XI} was obtained to 0.245 g-N/g-C. By substituting the N_{XI} into Eq. (4.1), the value of N_{XS} was found to be approximately 0.121 g-N/g-C. Fate of fecal nitrogen was summarized in the illustration of Fig. 4.7.

4.4 Conclusion

The characterization of fecal nitrogen was conducted: Composition of fecal nitrogen was found to be 75% biodegradable (NXS) and 25% originally inert type (NXI). Throughout the composting process, an inert type of nitrogen (NXIB) was reproduced by endogenous respiration (9% of fecal nitrogen). A total of 34% (NXI + NXIB) of fecal nitrogen remained as an inert type in the composted material, while 66% of fecal nitrogen was mineralized into ammonia. The contribution of nitrates to the nitrogen transformation was not found in this study.

References

Hotta S, Funamizu N (2007) Biodegradability of fecal nitrogen in composting process. Biores Technol 98(17):3412–3414. https://doi.org/10.1016/j.biortech.2006.10.045

Hotta S, Funamizu N (2009) Simulation of accumulated matter from human feces in the sawdust matrix of the composting toilet. Bio-Resour Technol 100(3):1310–1314. https://doi.org/10.1016/j.biortech.2008.07.044

Hotta S, Noguchi T, Funamizu N (2007) Experimental study on nitrogen components during composting process of feces. Water Sci Technol 55(7):181–186. https://doi.org/10.2166/wst.2007.143

Lopez Zavala MA, Funamizu N, Takakuwa T (2002) Characterization of feces for describing the aerobic biodegradation of feces. J Jpn Soc Civ Eng JSCE, No. 720/VII-25:99–105

Lopez Zavala MA, Funamizu N, Takakuwa T (2005) Biological activity in the composting reactor of the bio-toilet system. Biores Technol 96(7):805–812. https://doi.org/10.1016/j.biortech.2004.07.009

Chapter 5
Fate of Pathogens in Composting Process

Masahiro Otaki and Shinobu Kazama

Abstract Composting toilet has a characteristic that it has possibility to stock pathogens for a while when the patient uses it. Then health hazard aspects must be considered in the case of composting toilet more than water flush type. In this chapter, the typical and important pathogenic microbes relating to composting toilet are introduced. And the fate of model microbes which indicates pathogenic microbes in composting toilet was shown here. The decreasing rate of them depends upon the operating condition. We have to consider what appropriate conditions for enhancing reduction of pathogens are, and the quantitative microorganism risk assessment (QMRA) method is explained here to consider the acceptable risk when we handled the material such as sawdust. Based on QMRA consideration, the required reduction level is revealed, and the appropriate indicator microbes are suggested for monitoring high-tolerant pathogens in compost. In addition, the methods to estimate the inactivation mechanisms of microbes in compost are suggested in this chapter.

Keywords Appropriate indicator microbes · Pathogens · QMRA

5.1 Introduction

Health hazards relating to water have been a major topic for a long time. In particular, many have suffered from waterborne diseases due to pathogenic microorganisms. WHO reports that the disability-adjusted life year (DALY) due to diarrheal diseases which are typically waterborne reached almost 9% of the total

M. Otaki (✉)
Department of Human Environmental Science, Ochanomizu University,
2-1-1 Otsuka Bunkyo-ku, Tokyo 112-8610, Japan
e-mail: otaki.masahiro@ocha.ac.jp

S. Kazama
Department of Urban Engineering, The University of Tokyo,
7-3-1, Hongo, Bunkyo-ku, Tokyo 113-8656, Japan

N. Funamizu (ed.), *Resource-Oriented Agro-sanitation Systems*,
https://doi.org/10.1007/978-4-431-56835-3_5

Table 5.1 Typical waterborne pathogenic microorganisms

Category	Typical species
Helminths	*Ascaris lumbricoides, Necator americanus, Taenia saginata, Schistosoma mansoni*
Protozoans	*Entamoeba histolytica, Giardia lamblia, Cryptosporidium parvum, Cyclospora*
Bacteria	*Campylobacter jejuni, Pseudomonas aeruginosa, Shigella dysenteriae, Salmonella typhi, Vibrio cholerae*, Pathogenic *Escherichia coli*
Virus	Poliovirus, Echovirus, Enteroviruses, Rotavirus, Adenovirus, Norovirus

Made by authors based on the data from Haas et al. (1999)

disease burden and almost 40% of the water-, sanitation-, and hygiene (WSH)-related disease burden (Prüss-Üstün et al. 2008). The list of pathogenic microorganisms relating to waterborne diseases is shown in Table 5.1. A major source of these pathogenic microorganisms is human feces (see Table 5.1). Inappropriate sanitation facilities such as pit latrines without a slab or platform, hanging latrines, and bucket latrines cannot separate human feces from human contact (WHO; UNICEF 2012). On the other hand, flushing toilets can separate feces from the surrounding area, but receiving water bodies such as rivers, lakes, and groundwater, which have a potential of being sources of drinking water, are sometimes contaminated by the discharge from flushing toilets. In this case, appropriate treatment of such discharges or drinking water treatment must be considered (Table 5.2).

Composting toilets are categorized as non-water flush types, but the types keeping human feces in the reactor for a while. This has the potential to trap pathogens derived from infected persons, which raises the possibility of other users becoming infected when they handle the media, for example, exchange it to new one (Otaki et al. 2006). It is, however, different from other non-water-flush-type toilets, because it has some mechanisms to stop pathogens from activating due to the heating and drying effect caused by biological, physical, and chemical reactions.

Table 5.2 Concentration of enteric pathogens in feces

Organism	Concentration/(no./g)
Protozoan parasites	10^6–10^7
Enteric viruses	
Enterovirus	10^3–10^7
Rotavirus	10^{10}
Adenovirus	10^{12}
Enteric bacteria	
Salmonella spp.	10^4–10^{11}
Shigella	10^5–10^9
Ascaris	10^4–10^5
Indicator bacteria	
Coliform	10^7–10^9
Fecal coliform	10^6–10^9

Made by authors based on the data from Haas et al. (1999)

The effect created by stopping pathogens from activating is considered to be significantly influenced by the operating condition of the composting toilet. In this chapter, we describe the effect on microorganisms in composting toilets in various conditions. The required inactivation level will also be explained based on risk assessment discussions.

In this chapter, we focus on composting toilets that use sawdust as media though there are many types of materials that can be used in it. Sawdust is porous, which results in more air being contained and a larger surface area. It enhances the decomposing of feces in well-aired conditions and decreases the volume through the evaporation of water.

5.2 The Effect on Microorganisms in Composting Toilets Using Model Pathogens

One type of enteric bacteria called *Escherichia coli* is usually used as a model for pathogenic enterobacteria in water environments. Coliphage is usually used as a model virus, too. Here we used these model microbes for investigating the effect on microbes in composting toilets.

The effect on pathogens in sawdust can be changed by several factors such as temperature and moisture content. In order to reduce the pathogen risk, it is important to know the influence of other factors on them and to control the condition of the sawdust.

5.2.1 *Moisture Content, Temperature and pH*

The inactivation of microorganisms in sawdust followed a first-order reaction, expressed as follows (Burge et al. 1981; Otaki et al. 2006).

$$\ln(N/N_0) = -kt \tag{5.1}$$

where

N concentration of microorganisms at time t,
N_0 concentration of microorganisms at time 0,
k inactivation rate constant,
t retention time

The effect on *E. coli* as a model of pathogenic enterobacteria in several sawdust conditions was investigated (Kazama and Otaki 2011, 2012). The change in the concentration of *E. coli* in sawdust was determined by its growth on desoxycholate agar, and the inactivation rate constants were compared as seen in Fig. 5.1. The inactivation rate constant at 37 and 50 °C with 30% moisture content was twice as

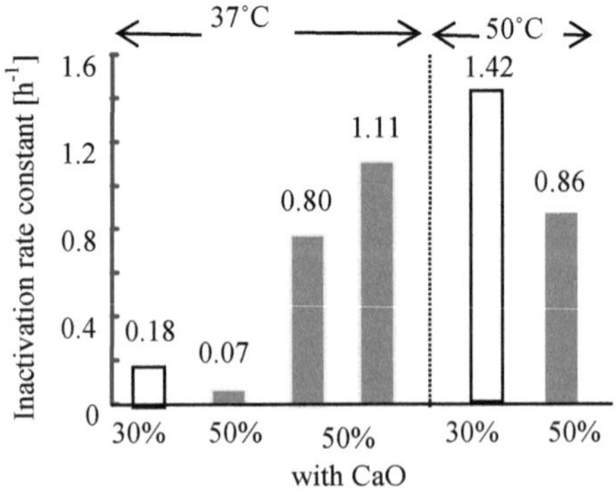

Fig. 5.1 Inactivation rate constants of *E. coli* in sawdust under the several conditions (edited from Kazama and Otaki (2011) with permission from the copyright holders, JSWE). White bars show the inactivation rate with 30% moisture conditions, gray bars show the inactivation rate constants with 50% moisture conditions

high as that with 50% moisture content, which means *E. coli* in low moisture conditions is more inactivate than that in high moisture conditions. The inactivation rate with 30 or 50% moisture content at 50 °C is significantly faster (10 times faster) than that at 37 °C. According to the results under these conditions, it was considered that high temperature conditions are more effective for inactivate *E. coli* than the low moisture conditions. The high pH conditions obtained by adding CaO significantly reduced the amount of *E. coli*, even in low temperatures with high moisture content.

5.2.2 *The Effect in Actually Operating Composting Toilets*

The condition (temperature, pH, etc.) of compost is known to change in the composting process (Hung 1993). Because the sawdust in a composting toilet is generally used for 6 months to 1 year before changing to new sawdust, the effect on pathogens would change over the duration of operating the toilet. The effect on *E. coli* or coliphage Qβ in sawdust obtained from actually operating composting toilets at several points in time was studied in order to estimate the relationship to changing factors (pH, moisture content, and electric current) (Kazama 2010). *E. coli* or Qβ was added to the sawdust, which was adjusted to 50% moisture content by addition of water after heating it at 70 °C for 3 h in order to reduce existing *E. coli.* The sawdust was just obtained from an actually operating composting toilet or non-treated sawdust. The temperature of the sawdust was set to 30 °C, which was observed in an actually operating composting toilet throughout the duration of use. The inactivation rate constants of *E. coli* and Qβ in treated sawdust and that of Qβ in non-treated sawdust have been compared in Fig. 5.2. The fate on the *E. coli* in the non-treated sawdust was not shown in this figure, because it was difficult to investigate it due to indigenous *E. coli.*

The inactivation rates of *E. coli* and Qβ in treated sawdust increased over 220 days. On the other hand, that of Qβ in non-treated sawdust increased once,

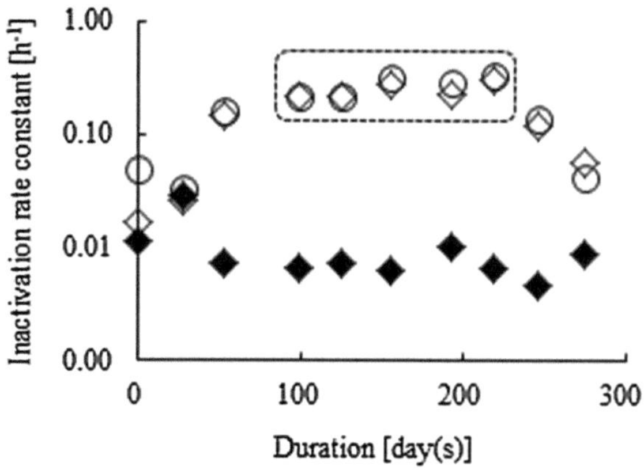

Fig. 5.2 Inactivation rate constants of *E. coli* and coliphage Qβ in sawdust (reprinted from Kazama 2010). Outlined circle; *E. coli* in treated sawdust, outlined diamond; Qβ in treated sawdust, filled diamonds; Qβ in non-treated sawdust. The data surrounded by broken line show the minimum inactivation rate constants calculated by the data under limit of quantization

but did not significantly change at any later stage. No significant relationship was observed between the inactivation rate constants and each factor (pH: 4.7–6.9 (treated sawdust) or 5.7 to 7.8 (non-treated sawdust), moisture content: 50% (treated sawdust) or 58–77% (non-treated sawdust), and electric current 0–4 ms) (Kazama 2010). It was thought that there were other factors that changed the effect on the *E. coli* and Qβ in treated sawdust, and in non-treated sawdust, the change of factors observed in this duration did not influence the effect on Qβ significantly and the inactivation rate was slow. The moisture content of non-treated sawdust fluctuated between 70 and 80% after 50 days; on the other hand, that of treated sawdust was set to 50%. Even though higher concentration of some factors in treated sawdust might cause higher inactivation rate in treated sawdust after 50 days, similar phenomenon did not appear in non-treated sawdust.

Both urine and feces were treated together in this composting toilet, which was thought to make the sawdust have a high moisture content and low temperature and be different from the normal composting process. Therefore, it was difficult to reduce the number of pathogens by methods such as high temperature, low moisture content, or high pH condition in the composting process. In this case, a high pH condition in addition to CaO is one solution to reduce pathogens easily and effectively.

It should also be noted that pathogenic viruses tend to have a stronger tolerance than pathogenic bacteria.

Coliphages, especially RNA coliphages with round shapes, such as Qβ and MS2, have been used as a model for pathogenic enteroviruses in the environment and water treatment processes to monitor the fate of fecal pathogenic viruses (Sobsey et al. 1995). The effect on several species of coliphages in different sawdust conditions were studied (Kazama and Otaki 2011; Kazama et al. 2011). The changes in the concentration of coliphages in sawdust were determined by plaque assay, and the inactivation rate constants were calculated.

5.2.3 Moisture Content, Temperature, and pH

Figure 5.3 shows the inactivation rate constants of Qβ in different sawdust conditions. The inactivation rate constant at 37 or 50 °C with 30% moisture content was higher than that with 50% moisture content. The inactivation rate constant with 30 or 50% moisture content at 50 °C is much higher than that with 30 or 50% at 37 °C. The high temperature and low moisture condition inactivated Qβ faster; however, the inactivation rates of Qβ were very low compared to that of *E. coli*. On the other hand, the inactivation rate of Qβ with a high pH condition was almost the same as that of *E. coli* and very efficient in reducing the number of Qβ.

5.2.4 The Fate of Several Species of Coliphages

There are several pathogenic enteroviruses, and potentially some of them having a higher tolerance than that of Qβ. The effect on four species of DNA (T4 and λ) and RNA coliphages (MS2 and Qβ) as model viruses was estimated, and the differences between them were assessed (Kazama and Otaki 2012).

Figure 5.4 shows the inactivation rate constants of coliphages at 30 and 50 °C. Every coliphage decreased more rapidly at high temperatures (Fig. 5.4b) than at low temperatures (Fig. 5.4a), because the inactivation rate constants at 50 °C were larger than at 30 °C. Low moisture content was also effective at decreasing coliphages. T4 coliphage had the strongest tolerance of all coliphages used in this experiment, and Qβ coliphage had the weakest tolerance.

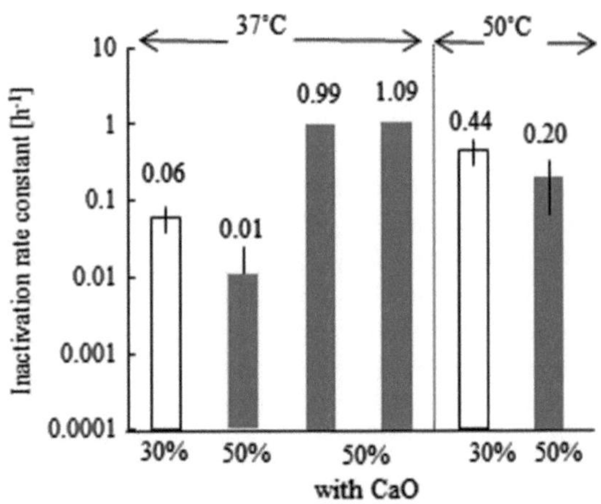

pH	6.65-6.90	9.64	10.20	6.50-6.70

Fig. 5.3 Inactivation rate contains of Qβ in sawdust under the several conditions. Edited from Kazama and Otaki (2011) with permission from the copyright holders, JSWE

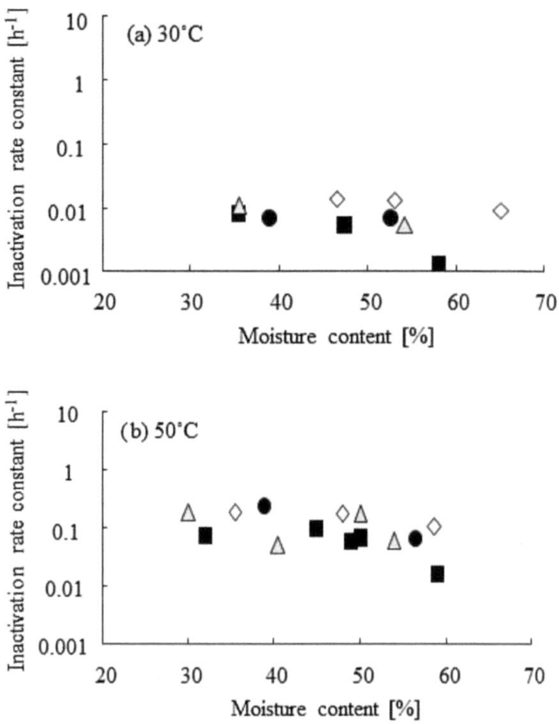

Fig. 5.4 Inactivation rate constants of coliphages with several moisture contents at 30 or 50 °C. Reprinted from Kazama et al. (2011) with permission from Elsevier

The number of days required for a safe level of infection risk (Nakagawa et al. 2006) was calculated using their inactivation rates. Figure 5.5 shows the relationship between temperature, moisture content, and the number of days required for a safe level of infection risk. T4 and λ (DNA phages) decreased slower than Qβ and MS2 (RNA phages). Their rates of decrease also depended on the sawdust temperature and moisture content. It was thought that the trend was that T4 coliphage had a stronger tolerance, especially in high moisture. According to the risk assessment estimation between T4 with Qβ, 15 days to 167 days were required for 12 log inactivation, which would guarantee a safe level of infection risk. Thus, the required number of days for safety was significantly different depending on the species and sawdust conditions. Most pathogenic enteroviruses have similar characteristics to Qβ or MS2, which are round in shape and have RNA genes. Therefore, their fate in sawdust was thought to be similar to that described here. Based on this assumption, although fewer days were required to decrease the risk, in the case that viruses have stronger tolerance to that seen for T4, longer or additional treatment may be necessary for safety.

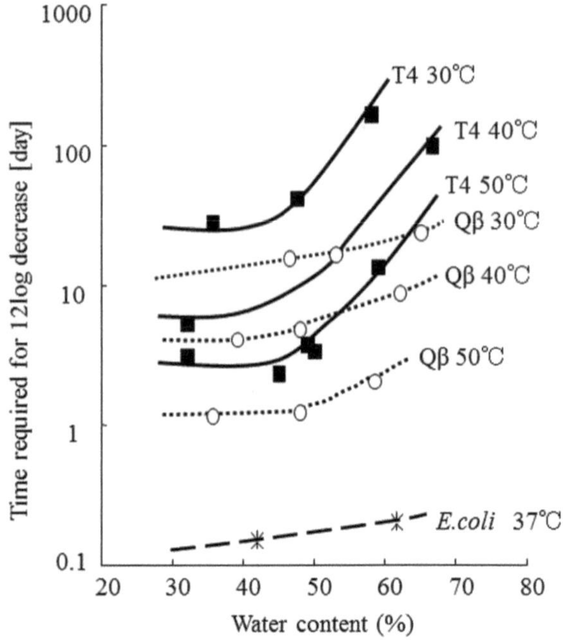

Fig. 5.5 Relationship among temperature, water content, and the days required for a safe level of infection risk. Reprinted from Kazama et al. (2011) with permission from Elsevier

5.3 Appropriate Operational Condition from the Viewpoint of QMRA

Here we describe quantitative microorganism risk assessment (QMRA) of composting toilets. QMRA generally consists of four steps: (1) hazard identification, (2) exposure assessment, (3) dose–response assessment and (4) risk characterization. Using these steps, we can estimate the infection risks to humans exposed to the pathogens released from the composting toilet. In addition, the required decreasing level of pathogens to ensure the safety of users can be calculated by QMRA. This information is useful to decide the appropriate operational condition of composting toilets from hygienic aspects.

The first step is hazard identification. In this step, the target microorganisms or symptoms should be identified according to the environment relating to the user of the composting toilet. In general, the target hazard is gastrointestinal pathogens as listed in Table 5.1 in the case of composting toilets. In the previous section, we described the tolerance of viruses and bacteria in the compost material. This indicated that viruses are thought to remain at higher levels in the material than the bacteria. We then focused on risks by pathogenic viruses such as enterovirus and rotavirus. These viruses are typical species that cause diarrhea. There is relatively sufficient information for risk assessment.

The second step is exposure assessment. In this step, possible exposure routes are assumed. In the use of composting toilets, pathogens are probably stored in the

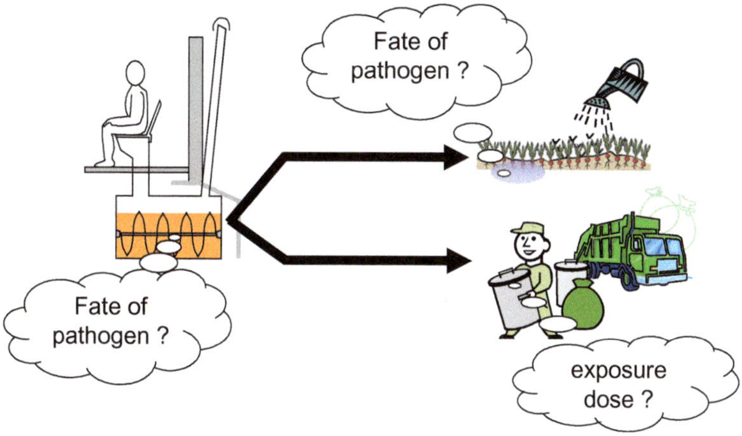

Fig. 5.6 New exposure routes of pathogens in composting toilet system. Reprinted from Otaki et al. (2007)

composting reactor for a while. It is then released into the environment with the compost material. Here, the dilution ratio by mixing compost material and the decreasing ratio (inactivated ratio) during that process that is maintained in the compost reactor have to be considered. After the composting reaction, the fate of pathogens and the possible exposure dose to humans also have to be considered. These routes of pathogens are significantly different from that of a flushing toilet. We need to take a significantly different viewpoint from flushing toilets when hygienic aspects are involved. New routes of pathogens have to be considered when a composting toilet system is used as shown in Fig. 5.6 (Otaki et al. 2007).

The typical exposure route might occur when the user changes the composting material, replacing it with new material. The timing of the change is based on the user conditions. Typically, it is partially replaced (e.g., one quarter is replaced), once every six months. When the user handles the material, there is a chance that the material may come into contact with hands or other parts of the body. It can occasionally come in the vicinity of the mouth and be ingested. Figure 5.7 shows the results of an experiment for measuring the amount of sawdust which was used as composting material that came into contact with hands with and without gloves. It showed that there are insignificant differences between wearing and not wearing gloves, and it depends on the water content of the material with a range of 0.01–1.0 g. These data can be used to estimate the amount ingested by humans when the composting toilet system is handled by humans.

The third step is dose–response assessment. Previous researchers have revealed the relationship between the dose of pathogens to human, in other words, the intake amount of pathogens, and the reaction of humans, i.e., the possibility of infection in the case of several pathogenic microorganisms. The relationship is expressed by two types of models: exponential models as Eq. (5.2) or beta poison models as Eq. (5.3).

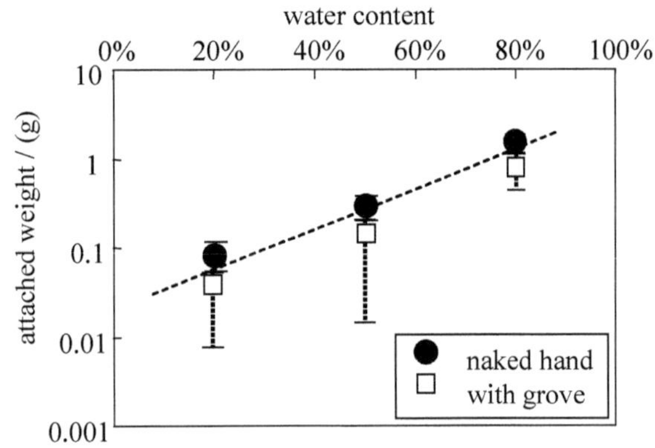

Fig. 5.7 Attached weight of composting toilet matrix onto hand. Reprinted from Nakagawa et al. (2006) with permission from the copyright holders, IWA Publishing

$$P(D) = \exp(-kD) \tag{5.2}$$

$$P(D) = 1 - (1 + (D/\beta)^{\alpha} \tag{5.3}$$

where D is intake dose. $P(D)$ is possibility of infection by intake dose D. k, α, β are parameters.

Haas et al. (1999) summarized the parameter values. Figure 5.8 shows the dose–response curve in several pathogenic microorganisms. As shown in the figure, rotaviruses are high risk at the same intake dose compared with other microorganisms.

The last step is risk characterization. In this step, the infection risk of target microorganisms is estimated based on the results in the previous three steps.

According to USEPA, the annual risk of infection by pathogenic microorganisms must be less than 10^{-4}. We set this value as the acceptable annual risk. Based on this value, the required level of decrease of pathogens in the composting toilet reactor using the exposure dose described above is calculated. The required level of decrease excluding about two pathogenic bacteria such as Salmonella and Shigella

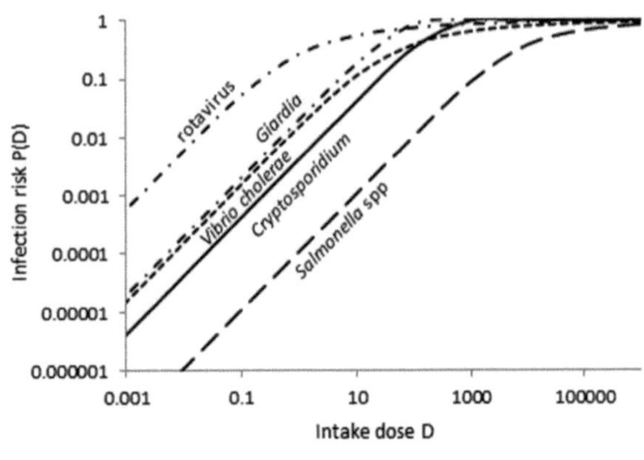

Fig. 5.8 Risk of infection at low dose

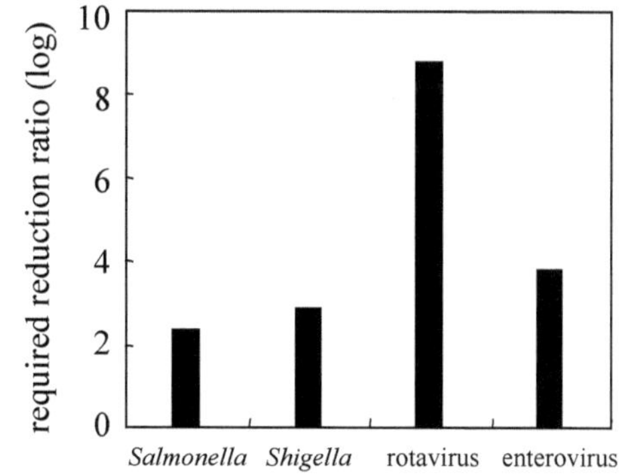

Fig. 5.9 Required reduction ratio for risk <10^{-4} in direct exposure. Reprinted from Nakagawa et al. (2003) with permission from the copyright holders, JSCE

and two pathogenic viruses such as rotaviruses and enteroviruses was estimated as shown in Fig. 5.9. We point out that a reduction of 10^{-9}, which excludes dilution by sawdust, was required for hygienic safety levels in the case of rotaviruses. The composting toilet must be operated in a manner that keeps the level of hygienic safety based on this estimation.

The key factor should be what operation is effective to reduce pathogens in composting toilets. *E. coli* and several coliphages were used as substitutes of pathogenic agents (Sobsey et al. 1995). They were fed to the system sampled from an operating composting toilet, and the change in concentration was measured as described in the previous chapter. Based on data regarding the inactivation rate under several conditions, QMRA was conducted and a diagram of operational conditions was suggested for the suppressed hygienic risk of <10^{-4}. Figure 5.10 shows the case in which the assumed pathogen was a rotavirus and it was assumed to follow the coliphage Qβ reduction rate, because both Qβ and rotaviruses are categorized as RNA viruses. In this figure, the appropriate operational zone was indicated below the curve, which is dependent on the retention time. This result is

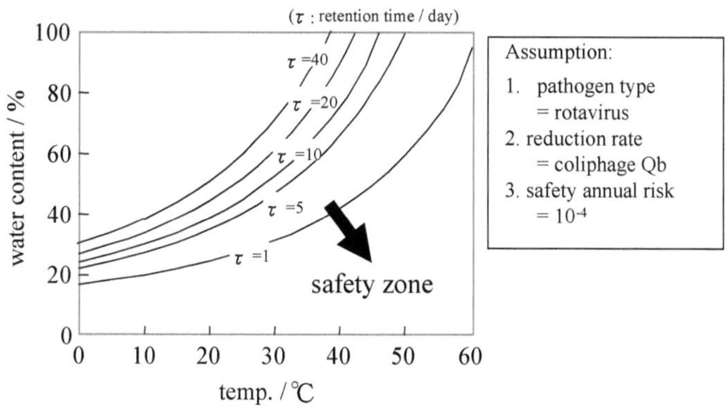

Fig. 5.10 Diagram of operational condition for hygienic risk <10^{-4}. Reprinted from Nakagawa et al. (2006) with permission from the copyright holders, IWA Publishing

considered to be an important guideline for appropriate operational conditions of composting toilets for hygienic safety (Nakagawa et al. 2006).

This suggestion has a weak point in that the reduction rate refers to different viral species. If reduction data in composting toilets using rotaviruses were available, more reliable risk assessment could be conducted and the best operational suggestion could be made.

5.4 Appropriate Microbial Index for Hygienic Safety

Total coliform or coliform groups are widely used as fecal indicators in water environments. It depends on the characteristics of them such as those below.

(1) Existence of a coliform group indicates that the water has fecal contamination, and a significant number of them are detected in the case of fecal contamination.
(2) The tolerance of coliform groups is usually higher than pathogenic bacteria. As such, the reduction of coliform groups indicates that there should be a greater reduction in pathogens.

However, this has no meaning in the case of composting, because many coliphages have higher tolerance than coliform bacteria (see Sect. 5.2). In other words, if the coliform is not detected in the composting media, there is still the possibility of viruses remaining there. Then we have to find other alternative microbial indices from the view point of a high tolerance in similar viruses.

The candidates are fecal streptococci, which is especially used as an indicator in seawater. This indicator has the following characteristics:

(1) It usually lives in the gastrointestine of humans and has less growing potential in the environment. It indicates fecal contamination.
(2) It has relatively high resistance to dry conditions, heat, and disinfectants.
(3) It has similar tendencies to the effect on coliphage in soil (Craig et al. 2002).
(4) The existence level is usually lower than coliform.

According to these characteristics, except for characteristic 4, it has high potential to be a good indicator in the composting toilet.

Figure 5.11 shows the detection level of fecal streptococci and other microbial indicators in actually operating composting toilets in Saitama prefecture, Japan. As shown in the figure, fecal streptococci were detected at relatively a high concentration level compared with other indices. The contained level of fecal streptococci in feces is generally considered to be lower than coliform groups. The higher level of detection in the actually operating composting toilet was the result of the high tolerance to its environment.

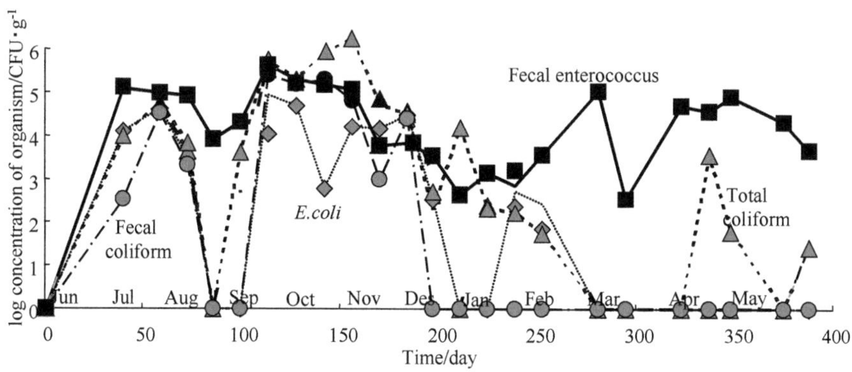

Fig. 5.11 Detected level of fecal enterococcus, total coliform, *E. coli,* and fecal coliform from the actually operated composting toilet from June 2006 to June 2007. Reprinted from Kazama and Otaki (2010) with permission from the copyright holders, JSCE

To compare the tolerance of fecal streptococci with the virus index, we conducted a laboratory-scale experiment using model fecal streptococci, i.e., *Enterococcus faecium*, and model viruses, i.e., coliphage T4 and λ. Figure 5.12 shows the concentration change of these microbes in composting toilets at various temperatures and water content conditions. As described in Sect. 5.3, the inactivation kinetics of these microbes is assumed to follow the first-order reduction. The rate constant values were then compared to understand the reduction rate of these microbes. As shown in the figures, the reduction rate of *E. faecium* was similar or less than other viral indices. In other words, enterococci have a high tolerance in compositing toilet materials in all conditions, except for low temperatures and high water content. In such conditions, all microbes can survive easily. We should therefore avoid operating the composting toilet in such conditions.

Figure 5.13 shows the inactivation rate constant of *E. faecium* and coliphage T4 in the composting toilet material with added CaO to increase the microbe's reduction rate (see Sect. 5.3). In this case, the inactivation rate of *E. faecium* was significantly enhanced by the addition of CaO even though T4 was not.

Consequently, enterococci can be used as a good indicator that shows the viral fate in composting toilets. But we have to look for other alternatives in some specific conditions, such as in low temperatures and high water content or with the addition of CaO.

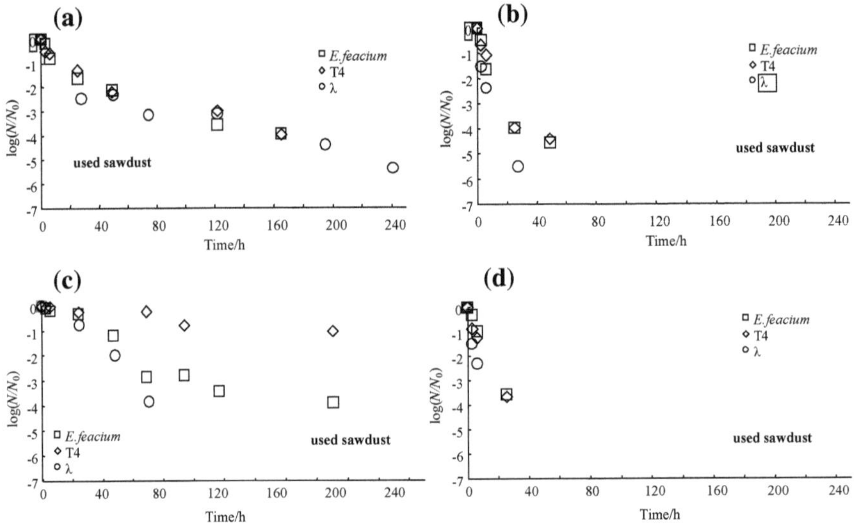

Fig. 5.12 Concentration change of *E. feacium*, coliphage T4 and λ in used sawdust with 30% water content at 37 °C (**a**), with 30% at 50 °C (**b**), with 50% at 37 °C (**c**), and with 50% at 50 °C (**d**). Reprinted from Kazama and Otaki (2010) with permission from the copyright holders, JSCE

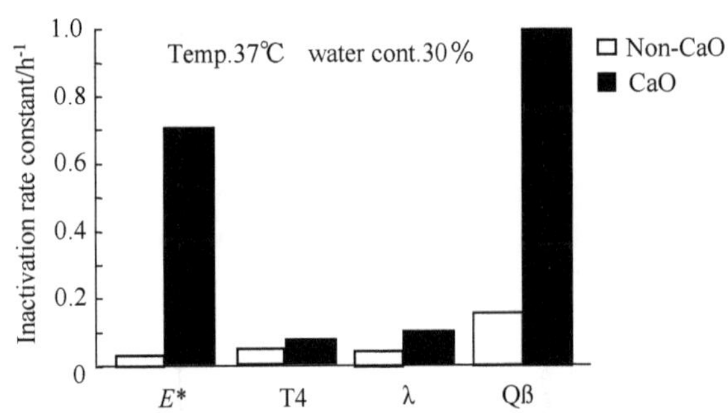

Fig. 5.13 Inactivation rate constants of microbes in sawdust with CaO and non-CaO at 37 °C in 30% water content. Reprinted from Kazama and Otaki (2010) with permission from the copyright holders, JSCE

5.5 New Methods for Estimating the Mechanisms of Microbe Inactivation

Some researchers have reported the effect on microorganisms in the compost materials, and operating conditions for maintaining safety are proposed according to these results. Most of all, researchers have used model pathogenic bacteria or viruses such as *E. coli* or coliphage, not the actual pathogenic microbes. The reasons why they have not used the actual pathogenic microbes were due to the inconvenience of handling and biosafety. The reliability of the effect on microbes in a composting toilet should be raised if the mechanisms of inactivation of microbes are explained with the reduction rate. The new methods for estimating the mechanisms of the inactivation of microbes are suggested here.

Usually, the reduction of bacteria in composting toilets was investigated by measuring the concentration change of target bacteria such as *E. coli*. The concentration was measured by the specific growing media which was specifically prepared for the target bacteria, called selective growing media. We tried to simultaneously use multiple growing media that had differing principles of detection. The measuring results from them were compared to estimations of the mechanisms of inactivation on the target bacteria.

Here, we explain the case of *E. coli* which is typically used as model pathogenic bacteria. We also used three types of growing media which were TSA (Tryptic soy agar, Difco. Co., Ltd), DESO (Desoxycholate agar, Eiken Chemical Co., Japan), and C-EC (Compact Dry EC, Nissui Pharmaceutical Co., Ltd. Japan). Compact Dry EC medium contains X-gluc and Magenta-GAL.

According to the detection principles of these three growing media, the damage to *E. coli* during inactivation can be estimated as shown in Table 5.3 (Kazama and Otaki 2012). The growth of *E. coli*, which has the capability of metabolizing proteins, can be detected by TSA. Non-detection by TSA means that *E. coli* might have damage relating to fatal conditions, such as metabolism function or nucleic acid damage. In the same way, the non-detection by DESO media means that it might have damage to the outer membrane or fatal damage, similar to the TSA case. In cases where some *E. coli* was detected by TSA, but not by DESO, this indicates that they have damage on the outer membrane, but not fatal damage. In cases where some was detected by TSA, but not by C-EC, this indicates that there might be some *E. coli* with damage to the enzyme activity, but this is not fatal. Table 5.4 summarizes the parts of the *E. coli* that are assumed to be damaged through detection differences between TSA, DESO, and C-EC.

Figure 5.14 shows the inactivation rate constants in three media during treatment of composting toilets for *E. coli*. No differences were seen at 37 °C with 30 and 50% water content, including in the case of high pH (CaO addition). As shown in Table 5.4, inactivation is mainly due to damage to nucleic acids or metabolic function. In other words, these conditions could fatally reduce *E. coli* at 50 °C with 30 and 50% water content. In short, high temperature conditions resulted primarily in damage to enzyme activity or membrane function.

Table 5.3 Assumed damages on *E. coli* during inactivation which result in undetection for each medium

Medium	Damages to *E. coli* which result in undetection
TSA	Nucleic acid and/or metabolism
DESO	Outer membrane and/or nucleic acid and/or metabolism
C-EC	Enzyme activity and/or nucleic acid and/or metabolism

Edited from Kazama and Otaki (2012) with permission from the copyright holders, JSWE

Table 5.4 Estimated parts of *E. coli* damages by the detection difference among three media

TSA	Media DESO	C-EC		Estimated parts which were damaged
×	×	×	→	Fatal damage such as nucleic acid and/or metabolism (and also enzymic activity and outer membrane)
O	O	×	→	Enzymic activity Not nucleic acid and outer membrane
O	×	O	→	Outer membrane Not nucleic acid and enzymic activity
O	×	×	→	Enzymic activity and outer membrane Not fatal

Edited from Kazama and Otaki (2012) with permission from the copyright holders, JSWE
O detected
× not detected

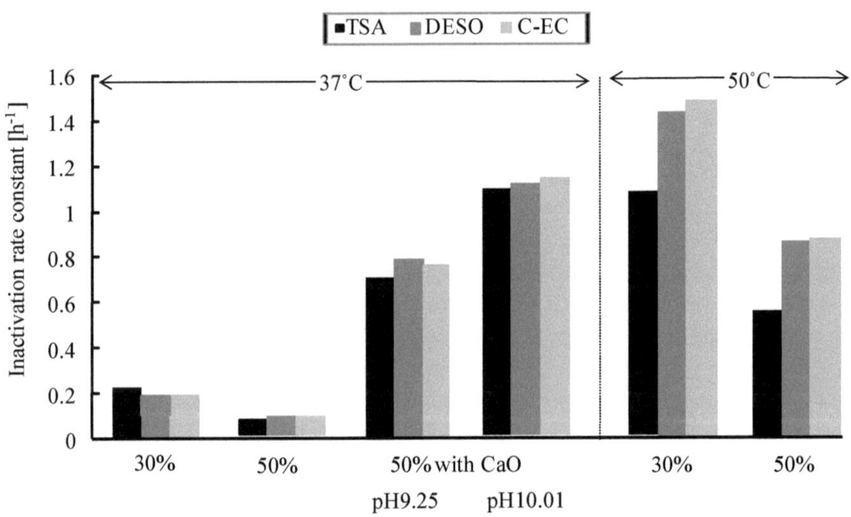

Fig. 5.14 Inactivation rate constants by TSA, DESO, and C-EC during the inactivation in various conditions. Reprinted from Kazama and Otaki (2011) with permission from the copyright holders, JSWE

5.6 Conclusions

The pathogenic risks from operating composting toilets are quite different from those in the case of flushing toilets. The pathogenic microorganisms might be reduced in the composting reactor according to the experiment results using model microbes such as *E. coli* and coliphage. It is thought that pathogenic viruses might have a higher tolerance than bacteria. As a result, more care was taken around pathogenic viruses than bacteria. The QMRA results also support that idea.

To assist hygienic safety, the optimal operational condition of composting toilets must be kept as shown in Fig. 5.10.

From the viewpoint of monitoring indicators, the coliform group is insufficient for estimating changes in viruses. Fecal enterococci are suggested to be better alternative indicators.

Understanding the mechanisms of inactivation might also help in considering the appropriate operational conditions to reduce the risks from various types of pathogenic microorganisms.

References

Burge WD, Colacicco D, Cramer WN (1981) Criteria for achieving pathogen destruction during composting. J Water Pollut Con F 53(12):1683–1690

Craig DL, Fallowfield HJ, Cromar NJ (2002) Comparison of decay rates of fecal indicator organisms in recreational coastal water and sediment. Water Supply 2(3):131–138

Haas CN, Rose JB, Gerba CP (1999) Quantitative microbial risk assessment. Wiley, USA

Hung RT (1993) The practical handbook of compost engineering. CRC Press, United States

Kazama S (2010) The inactivation mechanisms of bacteria and viruses in excrement treatment using sawdust. Ph.D thesis, Ochanomizu University (in Japanese)

Kazama S, Otaki M (2010) Fate of indicator bacteria for pathogenic viruses in composting toilet. J Jpn Soc Civil Eng Ser G 66(4):179–186. https://doi.org/10.2208/jscejg.66.179 (in Japanese)

Kazama S, Otaki M (2011) Mechanisms of bacteria and viruses inactivation in sawdust used in composting toilet. J Water Environ Technol 9(1):53–66. https://doi.org/10.2965/jwet.2011.53

Kazama S, Otaki M (2012) Inactivation mechanisms of *E. coli* in the sawdust used in composting toilet. J Water Environ Technol 10(4):363–374. https://doi.org/10.2965/jwet.2012.363

Kazama S, Tameike N, Nakagawa N, Otaki M (2011) A fate model of pathogenic viruses in a composting toilet based on coliphage inactivation. J Environ Sci 23(7):1194–1198. https://doi.org/10.1016/S1001-0742(10)60490-1

Nakagawa N, Yamagoshi K, Oe H, Otaki M (2003) Microbial risk assessment of the composting toilet. J Environ Syst Eng 748/VII-29:91–98 (in Japanese)

Nakagawa N, Oe H, Otaki M, Ishizaki K (2006) Application of microbial risk assessment on a residentially-operated bio-toilet. J Water Health 4(4):479–486

Otaki M, Nakagawa N, Akaishi F, Kubo K (2006) The fate of microorganisms in the composting toilet from the view point of hygienic risk. In: Proceedings of the 2nd international dry toilet conference. Tampere, Finland, vol 68, pp 16–19

Otaki M, Nakagawa N, Kazama S, Akaishi F, Tameike S (2007) Hygienic risk assessment and control in using composting toilet. In: Proceedings of the 5th international symposium on sustainable sanitation, pp 163–170

Prüss-Üstün A, Bos R, Gore F, Bartram J (2008) Safer water, better health. WHO Publishing, Geneva

Sobsey M, Battigelli DA, Handzel TR, Schwab KJ (1995) Male-specific coliphages as indicators of viral contamination of drinking water. AWWA Research Foundation, USA

WHO and Unicef (2012) Progress on drinking water and sanitation. WHO Publishing, Geneva

Chapter 6
Fate of Pharmaceuticals in Composting Process

Takashi Kakimoto and Yu Onoda

Abstract Composting toilet receives not only human excrement but also pharmaceuticals excreted in urine and/or feces. Since these compounds are physiologically active and its biodegradability seemed to be low, the fate and the effect of pharmaceuticals in composting process should be investigated. In our study, we focused on (1) the fate and the effect of antibiotics on the composting process of human excrement, (2) the factors affecting the degradation of antibiotics in composting reactor, and (3) the fate of pharmaceuticals other than antibiotics in composting process. Our results indicated that the amount of antibiotics expected from usually prescribed dose weaken the bacterial activity in composting reactor. Especially, effect of amoxicillin was strong. The composting reactor can treat antibiotics (amoxicillin, tetracycline, azithromycin). Though their degradation rates differed from antibiotics to antibiotics, the main factors affecting the degradation of antibiotics were found to be ammonia, phosphate, and hydroxyl ion. Since these concentrations in compositing toilet matrix were high in comparison with the other environmental media such as wastewater, this difference led the unique degradation mechanism in compositing toilet. As for pharmaceuticals other than antibiotics, laboratory-scale batch experiment showed that feces did not retard the degradation of pharmaceuticals, and among the pharmaceuticals examined, basic pharmaceuticals were treated faster than the acidic pharmaceuticals. But in the practically operated composting toilet, not only basic pharmaceuticals but also acidic and neutral ones were treated well. Detailed evaluation of the degradation mechanism and the degradation products would be necessary for future.

Keywords Abiotic degradation · Antibiotics · Bacterial activity
Composting process · Fate of pharmaceuticals

T. Kakimoto (✉)
Center for Environmental Science in Saitama, 914 Kamitanadare,
Kazo, Saitama 347-0115, Japan
e-mail: kakimoto.takashi@pref.saitama.lg.jp

Y. Onoda
Institute of Environmental Ecology, IDEA Consultants, Inc.,
1334-5 Riemon, Ooigawa-cho, Shida-gun, Shizuoka 421-0212, Japan

N. Funamizu (ed.), *Resource-Oriented Agro-sanitation Systems*,
https://doi.org/10.1007/978-4-431-56835-3_6

6.1 Fate of Pharmaceuticals in Human Body

Pharmaceuticals orally administered pass through the absorption; distribution; metabolism; and excretion process (Kato and Kamataki 2000). Schematic diagram of drug metabolism is shown in Fig. 6.1.

6.1.1 Absorption

The pharmaceuticals orally administered are mainly absorbed in upper small intestine. Although certain acidic pharmaceuticals can be absorbed in stomach, since the detention time in stomach is relatively short, and the surface area is much smaller than the intestine, it is rare for stomach to be the main absorption part of pharmaceuticals.

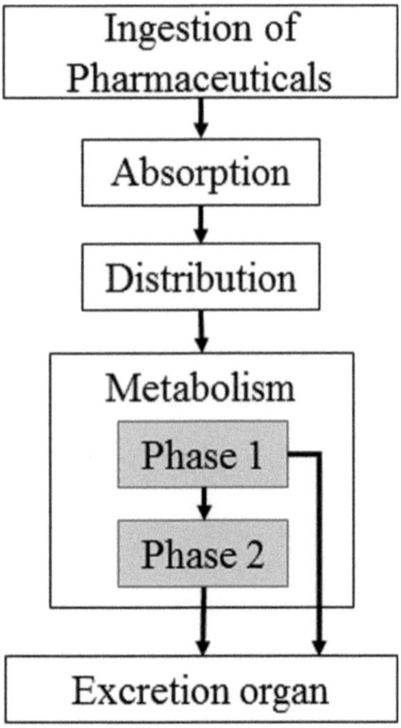

Fig. 6.1 Intake of pharmaceuticals and their absorption, metabolism, and excretion

6.1.2 Distribution

Pharmaceuticals absorbed are distributed by human circulation system. Most pharmaceuticals bind to plasma proteins in blood. Its ratio is called "percentage of protein binding." The binding of pharmaceuticals with protein is reversible, and only the free drugs (drugs without binding to protein) can pass through bio-membranes and show each pharmacological action at target organ.

6.1.3 Metabolism

Among the pharmaceuticals administered to human body, some kinds of pharmaceuticals are not absorbed and directly excreted (path A). Some kinds of the ones are also directly excreted due to its high hydrophilicity; however, most pharmaceuticals are hydrophobic and consequently metabolized into more hydrophilic form.

The pharmaceuticals absorbed are metabolized by phase 1 reaction. Among the phase 1 metabolites, a sort of pharmaceuticals that have enough hydrophilicity is excreted via urine and/or feces; however, most part of phase 1 metabolite is excreted after the phase 2 reaction.

Phase 1 reaction: Polar functional group such as –OH, –COOH, $-NH_3$, and so on is installed in this reaction. Due to this transformation, metabolites have less affinity for target site and result in less pharmacological activity and higher excretion ratio.

Phase 2 reaction: Polar functional group such as hydroxyl group, carboxyl group, amino group, thiol group in intact or metabolized pharmaceuticals is modified by the biological substances. This process is called phase 2 reaction or conjugation reaction. The phase 2 metabolite becomes more hydrophilic than the phase 1 metabolite.

6.1.4 Excretion

Pharmaceuticals are finally excreted as intact form or hydrophilic metabolites. In general, other than lung excretion, hydrophilic substances such as conjugates are easily excreted in comparison with hydrophobic substances. These pharmaceuticals and their metabolites are excreted into urine, bile, feces, breath, sweat, milk, and so on. One of the factors affecting the excretion pathway is pharmaceutical molecular weight. For instance, in the case of rat, the substance with more than 400 of molecular weight is preferentially excreted in bile (Feces). In contrast, low molecular weight conjugate is mainly excreted in urine. This is because glucuronide conjugate with high molecular weight is excreted in intestine via bile.

Table 6.1 shows the excretion pathway and the percentage of the unchanged drug and metabolite. Ordinary pharmaceutical dosage ranges from tens of mg to several thousands of mg a day. This consequently indicated that concentration of unchanged or metabolized pharmaceuticals in human excrement is calculated to be several mg day^{-1} levels. In general, it has been said that main excretion pathway of pharmaceutical is urine; however, as shown in Table 6.1, fecal excretion should not be neglected. Therefore, fate of pharmaceuticals and their metabolite is important in terms of stability of feces processing performance and management of pharmaceuticals in environment.

6.2 Effect of Pharmaceuticals in Composting Process of Feces

6.2.1 Single Exposure Effect

A wide variety of pharmaceuticals can enter the composting reactor. Pharmaceuticals administered to medical patients are generally excreted via urine and/or feces as a hydrophilic metabolite (Gilman et al. 2001). Among several pharmaceuticals, the dosage of antibiotics is generally high (e.g., the range is up to 6 g/day), and its excretion ratio as intact form is also high (up to 90% of dose) (Hirsch et al. 1999). This means that the excreted pharmaceuticals, especially antibiotics, may reduce microbial activity and may consequently lead to a delay in the composting process of feces in a composting toilet.

Since feces composting process is a kind of aerobic biodegradation process, we adopt the oxygen utilization rate (OUR) as an indicator of bacterial activity, and feces biodegradation test with and without pharmaceuticals were conducted in batch reactor.

Chemical structures of the pharmaceuticals examined in this study are listed in Table 6.2. They are the pharmaceuticals that seemed to be used for chronic disease and antibiotics. Among antibiotics, since penicillins, tetracyclines, macrolides, and fluoroquinolones antibiotics are frequently used, amoxicillin, tetracycline, azithromycin, levofloxacin were selected as representatives of them, respectively. Concentrations of the pharmaceuticals except antibiotics correspond to the 100-day equivalent load that were calculated based on their metabolic and excretion data. Initial concentration of each antibiotic corresponds to the 1-day load of them.

Figure 6.2 shows the effect of antibiotics on the composting process of feces. It was found that among the four kinds of antibiotics, penicillin group antibiotics amoxicillin showed the strongest effect on the composting process though amoxicillin concentration in composting reactor declined quickly. The maximum biodegradation rate (peak height of the OUR profiles) of amoxicillin was about 20%, and its peak area was apparently small in comparison with the control. This means that feces added to the reactor was not treated completely. Effect of other

Table 6.1 Metabolites of pharmaceuticals and its excretion pathway

	Unchanged	Metabolite	Route	References
Antibiotics				
Amoxicillin	60%		Urine	Todd and Benfield (1990)
Ampicillin			Urine	Haginaka and Wakai (1987)
Azithromycin	60%		Feces	Moffat et al. (2004)
Cefaclor	55.6–63.9%		Urine	Otomo et al. (1985)
Ciprofloxacin	70%		Both	Paton and Reeves (1988)
Norfloxacin	54%		Both	Paton and Reeves (1988)
Ofloxacin	85%		Both	Paton and Reeves (1988), Wong (1990)
Solcillin	44–53.9%		Urine	Hara and Mita (1969)
Tetracycline	80–90%		–	Hirsch et al. (1999)
Sulfadiazine	50%	22%	Urine	Andreasen et al. (1978)
Sulfamethoxazole	30–40%	75–80%	Urine	Welling et al. (1973), Rieder (1973)
Sulfamonomethoxine	14.1%	57%	Urine	Vree et al. (1990)
Blood lipid regulator				
Atorvastatin		89.4%	Feces	Ogami et al. (2005)
Simvastatin		72%	Feces	Vickers et al. (1990), Otawa et al. (1989)
Rosuvastatin	77%		Feces	Martin (2004)
Bezafibrate	50%	20%	Urine	Abshagen et al. (1979)
Fenofibrate	16%	50%	Both	Weil et al. (1990)
Gemfibrozil	5–70%		Urine	Zimetbaum et al. (1991), Malloy and Kane (2001)
Analgesics				
Ibuprofen		80%	Urine	Brocks and Jamali (1999)
Naproxen	1.1%		Urine	Andersen (1992)
Antihypertensives				
Atenolol	87%	12%	Both	Reeves et al. (1988)
Nadolol	11–14%		Urine	Takeuchi (1990)
Rosaltan	3.2–4.1%	6.7–7.9%	Feces	Nakajima et al. (1995)
Benazepril	0–1%	13–21%	Feces	Nakajima et al. (1991)
Captopril	50%	60%	Urine	Migdalof et al. (1984)
Amlodipine	9%	43%	Both	Beresford et al. (1988)

class of antibiotics (azithromycin, tetracycline, levofloxacin) was found to be weak in comparison with amoxicillin. The peak height of each OUR profiles is more than 60% of control, and bacterial respiration did not disappear. Thus, as far as this single exposure test shows, the effect of antibiotics but amoxicillin seems to be small practically. It was not clear that the reasons why the strength of antibacterial

Table 6.2 Pharmaceuticals examined in batch test

	Amitriptyline	Imipramine	Diclofenac	Ketoprofen
Antidepressant Analgesics	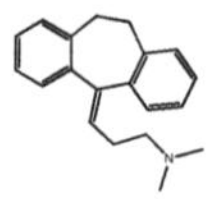	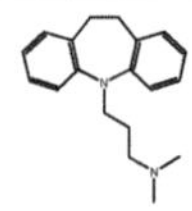	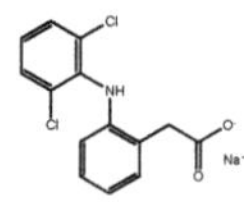	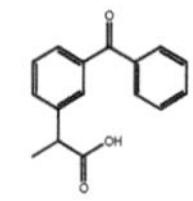
	Atenolol	Metoprolol	Bezafibrate	Carbamazepine
Beta-blocker Lipid regulator Antiepileptics	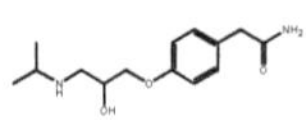	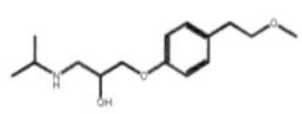	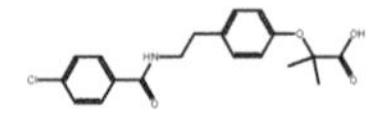	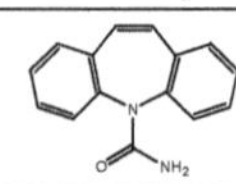
	Amoxicillin	Tetracycline	Azithromycin	Levofloxacin
Antibiotics	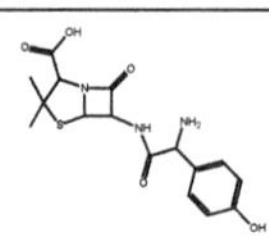	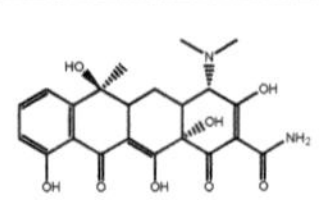	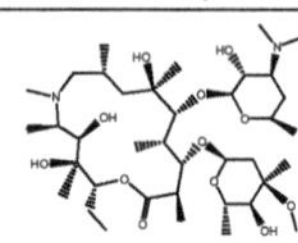	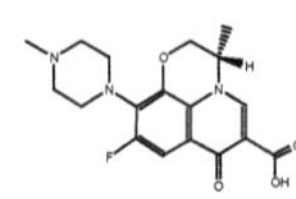

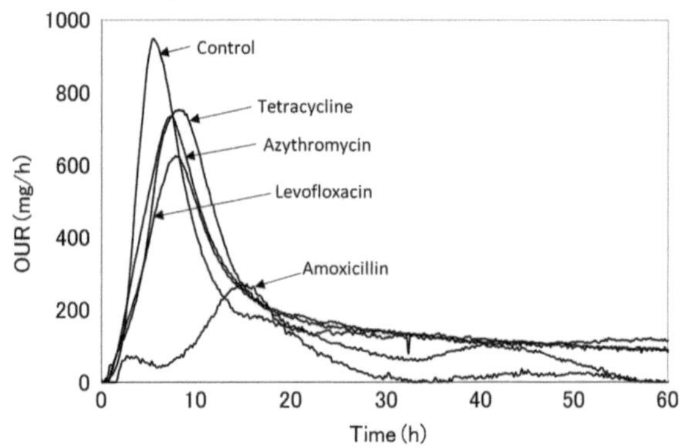

Fig. 6.2 Comparison of the effect of antibiotics on the composting process

activity varies depending on the type of antibiotics; however, antibiotics' mode of action, relationship between the antibacterial spectrum and bacterial species in composting matrix, and interactions between antibiotics and organic/inorganic matter (Ross and Riley 1992; Wetzstein et al. 1997) in composting reactor would affect the antibacterial activity.

In contrast, pharmaceuticals other than antibiotics did not show any significant effect on the bacterial respiration. The dosage to the composting reactor corresponded to the 100 days. Since some of these pharmaceuticals are for the chronic illness, 100-day equivalent dosage might be not enough for the long-term performance evaluation. However, these classes of pharmaceuticals are not likely to influence the composting process of feces.

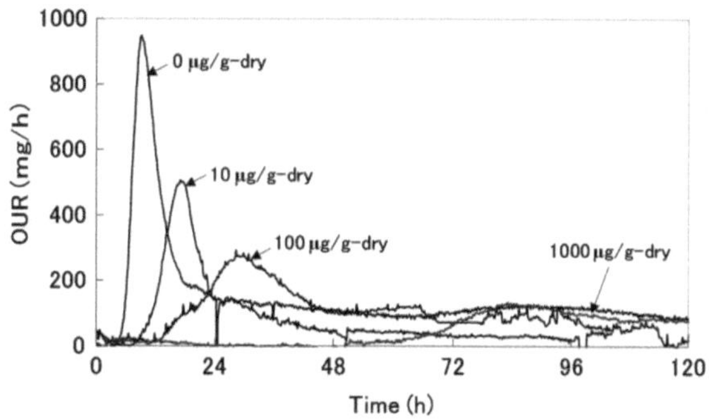

Fig. 6.3 Effect of amoxicillin on the composting process with variation of amoxicillin concentration. Operational condition: temperature 50 °C, *F/S* ratio 15%, initial moisture content (60%). Reprinted from Kakimoto et al. (2007), with permission from Elsevier

Figure 6.3 shows the effect of amoxicillin on the feces composting process with variation of initial amoxicillin concentration. Here, 100 µg/g-dry corresponds to the case where the one day load of amoxicillin is applied. This data indicate that even in the load of one-tenth (10 µg/g-dry), bacterial activity was significantly inhibited. In the case of 1000 µg/g-dry, which corresponds to the ten times load, there is no apparent peak. This means that the function of composting toilet was lost.

6.2.2 Reactivation of Amoxicillin-Dosed Composting Reactor by Intermittent Feeding of Feces

As seen in Fig. 6.3, the effect of amoxicillin is strong, and no automatic reactivation of bacteria was observed within 5 days from the start of the test. Therefore, it became important to find a way to reactivate the bacteria in the matrix. Loss of a processing function may be attributed to the following assumptions: (1) viable bacterial count that was responsible for the composting process decreased due to the amoxicillin dosage; (2) though amoxicillin concentration declined quickly, its degradation product may also have antibiotic effect.

It has been reported that the microbial flora in the composting toilet matrix consisted mainly of *Bacillus*, *Arthrobacter*, *Bifidobacterium,* and the *Halomonas* group (Imai 2003), and *Bifidobacterium* group of bacteria are a predominant species in intestinal microflora (Imabori and Yamakawa 1984). So it seemed to be possible to reactivate the bacteria in composting reactor by feeding of feces. Figure 6.4 shows the results of the intermittent feeding tests (10 and 1000 µg/g-dry). When the exposure of amoxicillin is 10 µg/g-dry, respiratory activity decreased in the first day, but from second day, OUR became high enough. This indicates that the feeding of feces can reactivate the composting process. After 96 h, peak height of OUR became small. This is not the effect of antibiotics but the effect of the excess moisture content of the matrix by addition of feces. In the case of 1000 µg/g-dry, unlike the 10 µg/g-dry case, intermittent feeding of feces could not reactivate the bacterial activity within a week although it eventually recovered.

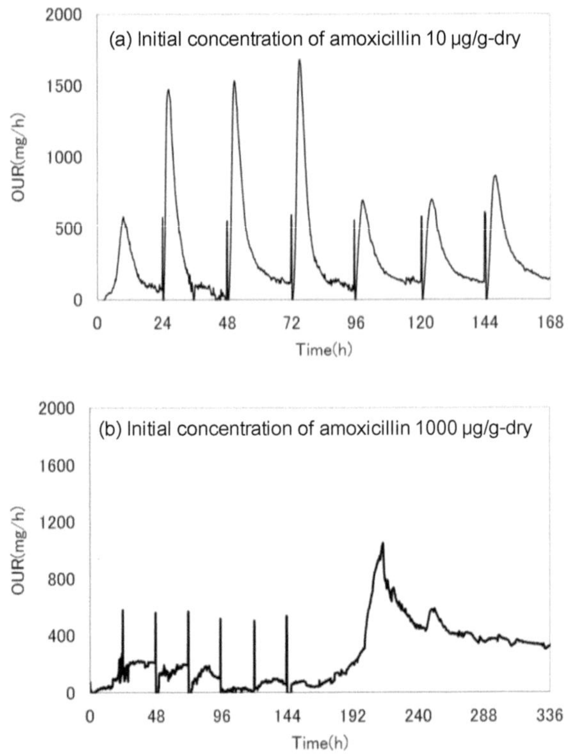

Fig. 6.4 OUR profiles of intermittent feces feeding test (every 24 h up to 144 h)

6.2.3 Role of Feces in the Composting Process

If the feces had acted simply as a bacterial carrier, it might be natural to think that not only the 10 μg/g-dry system but also the 1000 μg/g-dry system would show respiratory activity. These results implied that the feces acted as a substrate rather than as an actual bacterial carrier and that the differences of the time needed for the reactivation might be related to the difference of the initial viable bacterial counts and their activity. To confirm this, an experiment was performed using sterilized sawdust as a matrix. Since the bacteria living in the compost originally were killed by sterilization, no OUR peaks would be observed if the role of feces was simply a substrate for the bacteria in compost. Actually, OUR peak was not observed (Fig. 6.5). This result indicated that feces act as a bacterial substrate, and that the bacteria originally present in compost were different from the ones in feces.

Therefore, the initial bacterial count in toilet matrix should be kept high for rapid reactivation of the bacteria in the composting reactor against amoxicillin dosage. Although it depends on the excretion pathway of antibiotics, as for amoxicillin, urine separation would be an effective way for the stable operation of the composting process.

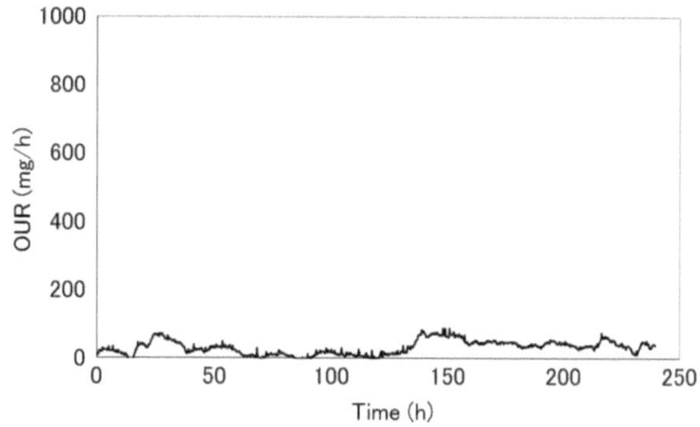

Fig. 6.5 OUR profile of the composting process using γ-ray sterilized matrix. Operational condition: temperature 50 °C, *F/S* ratio 15%, initial moisture content (60%). Reprinted from Kakimoto et al. (2007), with permission from Elsevier

6.3 Degradation of Antibiotics in Composting Process

In the Sect. 6.2, the effect of antibiotics on the composting process of feces was discussed, and it was found that although antibiotic concentration declined quickly, adverse effect of antibiotics on the feces decomposition process was observed. Therefore, elucidation of the fate and degradation mechanism of antibiotics in composting toilet seemed to be important for enabling the stable and sound operation of the composting type toilet.

6.3.1 Factors Affecting the Degradation of Antibiotics

Because composting process is a biological process, it might be natural to think that antibiotics are treated by bacteria not sensitive to the antibiotics. But the experimental results that compared the concentrations of four target antibiotics in toilet matrix and in the γ-ray sterilized one indicated that levofloxacin was persistent in composting process and that other antibiotics were treated. However, that was not biological process. Some literature says that chemical structure that expresses the antimicrobial activity can be decomposed by inorganic ions. For instance, beta-lactam ring in amoxicillin that shows the antibiotic effect can be degraded by ammonia, phosphate, hydroxyl ion. It has been reported that tetracyclines are inactivated by phosphate, ammonia, and hydroxyl ion (Kawabata et al. 1960; Tanaka and Nakamura 1995; Committee on analytical method of residual veterinary drugs 2000). Although there are a lot of reports about the stability of macrolides in acidic condition (Fiese and Steffen 1990), there are no reports about the stability of them in basic and high salinity conditions. Since urine contains phosphate and urea, phosphate, ammonia, and hydroxyl ion are present in high concentration in sawdust matrix. It is not apparent that azithromycin is sensitive to these ions; stability of azithromycin against the three ions was also investigated.

We monitored the concentration of three kinds of antibiotics (amoxicillin, tetracycline, azithromycin) in solutions with various phosphate, ammonia, and pH condition. This result indicated that amoxicillin and tetracycline were actually degraded by ammonia, phosphate, and hydroxyl ion. Azithromycin was degraded by phosphate and hydroxyl ion. As a result of the reaction kinetic analysis (Kakimoto and Funamizu 2007, 2006), amoxicillin and tetracycline decomposition in composting reactor was subject to the concentration of phosphate, ammonia, and pH level in the toilet matrix. As for azithromycin, though degradation profile could not be explained completely by the effect of pH and phosphate, their effect seemed to be large.

6.3.2 Antimicrobial Activity of Degradation Product

Since the main degradation reaction in compost and in the buffer solution (see Sect. 6.3.1) seemed to be the same, the final products would be also the same. Hence, to evaluate the antimicrobial activity of the degraded product in composting toilet matrix, we used the degradation product that was prepared in the pH 9.0–1.0 M of ammonium phosphate buffer stored in water bath (50 °C) for 5 days, for the minimum inhibitory concentration (MIC) test. In the MIC test, we prepared the fluid media (Mueller-Hinton media) containing 400 μg/L of antibiotics and the one containing the degradation product corresponding to the 400 μg/L of intact antibiotics. These two fluid media were diluted twice in sequence with inoculums of composting bacteria extracted from compost. After the incubation for 24 h, antimicrobial activity was evaluated by the presence or absence of turbidity. The result of MIC test showed that though 1.5 μg/L of amoxicillin shows the antimicrobial activity against the bacteria, the degradation product corresponding to 400 μg/L did not show any antimicrobial activity. Similar results were also confirmed in the case of tetracycline and azithromycin. Therefore, due to the unique environment of the toilet matrix, antibiotics can be inactivated.

6.4 Degradation of Other Pharmaceuticals in Composting Process

6.4.1 Effect of Feces Loading Ratio and Degradation Characteristics in Composting Process

Variable conditions of our composting toilet are organic loading ratio (*F*/*S* ratio), temperature, and water content, and they strongly affect the treatment performance (Lopez et al. 2004; Lopez and Funamizu 2005). During the biological treatment of xenobiotics such as pharmaceuticals, it is important to assess that the easily biodegradable organic matter in feces competitively interferes the degradation of pharmaceuticals or not. This study assessed the effect of coexisting organic matter in feces on the degradation of pharmaceuticals in composting process.

Figure 6.6 shows the OUR and the reduction of the pharmaceuticals in the composting process with different *F*/*S* levels (5, 15 and 20%) in batch system. The OUR profile indicated the degradation rate of feces, and it was apparent that the bacterial activity was high at the early stage of the process and beyond that small bacterial activity was observed. In the case of 5% *F*/*S* ratio, all pharmaceuticals (clofibric acid, pravastatin, bezafibrate, atenolol, metoprolol) showed almost the same reduction profiles, and their concentrations at 72 h were almost the same as the initial concentrations. However, in the case of 15 and 20%, the concentrations of the pharmaceuticals at 48 h were significantly lower for atenolol and metoprolol, and almost the same or slightly lower than those at 72 h in 5% *F*/*S* ratio for other acidic pharmaceuticals (clofibric acid, pravastatin, bezafibrate). Although the detailed mechanism of the change in the degradation rate was not clear, it was found that the easily biodegradable organic matter did not affect the degradation rate of the acidic pharmaceuticals. However, as the *F*/*S* ratio increased, degradation rate of basic pharmaceuticals increased.

6.4.2 Pharmaceuticals in Practically Operated Composting Toilet

We monitored the concentrations of target pharmaceuticals (Table 6.3) in composting toilet that is practically used by old couple. The target pharmaceuticals except ketoprofen and valsartan were manually added to the composting reactor once a week, and it was continued for about one year. The amount of each pharmaceuticals' dose to the composting reactor was determined in the maximum prescribed dose and metabolic data (Table 6.4). As for ketoprofen and valsartan, because the owner of the composting toilet frequently used them, we calculate the amount of dose to the composting reactor by hearting their amount of intake and metabolic data.

Figure 6.7 shows the concentration transition and decomposition rate of target compounds in composting toilet. We calculated the decomposition rate by dividing the amount of target compounds remaining in the reactor by the cumulative amount of those ones. For ketoprofen and valsaltane, we adopted the amount of intake told by the owner instead of cumulative amount of dose.

The monitoring data show that most pharmaceuticals except ketoprofen were decomposed in composting toilet. Especially, the concentrations of acetaminophen, bezafibrate, pravastatin, atenolol, and clarithromycin in compost samples were very low relative to those cumulative concentrations. This result indicates that these compounds had high degradability in composting processes. Concentrations of ibuprofen, mefenamic acid, and clofibric acid rose gradually until the 91st day. However, after the 91st day, when we reduced the additive amounts of them to

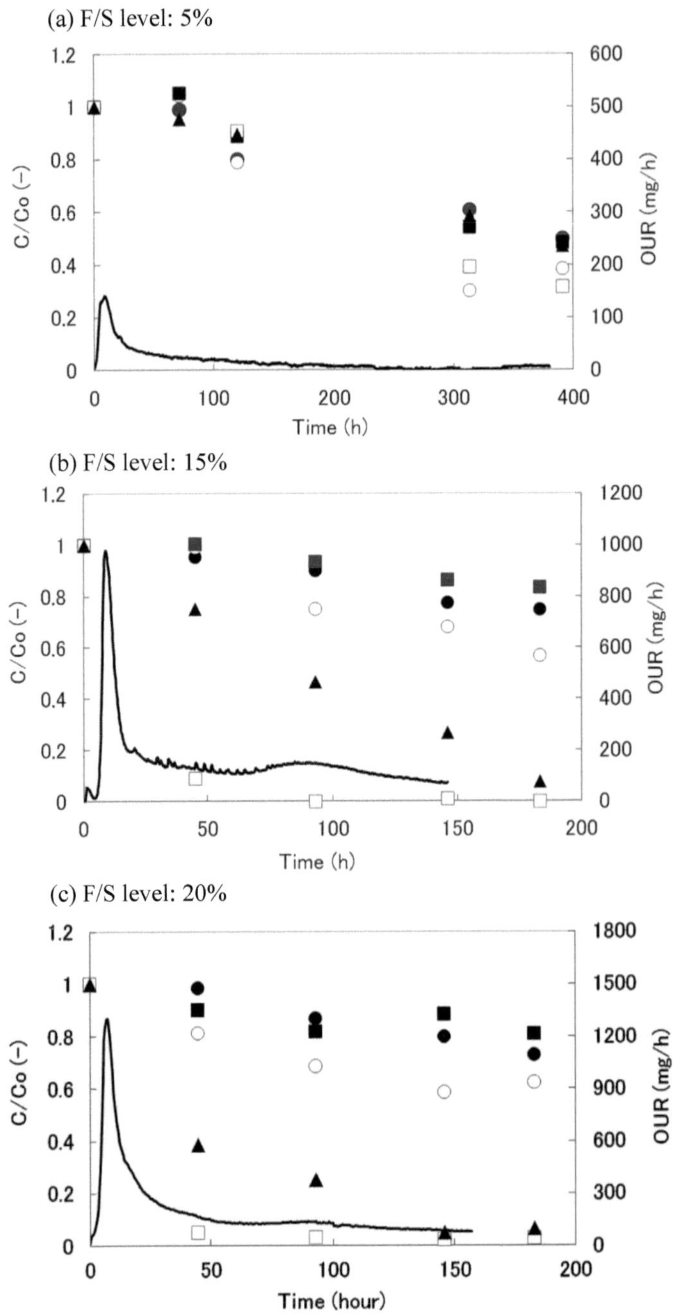

Fig. 6.6 Concentration of pharmaceuticals in batch test (filled circle: clofibric acid, circle: pravastatin, filled rectangle: bezafibrate, rectangle: atenolol, filled triangle: metoprolol)

Table 6.3 Target compounds for the experiment

Compounds	pKa	Medical use
Acetaminophen[a]	9.38	Antipyretic analgesics
Ibuprofen[a]	4.25	Antipyretic analgesics
Mefenamic acid[a]	4.2	Antipyretic analgesics
Ketoprofen[a]	3.9	Antipyretic analgesics
Indomethacin[a]	4.2	Anti-inflammatory agent
Bezafibrate[a]	3.4	lipid-lowering agent
Pravastatin sodium[a]	4.6	lipid-lowering agent
Clofibric acid[a]		Metabolites of lipid-lowering agent
Valsartan[a]	–	Antihypertensive agent
Carbamazepine[b]	–	Antiepileptic agent
Crotamiton[b]	–	Dermatological preparation
Propranolol[c]	9.45	Anti-arrhythmic drug
Atenolol[c]	9.6	Anti-arrhythmic agent
Metoprolol[c]	9.7	Anti-arrhythmic agent
Disopyramide[c]	8.36	Anti-arrhythmic agent
Sulpiride[c]	9.12	Psychoneurotic agent
Clarithromycin[c]	8.99	Antibacterial agent

[a]Acidic compounds
[b]Neutral compounds
[c]Basic compounds

one-fifth, their concentration was no longer rising. So, these compounds are degradable in composting process. But these pharmaceuticals would accumulate in the reactor when they are used as normal prescription due to their low degradability. Although indomethacin, carbamazepine, crotamiton, propranolol, metoprolol, disopyramide, and sulpiride tend to accumulate moderately, the decomposition rate of them was in the range of 40–60%. Valsartan was also accumulated moderately. As for ketoprofen, although the detection limit of the determination method in this study was higher than the calculated concentration of intact ketoprofen in the toilet matrix, we could monitor a slowly increasing trend of intact ketoprofen. Urinary excretion rates of intact ketoprofen and its metabolites are 2.7 and 60%, respectively, and main metabolites in urine were glucuronic acid conjugate that is easily deconjugated during biological treatment process such as activated sludge process. Thus, this result might imply that the ketoprofen metabolite (glucuronic acid conjugate) was deconjugated during the composting process, and the deconjugated ketoprofen has come to be detected.

Table 6.4 Load of target compounds in dry toilet

Compounds	Load of target compounds mg/week (until the 91st day)	Load of target compounds mg/week (after the 91st day)
Acetaminophen	300	300
Ibuprofen	1000	200
Mefenamic acid	2000	400
Ketoprofen[a]	–	–
Indomethacin	2	2
Bezafibrate	1000	200
Pravastatin sodium	10	10
Clofibric acid	1000	200
Valsartan[a]	–	–
Carbamazepine	40	40
Crotamiton	7	7
Propranolol	2	2
Atenolol	300	300
Metoprolol	40	40
Disopyramide	900	900
Sulpiride	600	600
Clarithromycin	10	10

[a]Ketoprofen and Valsartan were administered to the owner. Therefore, manual input was not conducted

We monitored the 17 kinds of pharmaceuticals that consist of 9 acidic, 2 neutral, and 6 basic pharmaceuticals in practically operated composting toilet for a year. As a result, though it was difficult to classify the degradability depending on the physicochemical properties of selected pharmaceuticals, it was shown that the processing function of feces was kept well even if pharmaceuticals including antibiotics are added and that composting toilet can degrade the pharmaceuticals. However, challenges still remain in terms of the management of pharmaceuticals and its degradation product remaining in the toilet matrix.

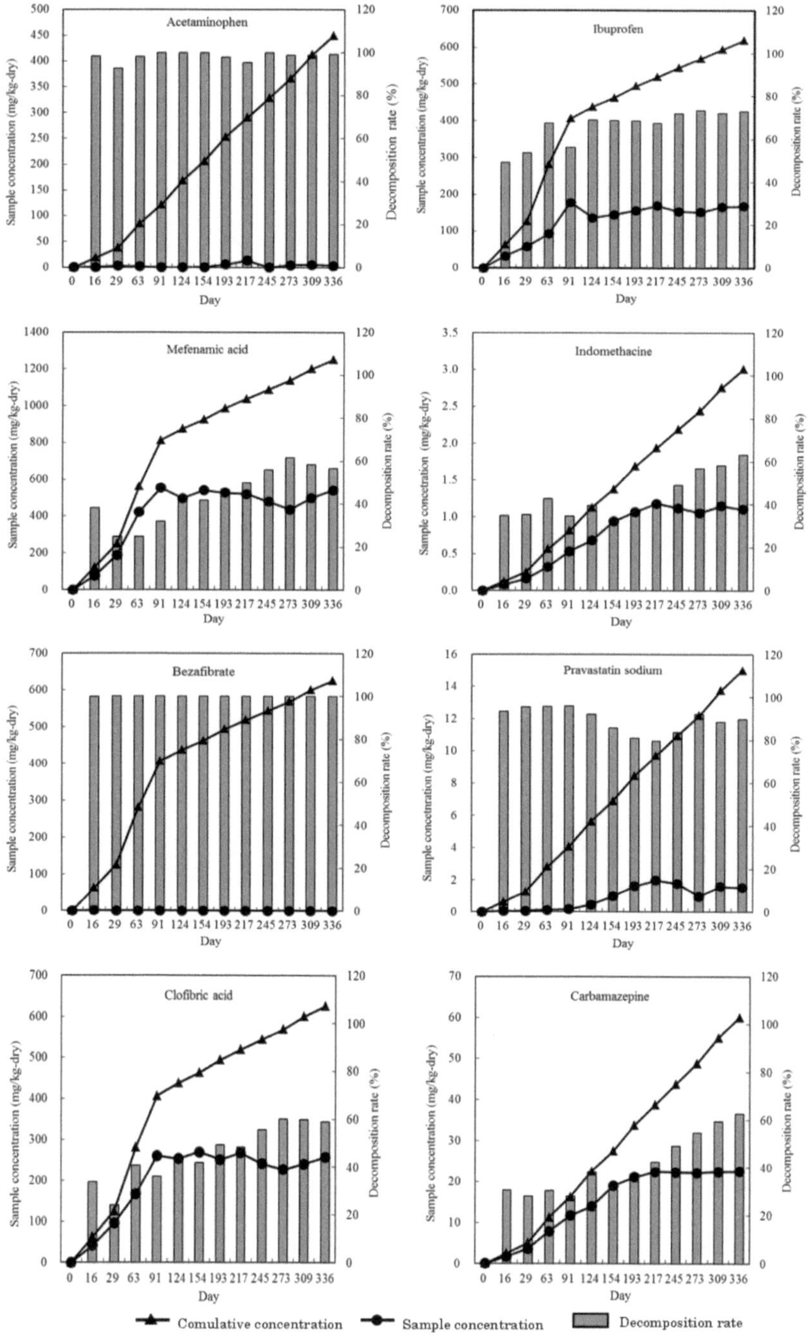

Fig. 6.7 Decomposition of pharmaceuticals during composting process

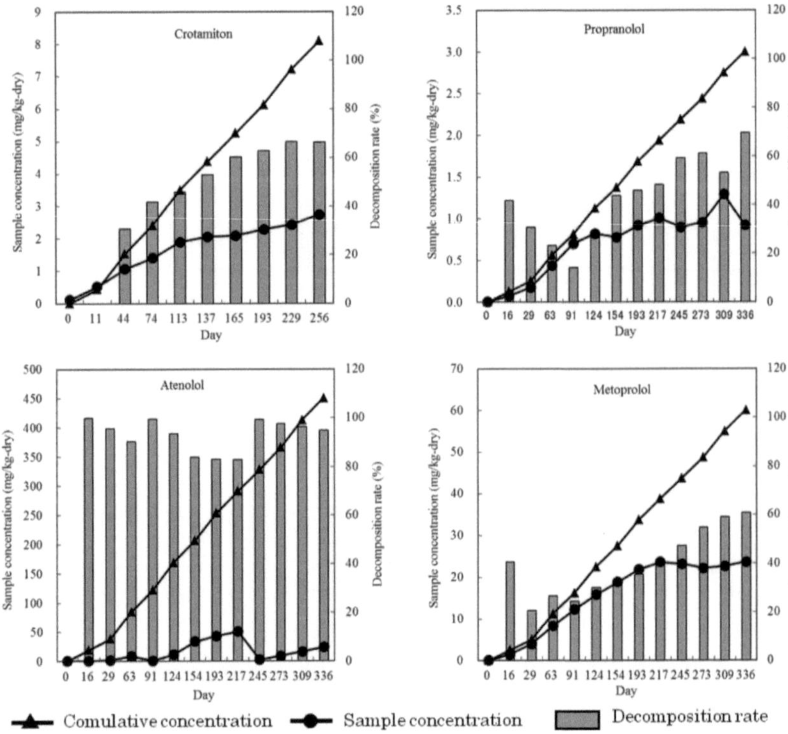

Fig. 6.7 (continued)

References

Abshagen U, Bablok W, Koch K, Lang PD, Schmidt HA, Senn M, Stork H (1979) Disposition pharmacokinetics of bezafibrate in man. Eur J Clin Pharmacol 16:31–38. https://doi.org/10.1007/BF00644963

Andersen JV (1992) Simultaneous quantitative determination of naproxen, its metabolite 6-O-desmethylnaproxen and their five conjugates in plasma and urine samples by high-performance liquid chromatography on dynamically modified silica. J Chromatogr 577:325–333

Andreasen F, Elsborg L, Husted S, Thomsen O (1978) Pharmacokinetics of sulfadiazine and trimethoprim in man. Europ J Pharmacol 14:57–67. https://doi.org/10.1007/BF00560259

Beresford AP, McGibney D, Humphrey MJ, Macrae PV, Stopher DA (1988) Metabolism and kinetics of amlodipine in man. Xenobiotica 18(2):245–254. https://doi.org/10.3109/00498258809041660

Brocks SR, Jamali F (1999) The pharmacokinetics of ibuprofen in animals and humans. In: Ibuprofen: a critical bibliographic review. Taylor and Francis, UK, pp 87–142

Committee on analytical methods of residual veterinary drugs (2000) Analytical method of veterinary drugs in livestock and marine products: part 7. Food Sanitation Res 50(10):7–18

Fiese EF, Steffen SH (1990) Comparison of the acid stability of azithromycin and erythromycin A. J Antimicrob Chemother 25(Suppl. A):39–47

Gilman AG, Hardman JG, Limbrid LE (2001) Goodman & Gillman's the pharmacological basis of therapeutics, 10th edn. McGraw-Hill, New York

Hara S, Mita T (1969) Absorption and excretion of solcillin. J New Remedies Clin 18(11):1602–1641

Haginaka J, Wakai J (1987) Liquid chromatographic determination of ampicillin and its metabolites in human urine by postcolumn alkaline degradation. J Pharm Pharmacol 39:5–8

Hirsch R, Ternes T, Haberer K, Kratz K (1999) Occurrence of antibiotics in the aquatic environment. Sci Total Environ 225:109–118. https://doi.org/10.1016/S0048-9697(98)00337-4

Imabori K, Yamakawa T (1984) Dictionary of biochemistry. Tokyo Kagaku Dozin, Tokyo

Imai Y (2003) Verification analysis of microflora in biotoilet. Bachelor course thesis of Hokkaido University, Japan

Kakimoto T, Funamizu N (2006) Effect of phosphate, ammonia and pH on the degradation of antibiotics in the composting toilet. Environ Eng Res 43:429–437

Kakimoto T, Funamizu N (2007) Factors affecting the degradation of amoxicillin in composting toilet. Chemosphere 66(11):2219–2224. https://doi.org/10.1016/j.chemosphere.2006.09.060

Kakimoto T, Osawa T, Funamizu N (2007) Antibiotic effect of amoxicillin on the feces composting process and reactivation of bacteria by intermittent feeding of feces. Biores Technol 98(18):3555–3560. https://doi.org/10.1016/j.biortech.2006.11.029

Kato and Kamataki (2000) Drug-metabolism—as a medical pharmacy and toxicology, 2nd edn. Tokyo Kagaku Dojin, Tokyo

Kawabata T, Sakaguchi G, Nakamura Y, Akano T (1960) Factors affecting the decomposition of oxytetracycline and chlortetracycline. J Antibiot Ser A 13(3):180–185

Lopez Zavala MA, Funamizu N, Takakuwa T (2004) Modeling of aerobic biodegradation of feces using sawdust as a matrix. Water Res 38:1327–1339. https://doi.org/10.1016/j.watres.2003.10.028

Lopez Zavala MA, Funamizu N (2005) Effect of moisture content on the composting process in a biotoilet system. Compost Sci Util 13:208–216

Malloy MJ, Kane JP (2001) Agents used in hyperlipidemia. In: Katzung BG, Norwalk C (eds) Basic and clinical pharmacology. Mc-Graw-Hill Publisher, New York

Martin PD (2004) Metabolism, excertion, and pharmacokinetics of rosuvastatin in health adult male volunteers. Jpn Pharmacol Ther 32:S18–S19

Migdalof BH, Antonaccio MJ, McKinstry DN, Singhvi SM, Lan SJ, Egli P, Kripalani KJ (1984) Captopril: pharmacology, metabolism, and disposition. Drug Metab Rew 15(4):841–869. https://doi.org/10.3109/03602538409041080

Moffat AC, Osselton MD, Widdop B (2004) Clarke's analysis of drugs and poisons. Pharmaceutical Press, London

Nakajima M, Kanamaru M, Uno K, Yoshimoto T (1991) Phase I study of CGS 14824A (Benazepril HCl) A new angiotensin converting enzyme inhibitor. (1st report). Single oral administration in healthy volunteers. J Clin Ther Med 7(5):949–974

Nakajima M, Uematsu T, Kanamaru M, Takayama F, Kamei K (1995) Phase I study of MK-954, a new angiotensin II receptor antagonist. Results of single oral administration. Jpn J Clin Pharmacol Ther 26(3):671–684. doi:https://doi.org/10.3999/jscpt.26.671

Otomo M, Ito M, Iida M, Yasuda M, Sonoyama T, Matsuda S (1985) Phase I clinical studies of S6472 (sustained release preparations of cefaclor). J Antibiot 38(3):822–832. https://doi.org/10.11553/antibiotics1968b.38.822

Ogami K, Kouke A, Komori K, Suzuki Y, Yasojima M (2005) Hitoyou iyakuhin butsuri kagaku teki jyouhousyu (Physical and chemical information collection of medical products for human use). Gihodo, Tokyo

Otawa M, Uchiyama N, Saito Y, Suzuki A, Tanno C, Nakayama N, Goto Y (1989) Phase I studies on MK-733, an inhibitor of HMG-CoA reductase. I: Single dose and multiple dose studies in healthy volunteers. J Clin Ther Med 5(6):1123–1140

Paton JH, Reeves DS (1988) Fluoroquinolone antibiotics microbiology, pharmacokinetics and clinical use. Drugs 36:193–228. https://doi.org/10.2165/00003495-198836020-00004

Reeves PR, McAinsh J, McIntosh DA, Winrow MJ (1988) Metabolism of atenolol in man. Xenobiotica 8(5):313–320. https://doi.org/10.3109/00498257809060956

Rieder J (1973) Metabolism and techniques for assay of trimethoprim and sulfamethoxazole. J Infect Dis 128:S567–S573

Ross DL, Riley CM (1992) Physicochemical properties of the fluoroquinolone antimicrobials. III. Complexation of lomefloxacin with various metal ions and the effect of metal ion complexation on aqueous solubility. Int J Pharm 87:203–213. https://doi.org/10.1016/0378-5173(92)90244-V

Takeuchi Y (1990) Drugs for abnormal pulse. The pharmacokinetics of antiarrhythmic drugs. J Pract Pharm 41(7):987–992

Tanaka N, Nakamura S (1995) Broad outline of antibiotics, 4th edn. Tokyo University Press, Tokyo

Todd PA, Benfield P (1990) Pharmacokinetic properties and therapeutic use, amoxicillin/ clavulanic acid an update of its antibacterial activity. Drugs 39(2):264–307

Vickers S, Duncan CA, Vyas KP, Kari PH, Arison B, Prakash SR, Ramjit HG, Pitzenberger SM, Stokker G, Duggan DE (1990) In vitro and in vivo biotransformation of simvastatin, an inhibitor of HMG CoA reductase. Drug Metab Dispos 18(4):476–483

Vree TB, Kolmer EW, Martea M, Bosch R, Shimoda M (1990) High-performance liquid chromatography of sulphadimethoxine and its N1-glucuronide, N4-acetyl and N4-acetyl-N1-glucuronide metabolites in human plasma and urine. J Chromatogr 526:119–128

Weil A, Caldwell J, Strolin-Benedetti M (1990) The metabolism and disposition of 14C-fenofibrate in human volunteers. Drug Metab Dispos 18(1):115–120

Welling PG, Craig WA, Amidon GL, Kunin CM (1973) Pharmacokinetics of trimethoprim and sulfamethoxazole in normal subjects and in patients with renal failure. J Infect Dis 28:S556–S566

Wetzstein HG, Schmeer N, Karl W (1997) Degradation of the fluoroquinolone enrofloxacin by the brown rot fungus Gloeophyllum striatum: identification of metabolites. Appl Environ Microbiol 63:4272–4281

Wong FA (1990) The metabolism of ofloxacion in humans. Drug Metab Dispos 18(6):1103–1104

Zimetbaum P, Frishman WH, Kahn S (1991) Effects of gemfibrozil and other fibric acid derivatives on blood lipid and lipoproteins. J Clin Pharmacol 31:25–37. https://doi.org/10.1002/j.1552-4604.1991.tb01883.x

Chapter 7
Fate of Water in Composting Toilet

Ryusei Ito, Shingo Ogawa and Naoyuki Funamizu

Abstract In composting toilets using sawdust as a matrix, drying the mixture of sawdust and urine is one of the most important processes. This is because the moisture content of the mixture is an important factor in its biodegradation process and must be kept at about 65%. In order to operate the toilet stably, appropriate design and operation guidelines are necessary, but this requires the drying characteristics of the sawdust matrix. Therefore, in this study, we studied the drying characteristics of sawdust using two devices of different sizes, laboratory-scale and full-scale composting reactors, and made simple design and formulation of operational guidelines from the viewpoint of drying. As a result, it was found that the drying speed of sawdust mixture can be described by the dry model with modified vapor pressure at surface. Moreover, from the laboratory-scale drying experiments, it was found that there is a possibility of obtaining the drying rate of the toilet on a full scale by considering the flow of dry air. Drying can be accelerated by mixing, and its effect was increased as the temperature of the reaction tank increased. From these results, it became possible to design composting toilets and to formulate operational guidelines.

Keywords Design and operating guidelines · Drying of sawdust matrix
Vapor pressure and dry model

R. Ito (✉) · S. Ogawa
Graduate School of Engineering, Hokkaido University,
Kita13-Nishi 8, Kita-Ku, Sapporo City 060-8628,
Hokkaido, Japan
e-mail: ryuusei@eng.hokudai.ac.jp

N. Funamizu
Graduate School of Global Food Resources, Hokkaido University,
Kita-9, Nishi-9, Kita-Ku, Sapporo 060-8589, Japan

N. Funamizu
Research Institute of Humanity and Nature, Motoyama Kamigamo,
Kita-Ku, Kyoto 603-8047, Japan

N. Funamizu (ed.), *Resource-Oriented Agro-sanitation Systems*,
https://doi.org/10.1007/978-4-431-56835-3_7

7.1 Introduction

In composting reaction, if the moisture content of matrix is higher than the optimum value, oxygen supply is not performed, and then anaerobic fermentation occurs which causes malodor generation (Lopez and Funamizu 2005). Conversely, if the moisture content is too low, the rate of biodegradation reaction will be slow. Therefore, it is necessary to keep the moisture content within the sawdust matrix at 50–65% (Horisawa et al. 2000). Since the water content of matrix is determined by the water balance in the reaction tank, supply of water by using toilet and removal of water by drying are important factors. This supply is determined by the number of times of use, and the removal of water is determined by the drying rate. Therefore, we examined the effect of such drying conditions as temperature and relative humidity of air on drying rate of sawdust matrix.

7.2 Experiment

7.2.1 Laboratory-Scale Test

Figure 7.1 shows a schematic diagram of laboratory-scale experimental equipment. Sawdust which was sieved from 0.3 to 2.36 mm was used. Sawdust moistened to a moisture content of 70% was placed in a drying container and kept at a constant temperature with a water bath. The drying area in this drying container was 91.5 cm^2. Air conditioner and air fan circulated constant temperature and humidity air. The flow condition was kept as a laminar flow with $Re = 752$. The temperature of the sawdust surface was measured at 2 cm from the surface. Drying experiments were conducted under the experimental conditions summarized in Table 7.1. Drying speed during the experiment was calculated from the change of the weight of the vessel.

7.2.2 Full-Scale Test Condition of the Butch Test

Figure 7.2 shows a schematic diagram of an experimental system using a single-use full-scale composting toilet. The sawdust used in the drying experiment was the same as those mentioned in the previous section. The dry area of the toilet in the reaction tank was 1690 cm^2. For heating the reaction tank, a heating wire was used and the heater was controlled to a predetermined temperature. The constant temperature and humidity air were introduced and circulated using a funnel having a diameter of 90 cm to the center of the upper surface of the toilet from air conditioner. Reynolds number of circulating air on the upper surface of sawdust was $Re = 1462$. As the temperature of the sawdust surface, the temperature measured at 2 cm from the surface was used. The experimental conditions are shown in Table 7.2.

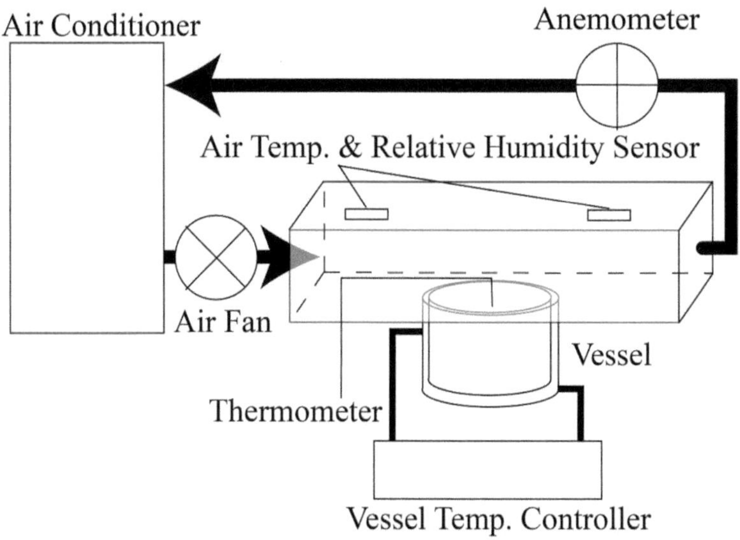

Fig. 7.1 Schematic diagram of laboratory-scale experimental system [reprinted from Ito et al. (2006)]

Table 7.1 Experimental conditions for laboratory-scale test

Run	Vessel temperature (K)	Air temperature (K)	Relative humidity of air (%)
L-1-1	293	333	20
L-1-2	303		
L-1-3	313		
L-1-4	323		
L-1-5	333		
L-2-1	323	293	55
L-2-2		303	
L-2-3		313	
L-2-4		323	
L-2-5		333	
L-3-1	323	298	50
L-3-2			60
L-3-3			70
L-3-4			80
L-3-5			90
L-4-1	338	308	50
L-4-2			60
L-4-3			70
L-4-4			80
L-4-5			90

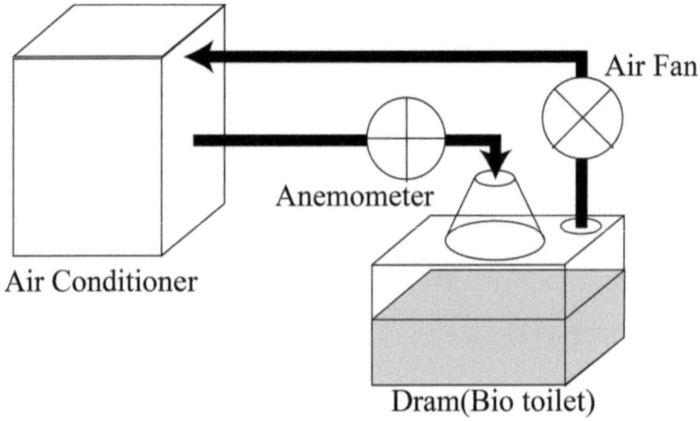

Fig. 7.2 Schematic diagram of an experimental system using a single-use full-scale composting toilet [reprinted from Ito et al. (2006)]

Table 7.2 Experimental conditions for full-scale test

Run	Vessel temperature (K)	Air temperature (K)	Relative humidity of air (%)
F-1-1	313	297	45
F-1-2	323		
F-1-3	333		
F-2-1	313	297	55
F-2-2	323		
F-2-3	333		
F-3-1	313	297	80
F-3-2	323		
F-3-3	333		

Since the full-scale compost-type toilet has large reaction tank, the drying progresses from the surface, and the distribution of moisture content is formed in the sawdust layer. In order to prevent this moisture distribution and to make water content in the sawdust layer uniform, there is a mechanism to mix sawdust matrix at regular intervals. Therefore, the effect of mixing frequency on drying rate was also evaluated. The experimental conditions are summarized in Table 7.3.

Table 7.3 Experimental conditions on the effect of mixing

Run	Vessel temperature (K)	Air temperature (K)	Relative humidity of air (%)	Mixing frequency (times/day)
F-4-1	313	297	45	3, 4, 6, 12, 24
F-4-2	323			
F-4-3	333			
F-5-1	313	297	55	3, 4, 6, 12, 24
F-5-2	323			
F-5-3	333			
F-6-1	313	297	80	3, 4, 6, 12, 24
F-6-2	323			
F-6-3	333			

7.3 Summary of Test Results

7.3.1 Drying Rate of Sawdust Matrix in the Laboratory-Scale Test

Figure 7.3 shows the drying characteristic curves of sawdust at various humidity conditions. When the moisture content of sawdust matrix is high, the drying rate is constant, and this high moisture condition is called as the constant drying rate zone. Where the water content is low, the drying rate decreases with decreasing moisture content.

It is well known that the drying rate is proportional to the difference between air vapor pressure, VP_A, and drying surface vapor pressure, VP_S (the society of chemical engineers, Japan 1958). In the constant drying rate zone, the vapor pressure of drying surface, VP_S, is equal to the saturated vapor pressure which

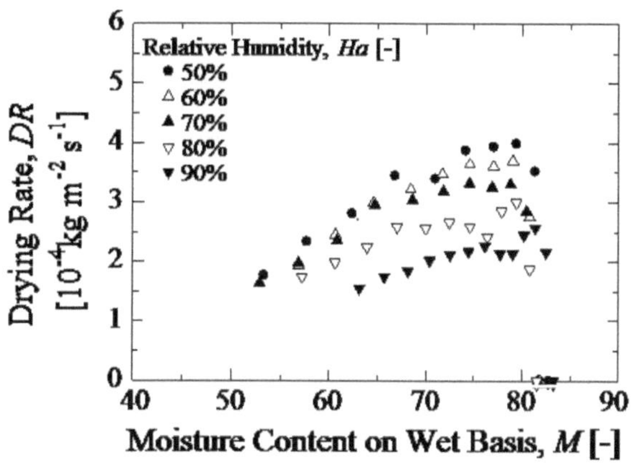

Fig. 7.3 Drying characteristic curves obtained from laboratory-scale test [reprinted from Ito et al. (2006)]

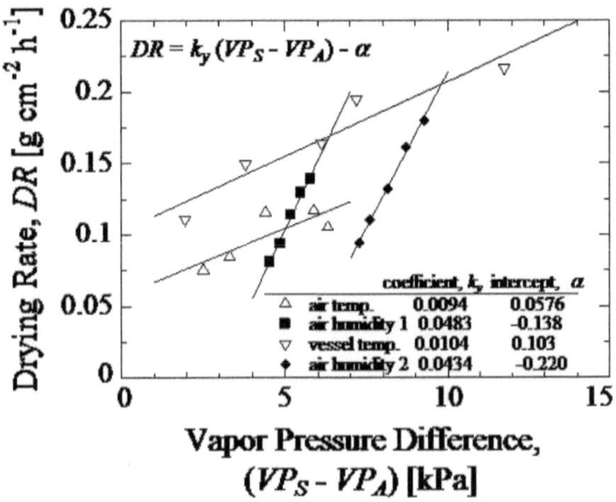

Fig. 7.4 Relationship between drying rate and vapor pressure difference, $VP_S–VP_A$, from laboratory-scale test data [reprinted from Ito et al. (2006)]

depends on drying surface temperature. Figure 7.4 is the relationships between drying rate in the constant drying zone and vapor pressure difference, $VP_S–VP_A$, showing that there is a linear relationship between drying rate and vapor pressure difference, but the straight lines are different for each experiment series. This may be because the temperature of the sawdust surface used for calculating saturated vapor pressure was different from the true temperature, depending on experimental conditions. Therefore, we examined the possibility that the vapor pressure of drying surface of the sawdust layer can be corrected by humidity of air, and the following formula was studied.

$$\mathrm{DR} = k\left(\mathrm{VP}'_{\mathrm{S}} - \mathrm{VP}_{\mathrm{A}}\right), \mathrm{VP}'_{\mathrm{S}} = \mathrm{VP}_{\mathrm{S}} + f(\mathrm{Ha}) \tag{7.1}$$

where k is mass transfer coefficient (s m^{-1}), DR is drying rate (kg m^{-2} s^{-1}), VP_S is saturated vapor pressure (Pa) calculated by using drying surface temperature, VP'_S is modified vapor pressure (Pa) at drying surface, VPA is vapor pressure (Pa) of air, f(Ha) is the vapor pressure correction value (Pa) which is a function of the relative humidity of air, and Ha is relative humidity of air.

The formula of f(Ha) was obtained by using experimental data as $f(\mathrm{Ha}) = -0.0759\mathrm{Ha} + 14.6$, and every drying data are correlated by Eq. (7.1) as shown in Fig. 7.5.

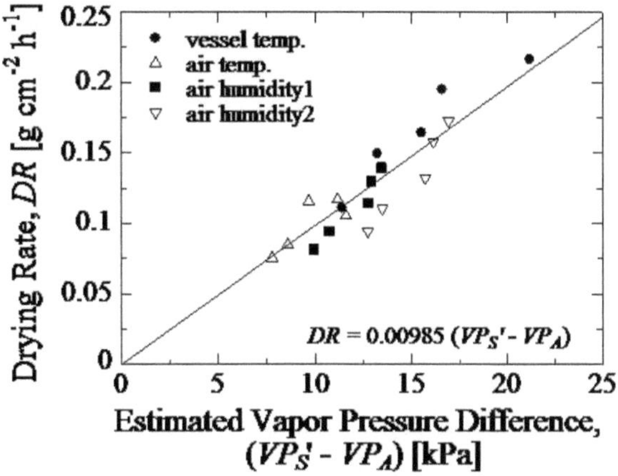

Fig. 7.5 Relationship between drying rate and modified vapor pressure difference, VP'$_S$–VP$_A$, from laboratory-scale test data [reprinted from Ito et al. (2006)]

7.3.2 *Drying Rate of Sawdust Matrix in the Full-Scale Test and Comparison with Data from Laboratory-Scale Test*

Figure 7.6 is the correlation results of drying rate by using Eq. (7.1), showing that even in the full-scale toilet, drying rate is estimated by Eq. (7.1). The value of mass transfer coefficient in Eq. (7.1) for the laboratory-scale data is estimated as 2.74×10^{-5} (kg m^{-2} s^{-1} K Pa^{-1}), and 1.12×10^{-5} (kg m^{-2} s^{-1} K Pa^{-1}) is obtained from the full-scale test. This difference is thought to be due to the different flow velocities of circulating air. Therefore, all the data from laboratory-scale and full-scale were organized based on the dimensionless formula of mass transfer (Robert 1968), and finally, the following formula was obtained:

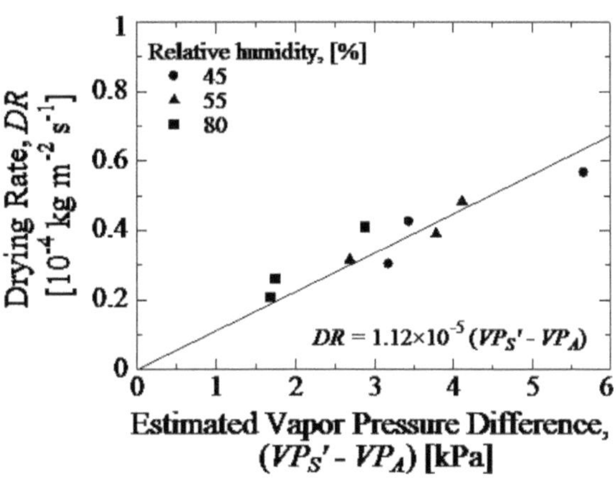

Fig. 7.6 Relationship between drying rate and modified vapor pressure difference, VP'$_S$–VP$_A$, from full-scale test data [reprinted from Ito et al. (2006)]

$$\text{Sh} = 0.664\, Re^{1/2} Sc^{1/3} \tag{7.2}$$

where Sh is Sherwood number, *Re* is Reynolds number, and *Sc* is Schmidt number.

7.3.3 *Effect of Mixing Frequency on Drying Rate of Full-Scale Toilet*

Figure 7.7 shows the relationship between drying rate and mixing frequency. The drying rate linearly increases with mixing frequency. This is because the water vapor contained in the voids space in the sawdust matrix layer moves to the surface by mixing, and the water vapor is discharged into the circulating air. The relationship between the drying rate and mixing frequency is described by the following equation:

$$\text{DR} = aM + \text{DR}_0 \tag{7.3}$$

where a is a coefficient, M is the number of mixing per day, and DR_0 is the drying rate without mixing.

The coefficient a is expressed by reactor temperature as shown in Fig. 7.8. Since the value of DR_0 can be estimated by Eqs. (7.1) and (7.2), the drying rate of full-scale toilet with mixing mechanism can be obtained from Eq. (7.3) based on mixing frequency and reaction vessel temperature.

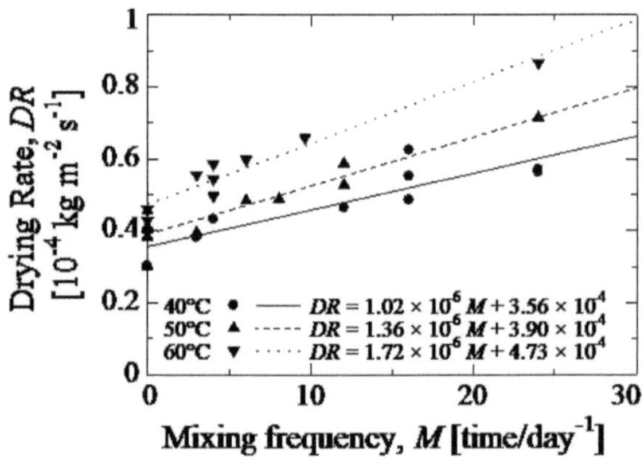

Fig. 7.7 Relationship between drying rate and mixing frequency from full-scale test data [reprinted from Ito et al. (2006)]

Fig. 7.8 Coefficient a in Eq. (7.3) [reprinted from Ito et al. (2006)]

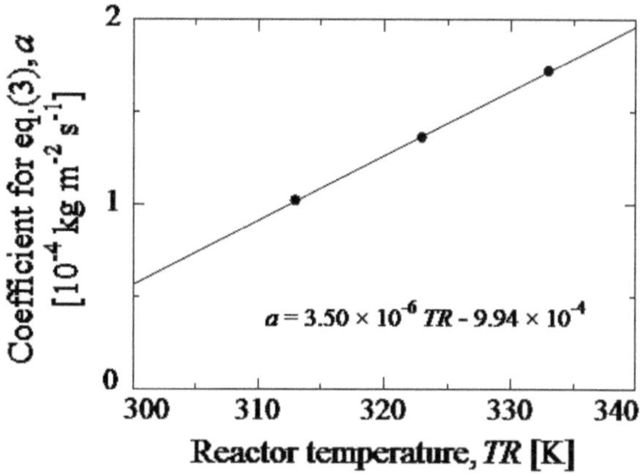

7.4 Conclusion

In order to formulate guidelines for the design and operation conditions of the composting toilet, the drying characteristics of the sawdust matrix were studied. As a result, it was found that the drying rate can be described by using dry model and it can be obtained by correcting the temperature of the sawdust layer surface from operating conditions. Moreover, from the laboratory scale drying experiments, it was found that there is a possibility of obtaining the drying rate of the toilet on a real scale by considering the flow of dry air. Further, drying can be accelerated by stirring, and its effect becomes large as the temperature of the reaction tank increases. From these results, it became possible to design compost-type toilets and formulate operational guidelines.

References

Horisawa S, Tamai Y, Sakuma Y, Doi S, Terazawa M (2000) Effect of moisture content of a wood matrix on a small scale biodegradation system for organic waste. J Wood Sci 46:317–321. https://doi.org/10.1007/bf00766223

Ito R, Ogawa S, Funamizu N (2006) Drying characteristics of sawdust matrix in the dry composting toilet. Environ Eng Res 43:437–442. https://doi.org/10.11532/proes1992.43.437 (in Japanese)

Robert ET (1968) Mass transfer operations, 2nd edn. McGraw-Hill, New York, pp 54–55

The society of chemical engineers, Japan (1958) Chemical engineering handbook, ver. 2. Maruzen, Tokyo (in Japanese), pp 539–541

Zavala MAL, Funamizu N (2005) Effect of moisture on the composting process in the bio-toilet system. Compost Science & Utilization 13(3-summer):208–216. https://doi.org/10.1080/1065657x.2005.10702242

Chapter 8
Design of Composting-Type Toilet

Miguel Ángel López Zavala

Abstract Based on results of experimental research presented in previous chapters, criteria for the proper design and operation of composting toilet were established. In addition, size of the composting reactor was determined attending the already established criteria and based on results of drying tests conducted in a laboratory-scale drying device designed by us. Design procedure considered the daily water loading, due to contributions of urine and feces, as the main criterion. Organic loading, in terms of feces/sawdust ratio (*F*/*S*), resulted to be a non-governing factor for the design, because it was very low. Even though mixing frequency may enhance the biodegradation and drying rates in the composting reactor, the design was conducted regarding no mixing conditions due to the evaluation of its effect on such rates has not been studied yet. However, mixing frequency was considered an important factor for operating the system. The establishment of operation criteria led to the formulation of an operation scheme where three main zones are distinguished: (i) "green zone" where the best composting performance is expected; (ii) "yellow zone" where biodegradation can be conducted but performance is not the most efficient; (iii) other zones where operation of the bio-toilet is not recommended because odor problems and human health risk will develop, high operation cost for concept of excessive mixing may result, or in the worst case, biodegradation of feces will not occur.

Keywords Drying rates · Drying surface · Drying tests · Operation scheme
Organic loading · Water loading

M. Á. López Zavala (✉)
Tecnológico de Monterrey, Water Center for Latin America and the Caribbean, Av. Eugenio Garza Sada Sur no. 2501, Col. Tecnológico, C.P. 64849 Monterrey, N.L., Mexico
e-mail: miganloza@itesm.mx

N. Funamizu (ed.), *Resource-Oriented Agro-sanitation Systems*,
https://doi.org/10.1007/978-4-431-56835-3_8

8.1 Introduction

The proper design of composting toilet implies that the system fulfills criteria such as safety, functionality, economy, aesthetics, and social and environmental affordability. Attending the criteria of functionality and social and environmental affordability, the composting toilet must be designed with the aim of accelerating decomposition of human excreta, optimizing efficiency, and minimizing any potential environmental or nuisance problems (odor). These objectives must couple safeness, low manufacturing and operational costs, and aesthetics, to become an affordable option for the users. Thus, the design process of the bio-toilet will consist of determining the size of the composting reactor that ensures the achievement of the criteria mentioned previously.

There are five important factors or aspects with which we must deal at the time of establishing criteria for the proper design and operation of the bio-toilet, and these are degradation of organic matter, water balance, nutrients (mainly nitrogen processes), pathogens inactivation, and fate of micro-pollutants.

Based on the results achieved in previous studies on such issues, an establishment of criteria for the proper design and operation is the aim of this paper. Micro-pollutants' degradation is not discussed in this paper. Because bio-toilet deals with the evaporation of the water contained in human excreta, theory and concepts of drying of solids are discussed and the results of drying tests conducted in a laboratory-scale drying device are used in the design process; however, results of drying tests, conducted in a laboratory-scale drying device designed by us.

8.2 Considerations for the Operation

As mentioned above, the composting toilet is managed with the aim of accelerating decomposition, optimizing efficiency, and minimizing any potential environmental or nuisance problems (odor). The conditions under which the composting reactor must be operated to meet these goals are given by the results of the research presented in previous chapters, especially related to moisture content and temperature. Thus, criteria for the proper operation of the system under the current configuration are summarized as follows:

(a) Continuous thermophilic aerobic conditions and organic loading are not limiting factors for the composting process in the bio-toilet system. However, constraints associated with high moisture contents may affect the performance of the composting process if organic loads beyond the F/S = 25% are planned to be used.
(b) Optimum moisture control and management in the composting reactor of the bio-toilet system should take into account not only high performance on biodegradation of feces, but also problems of odor and anaerobic emissions, and necessities of maintenance and services, associated with frequency of

sawdust replacement. These requirements could be achieved by keeping moisture contents at 60% or little higher, but avoiding levels near or higher than 65%, i.e., moisture levels near, but lower than the critical moisture. Very low moisture contents, less than 50%, should also be avoided to ensure proper environment for microorganisms and consequently faster and complete stabilization of organic matter contained in feces.

(c) Composting in the bio-toilet system is characterized by different biological response of microorganisms depending on the moisture content under which the process is conducted. Low moisture contents (<65%) ensured aerobic degradation of feces, whereas high moisture levels ($\geq 65\%$) caused both aerobic and anaerobic decomposition.

(d) Because anaerobic conditions occurred at high moisture contents ($\geq 65\%$), microorganisms' activity generated odor and VFA emissions. In addition, simultaneous aerobic and anaerobic processes at high moisture levels caused the increase of sulfate concentrations and formation of nitrites in the sawdust matrix, even though the composting process was conducted at thermophilic temperatures. At low moisture contents, anaerobic emissions, nitrification products, and increase of sulfate concentrations were not detected.

(e) Critical moisture was the frontier moisture content which defined either the highest degradation rates or the beginning of odor and anaerobic emissions, increases of sulfate concentrations, and the nitrification process under thermophilic conditions. Thus, critical moisture may be adopted as a simple physical operation parameter.

(f) Mesophilic and thermophilic microorganisms showed different response to the temperature; additionally, results suggest that the optimum temperature from the viewpoint of feces biodegradability is within the thermophilic range, nearly to 60 °C.

(g) At 70 °C, the activity of biomass was very low. It was clear that enzymatic activity of microorganisms diminished remarkably at this high temperature. At temperatures lower than 60 °C, biodegradation rates of feces slowed following a pattern described by the Arrhenius equation. In the mesophilic range, hydrolyzability of organic matter was remarkably dependent on the temperature at which the composting process was conducted.

(h) Temperatures over than 45 °C (thermophilic range) were more effective for inhibiting the coliforms' colony formation and bacteriophages' plaque formation, consequently for pathogens' inactivation.

(i) Coliform bacteria colony formation at 60 °C was inhibited in only 1.17 h, whereas at 50 and 45 °C the inactivation occurred after 8 and 24 h, respectively. On the other hand, at temperatures lower than 45 °C, it took much longer time to obtain 6-log reduction of CFU. The decay process of bacteriophage also showed the same profile.

(j) Temperature distribution in the bio-toilet system was not uniform, even just after mixing the sawdust matrix. The non-uniformity in temperature distribution causes reducing overall performance of destruction of pathogens. Results of risk assessment showed that reduction of infection risk to an acceptable level

(1×10^{-4} per year) during the compost withdrawal is achieved by i) by mixing the sawdust mixing 20 times per day during 2 days or (ii) by mixing 15 times per day during 3 days after the last using-event of the bio-toilet in current operating bio-toilet. Eliminating low-temperature zone is more effective to reduce reaction time to obtain affordable infection risk, especially in low mixing frequency such as 2 times per day than increasing high-temperature volume.

(k) Low pH (6–7) should be kept in the composting reactor of the bio-toilet system to ameliorate the high nitrogen losses observed. During feces' degradation conducted at 55 °C, T-N losses in the form of NH_3 were on the order of 94%, regardless of the organic loading. Experiments with urine confirmed that ammonification of organic nitrogen occurred in the composting reactors at high and low temperatures. At 40 and 60 °C, lower T-N reductions were observed, but the ammonia releases from the composting reactors were approximately 97.5% of T-N reductions. Unlike, at 20 and 30 °C, T-N reductions were approximately 81.5 and 89.1%, respectively, whereas ammonia losses resulted to be about 57% of T-N reductions.

(l) Nitrification products were not detected in all range of temperatures evaluated, except when high moisture content were kept. High salinity, especially high concentrations of Cl, seems to be the main inhibitory factor for the nitrification process in the composting reactor. Therefore, alternative methods to conserve nitrogen in the sawdust matrix, for instance mineralization of NH_4 in the form of struvite crystals, should be evaluated to enhance the agricultural quality of the compost generated, and to reduce the human health risk and air pollution problems due to ammonia emissions.

In the practice, above optimum temperature and moisture conditions in the composting reactor can be achieved by (i) monitoring continuously the composting process; (ii) manipulating the temperature controller of the system, the mixing frequency, and the water amount for cleaning purposes. pH control is more difficult to achieve, but the addition of sulfates containing, for instance magnesium, may help to reduce pH or enhance the formation of struvite crystals at higher pH.

Based on above considerations, an operation scheme for the composting toilet was prepared (Fig. 8.1). In this scheme, three parameters were considered: temperature, moisture content, and mixing frequency. The effect of mixing frequency on human health risk was studied by Nakata et al. (2003); however, the effect on moisture content reduction due to drying process and on biodegradation rates has not been studied yet; therefore, the operation zones associated with mixing frequency are not conclusive, so that they are delimited by dashed lines. The white-dotted zone (green zone) denotes the conditions where the best performance of the bio-toilet is expected. In the shadow-dotted zone (yellow zone), the composting process can be conducted, but performance is not the most efficient. In the non-dotted zones, operation of the bio-toilet is not recommended because odor problems and human health risk will come, high operation cost for concept of excessive mixing, or in the worst case, biodegradation of feces will not occur because microorganisms will die due to adverse environmental conditions.

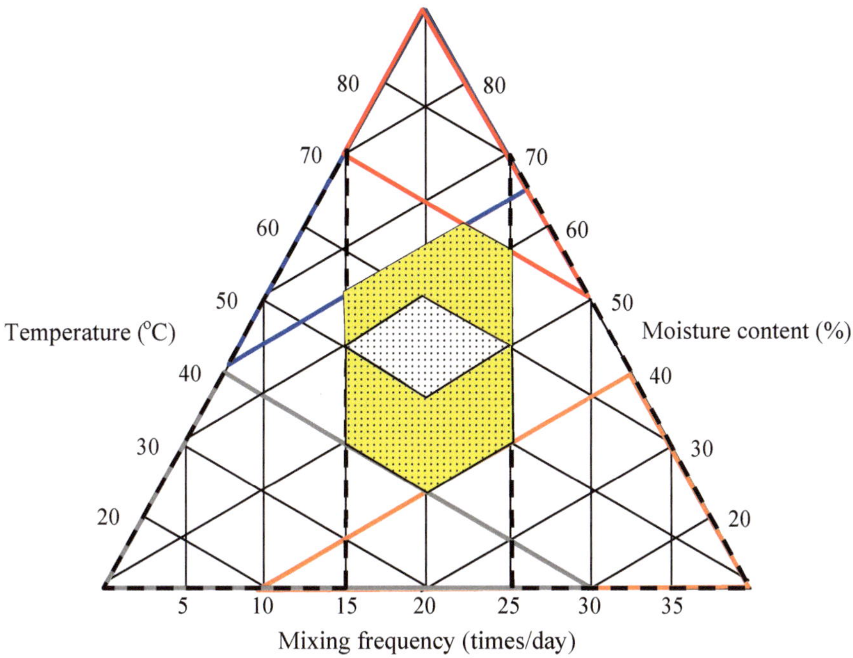

Fig. 8.1 Criteria for proper operation of the bio-toilet system (reprinted from Zavala and Funamizu (2006) with permission from the copyright holders, IWA Publishing)

8.3 Considerations for the Design

The size of the composting reactor is determined by conditions such as:

(a) Water loading rate due to daily contributions of urine, water contained in feces, and water for cleaning the toilet bowl.
(b) Drying rate, i.e., evaporation rate of water contained in urine and feces and cleaning water.
(c) Organic loading rate due to daily feeding of feces and toilet paper; i.e., feces/sawdust ratio (*F*/*S*).
(d) Operating conditions described above.
(e) Mixing frequency.

The water loading affects the water balance in the bio-toilet system. The average daily feces and urine production rates per capita per day found in the literature are approximately 130 g (wet basis) for feces and 1200 mL for urine (Almeida et al. 1999; Del Porto and Steinfeld 1998). The water content of feces is approximately 82% (Zavala et al. 2002). Thus, the daily water loading rate of human excreta totaled 1307 mL per capita per day (or 1325 g per capita per day, if the density of urine is 1.015 g/cm^3). Because no studies have been conducted to evaluate the amount of water needed for cleaning the bio-toilet, but regarding that the amount of water for cleaning the bio-toilet is relatively low, cleaning water may be considered negligible for design purposes. On the other hand, the design of the composting reactor must ensure the evaporation of the water loaded in one day, in such a way the bio-toilet is capable of receiving the next-day water loading. If the water content of feces is 82%, the daily organic loading in dry basis due to feces is 23.5 g per capita per day. Compared with the water loading rate, the organic loading rate is very low. Therefore, as mentioned previously, the size of the composting reactor will be mainly determined by the water loading rate.

Even though toilet paper is biodegradable organic material, its decomposition is quite slow compared with feces' biodegradation because toilet paper is a cellulosic material that is degraded by actinomycetes and/or fungi, microorganisms that are not dominant in the composting reactor of the bio-toilet system. Therefore, for design purposes, toilet paper may be considered as a part of the bulking matrix (sawdust).

Mixing in the bio-toilet enhances the incorporation of toilet wastes and air into the sawdust matrix, and the homogenization of the sawdust matrix in terms of organic matter, microorganisms, water content, and temperature distribution. Thus, the direct benefits of mixing are acceleration of toilet wastes' decomposition, increase of drying rates that will lead to smaller size of the composting reactor, enhancement of pathogens inactivation, and reduction of odors and possible anaerobic emissions. In general, mixing will improve the performance of the bio-toilet system. Nakata et al. (2004) evaluated the mixing frequency effect on the reduction of health risk when compost is withdrawn from the composting reactor. However, an evaluation of mixing frequency on drying rates has not been conducted yet; therefore, for design purposes, mixing may be considered as a "safety factor." Thus, the size of the composting reactor may be estimated regarding no mixing conditions.

Drying in the composting reactor is an important process to maintain a proper water balance in the sawdust matrix and, consequently, for the design of the system. The drying process was evaluated in the bio-toilet system (Zavala and Funamizu 2006). The procedure adopted for conducting the drying tests and their results are not reported in this paper. In their design procedure, the surface area required to evaporate the water loads was estimated by using concepts of drying theory; these concepts were also initially applied to determine the mass and volume of sawdust requirements; however, these parameters were then revised to consider the accumulation of stabilized solids. This accumulation process finally governed the estimation of mass and volume of sawdust required.

Mixing will enhance the drying process; consequently, it may be thought that dimensions of the composting reactor may be reduced. However, because the effect of mixing in the drying process is difficult to evaluate with precision, mixing may result as being more important for operation rather than for design purposes. Thus, the mixing effect was not regarded in the design procedure. In addition, in this study, the shape of the composting reactor is considered to be parallelepiped; however, the current configuration is a semicircle in the bottom due to the shape of the mixing mechanism utilized. The dimensions determined here correspond to non-mixing conditions; therefore, these may change depending on the type and shape of the mixing mechanism.

It was established above that the composting process in the bio-toilet should be conducted at 60 °C and 60% moisture content for good performance. Based on these recommendations and regarding the results of the drying tests, the size of the composting reactor may be determined as follows.

8.3.1 Estimation of the Drying Surface of the Composting Reactor (A)

The drying surface A can be expressed as follows:

$$A = \frac{M_s(W_1 - W_2)}{t\mathrm{DR_m}} \text{ and } \mathrm{DR_m} = \frac{\mathrm{DR_1} - \mathrm{DR_2}}{\ln\left(\frac{\mathrm{DR_1}}{\mathrm{DR_2}}\right)} \tag{8.1}$$

where $\mathrm{DR_m}$ is the logarithmic average of the rate $\mathrm{DR_1}$, at moisture content W_1, and $\mathrm{DR_2}$ at W_2.

W_1 and W_2 are the initial and final moisture contents, respectively.

t is time interval, while the moisture content changes from its initial value W_1 to its final value W_2.

The water content is, by definition, the mass of water divided by the mass of dry solids

$$W = \frac{M_w}{M_s} \tag{8.2}$$

Thus, Eq. (8.1) can be transformed into

$$A = \frac{(M_{w_1} - M_{w_2})}{t\mathrm{DR_m}} \tag{8.3}$$

The moisture content in the composting reactor should not be lower than 50% to ensure proper environmental conditions for microorganisms responsible of feces degradation, as discussed above. Thus, the moisture content of the sawdust matrix should change from 60 to 50% when drying occurs and from 50 to 60% due to contributions of human excreta. This 10% of moisture content variation must ensure the evaporation of total water contained in excreta. Thus,

$$M_{w_1} = \text{Mass of urine} + \text{mass of water in feces} + \text{mass of water for cleaning}$$

It was established above that the daily water loading rate of human excreta totalized 1307 ml per capita per day (or 1325 g per capita per day, if the density of urine is 1.015 g/cm^3). Neglecting the contribution of water for cleaning results in

$$M_{w_1} = 1325 \text{ g per capita}$$

Because all water contained in excreta must be evaporated during the drying period

$$M_{w_2} = 0 \text{ g}$$

In order to receive the water contribution of the next day, the composting reactor must evaporate the daily water contribution in 1 day; thus, time of drying will be

$$t = 24 \text{ h}$$

Figure 8.2 shows the results of drying tests conducted. Two drying stages were observed, as described by the drying theory, the falling-rate and the constant-rate. The constant drying period was found at the moisture more than approximately 65% (or 1.85 g g^{-1} ds), and the corresponding drying rate was about 0.137 g cm^{-2} h^{-1}. As seen in the falling rate period, drying rate is almost linear to the moisture content. This result is important because the calculations during the design process due to the operation of the composting reactor of the toilet should be conducted in this drying period, nearly to 50% (or 1.00 g g^{-1} ds) and 60% (or 1.25 g g^{-1} ds) moisture content.

From Fig. 8.2, the drying rates at 60% (DR_1) and 50% (DR_2) moisture contents are

$$DR_1 = 0.123 \text{ g cm}^{-2}\text{h}^{-1} \text{ and } DR_2 = 0.088 \text{ g cm}^{-2} \text{ h}^{-1}$$

Substituting values, the logarithmic average DR_m yields

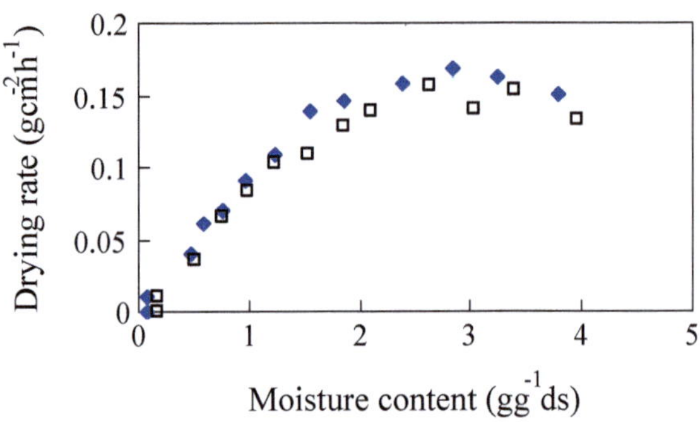

Fig. 8.2 Experimental drying rate profiles

$$\mathrm{DR_m} = \frac{0.123 - 0.088}{\ln\left(\frac{0.123}{0.088}\right)} = 0.1045\ \mathrm{g\ cm^{-2}h^{-1}}$$

Thus, the drying surface is estimated substituting all these values in Eq. (8.3)

$$A = \frac{(1325 - 0)}{24(0.1045)} = 528.2\ \mathrm{cm^{-2}}\text{per capita}$$

8.3.2 *Estimation of the Sawdust Matrix Volume (V_T)*

For design purposes, the volume of the sawdust matrix must ensure that (i) moisture content becomes approximately 60% when excreta is totally added to the system in only one event, and (ii) water contained in excreta is totally evaporated in 24 h; i.e., moisture content of the sawdust matrix becomes approximately 50% one day after the excreta are fed.

Rearranging Eq. (8.2) and regarding that M_s is the mass of dry sawdust (M_{ds}) plus the mass of dry feces (M_{df}) yields

$$M_s = \frac{M_w}{W} = M_{ds} + M_{df} \tag{8.4}$$

The mass of dry feces generated per person per day was determined above as 23.5 g; therefore, the mass of dry sawdust required to have a W of 60% (or 1.5 g g^{-1}) is estimated as follows:

$$M_{ds} = \frac{M_w}{W} - M_{df} = \frac{1325}{1.5} - 23.5 = 859.8\ \text{g per capita}$$

With these results, the *F/S* ratio at which the composting reactor of the toilet will work is estimated

$$\frac{F}{S} = \frac{23.5}{859.8} 100 = 2.7\%$$

As seen, the *F/S* ratio is very low; consequently, it does not govern the design of the composting reactor.

The average density of the sawdust used in the bio-toilet under operation is about 0.19 g/cm^3 (Horisawa et al. 2000; Terazawa et al. 1995); therefore the volume of dry sawdust is

$V = \frac{M_{ds}}{\delta_s} = \frac{859.8}{0.17} = 5057.6\ \mathrm{cm^3}$ per capita (or 5.1 l per capita, approximately).

The daily use of the bio-toilet causes the accumulation of biologically inert material in the composting reactor. This material occupies volume in the reactor that must be considered during the design process. It was mentioned above that 44% of solids remain in the composting reactor after degradation. Thus, the masses of solids that will accumulate in 6 months and in 1 year are

in six months: $M_{S_{ac}} = (M_{df})(0.44)(182.5\ \text{days}) = (23.5)(0.44)(182.5) = 1887.0$ g per capita

in one year: $M_{S_{ac}} = (M_{df})(0.44)(365\ \text{days}) = (23.5)(0.44)(365) = 3774.1$ g per capita.

Because of the high porosity of the sawdust (78% approximately; Horisawa et al. 2000), the stabilized solids will fill up partially the pores of sawdust. When it happens, physical properties of the sawdust matrix such as density, porosity, air and water retention, and water storage capacity will also change affecting consequently the performance of the composting process. To avoid this, additional volume for remaining solids should be considered. The density of stabilized organic solids is unknown; however, studies of physical properties conducted on fine-texture mull soils report bulk densities of 0.74 g cm^{-3} (Wall and Heiskanen 2003). Adopting this value as the density of the remaining solids in the composting reactor, the additional volume to be considered in the design is as follows:

if the compost is withdrawn every six months: $V_{S_{ac}} = \frac{M_{S_{ac}}}{\delta_{S_{ac}}} = \frac{1887.0}{0.74} = 2550.0\ \text{cm}^{-3}\text{per capita}$.

if the compost is withdrawn annually: $V_{S_{ac}} = \frac{M_{S_{ac}}}{\delta_{S_{ac}}} = \frac{3774.1}{0.74} = 5100.1\ \text{cm}^{-3}\text{per capita}$.

As seen, if the compost is withdrawn every six months, the volume of solids accumulated in the reactor will be almost the half of that of sawdust. But, if the compost is withdrawn every year, the volume of solids will equal that of sawdust is during one year the volume of solids accumulated equals that of sawdust. Thus, the total volume for composting matrix will be

if the compost is withdrawn every six months:

$$V_T = V + V_{S_{ac}} = 5057.6 + 2550.0 = 7607.6\ \text{cm}^{-3}(\text{or } 7.6\text{l})\text{per capita}$$

if the compost is withdrawn annually:

$$V_T = V + V_{S_{ac}} = 5057.6 + 5100.1 = 10,157.7\ \text{cm}^{-3}(\text{or } 10.2\text{l})\text{per capita}.$$

8.3.3 Dimensions of the Composting Reactor

Using these values and the drying surface determined above, the thickness of the composting matrix (regarding a rectangular reactor) at the time of withdrawing the compost (H_s) is

if the compost is withdrawn every six months: $H_s = \frac{V_T}{A} = \frac{7607.6}{528.2} = 14.4$ cm
if the compost is withdrawn annually: $H_s = \frac{V_T}{A} = \frac{10157.7}{528.2} = 19.2$ cm

These dimensions were estimated regarding that no mixing is conducted. Therefore, the shape of the mixing mechanism or the mixing system may affect such dimensions. In addition, free space for air circulation will also contribute with the total depth of the composting reactor. 10 cm may be reasonable. Thus, the total depth of the reactor could be

if the compost is withdrawn every six months: $H_T = H_s + H_f = 14.4. + 10 = 24.4$ cm
if the compost is withdrawn annually: $H_T = H_s + H_f = 19.2. + 10 = 29.2$ cm

Mixing will enhance the drying process; consequently, it may be thought that dimensions of the composting reactor may be reduced. However, because effect of mixing in drying process is difficult to evaluate with precision, mixing may result more important for operation rather than for design purposes. Thus, mixing effect was not regarded in the design procedure.

The dimensions of a rectangular composting reactor required for 1–20 users are summarized in Table 8.1. In addition, in this study the shape of the composting reactor is considered to be parallelepiped; however, the current configuration is a semicircle in the bottom due to the shape of the mixing mechanism utilized. The dimensions here determined correspond to non-mixing conditions; therefore, these may change depending on the type and shape of the mixing mechanism.

Table 8.1 Dimensions of the composting reactor of the composting toilet for 1–20 users

Dimension	Number of users											
	1	2	3	4	5	6	7	8	9	10	15	20
Area (dm^2)	5.3	10.6	15.8	21.1	26.4	31.7	37.0	42.3	47.5	52.8	79.2	105.6
Width (cm)	18.8	26.5	32.5	37.5	42.0	46.0	49.6	53.1	56.3	59.3	72.7	83.9
Length (cm)	28.1	39.8	48.8	56.3	62.9	68.9	74.5	79.6	84.4	89.0	109.0	125.9
High[a]	24.4	24.4	24.4	24.4	24.4	24.4	24.4	24.4	24.4	24.4	24.4	24.4
Volume[a] (L)	12.9	25.8	38.7	51.6	64.4	77.3	90.2	103.1	116.0	128.9	193.2	257.8
High[b]	29.2	29.2	29.2	29.2	29.2	29.2	29.2	29.2	29.2	29.2	29.2	29.2
Volume[b] (L)	15.4	30.8	46.3	61.7	77.1	92.5	108.2	123.4	138.8	154.2	231.4	308.5
M_{ds} (kg)	0.86	1.72	2.58	3.44	4.30	5.16	6.02	6.88	7.74	8.60	12.90	17.20

[a]Sawdust withdrawn every six months
[b]Sawdust withdrawn every year
L 1.5 W
M_{ds} Mass of dry sawdust

References

Almeida MC, Butler D, Friedler E (1999) At-source domestic wastewater quality. Urban Water 1(1):49–55. https://doi.org/10.1016/S1462-0758(99),00008-4

Del Porto D, Steinfeld C (1998) The composting toilet system book. Center for Ecological Pollution Prevention (CEPP). Concord Massachusetts

Horisawa S, Tamai Y, Sakuma Y, Doi S, Terazawa M (2000) Effect of moisture content of a wood matrix on a small scale biodegradation system for organic solid waste. J Wood Sci 46(4): 317–321. https://doi.org/10.1007/BF00766223

Nakata S, Zavala MAL, Funamizu N, Otaki M, Takakuwa T (2003) Temperature effect on pathogens decline in the bio-toilet system. In: Proceedings of the 1st international dry toilet conference: dry toilet 2003. Tampere, Finland

Nakata S, Zavala MAL, Funamizu N, Takakuwa T (2004) Pathogens risk assessment as a tool for deciding the proper operation of the bio-toilet's composting reactor. Hokkaido University, Japan, Graduate School of Enginnering

Terazawa M, Tamai Y, Horisawa S, Yamashita K (1995) Ability of sawdust as an artificial soil. Biodegradation of non-lignocellulosic substances II. Division of Environmental Resources, Graduate School of Agriculture, Hokkaido University, Japan, 77–84

Wall A, Heiskanen J (2003) Water retention characteristics and related physical properties of soil on afforested agricultural land in Finland. For Ecol Manage 186(1–3):21–32. https://doi.org/10.1016/S0378-1127(03)00239-1

Zavala MAL, Funamizu N (2006) Design and operation of the bio-toilet system. Wat Sci Tech 53(9):55–61. https://doi.org/10.2166/est.2006.277

Zavala MAL, Funamizu N, Takakuwa T (2002) Characterization of feces for describing the aerobic biodegradation of feces. J Environ Syst Eng JSCE 720/VII-25:99–105

Part III
Technologies for Urine

Chapter 9
Components of Pure Fresh Human Urine and Their Fate in Storage Process

Takashi Kakimoto, Hitoshi Shibuya, Haruki Suzuki, Shinya Hotta and Naoyuki Funamizu

Abstract Separation of urine at source followed by on-site storage and adequate treatment has become a basic idea for more sustainable alternatives in management of household wastewater. Compared to interests in phosphate, nitrogen, and micro-pollutants, discussion about organic matter seemed to be not enough especially on the changes in the character of urine during storage process. Fresh human urine was collected and concentrations of 40 amino acids, 8 organic acids, 4 nitrogen, and 6 inorganic ion species. Molecular weight distribution of organic matter was measured. Electrochemical characteristics of urine were also investigated by cyclic voltammetry. Storage test in batch system at 30 °C for 90 days were conducted to investigate their fate. Decline of pharmaceuticals was observed in storage test. Most of the organic matter except urea decomposed to smaller molecule within 100 Da. Pharmaceuticals (Tetracycline and Metoprolol) remarkably decreased during the test. Fate of nitrogen was mostly described by urea hydrolysis whereas declines of creatinine and uric acid were also found.

T. Kakimoto (✉)
Center for Environmental Science in Saitama, 914 Kamitanadare, Kazo, Saitama 347-0115, Japan
e-mail: kakimoto.takashi@pref.saitama.lg.jp

H. Shibuya · H. Suzuki
Division of Laboratory and Transfusion Medicine, Hokkaido University Hospital, Kita-14, Nishi-5, Kita-ku, Sapporo 060-8648, Japan

S. Hotta
Research and Development Center, Asahi Kasei Corporation, Kawasaki-ku, Kawasakishi 210-0863, Japan

N. Funamizu
Graduate School of Global Food Resources, Hokkaido University, Kita-9, Nishi-9, Kita-ku, Sapporo 060-8589, Japan

N. Funamizu
Research Institute of Humanity and Nature. Motoyama Kamigamo, Kita-ku, Kyoto 603-8047, Japan

N. Funamizu (ed.), *Resource-Oriented Agro-sanitation Systems*,
https://doi.org/10.1007/978-4-431-56835-3_9

Approximately 30% of initial phosphate ion seemed to be precipitated because of crystallization with Mg and Ca ions which levels decreased under detection limit. Results of cyclic voltammetry remarkably showed changes of urine characteristics on the process of storage.

Keywords Organic matter · Molecular weight distribution · Organic nitrogen Phosphate precipitation · Urine storage

9.1 Introduction

Separation of urine at source has been a basic idea for more sustainable management of wastewater in household (Hellstrom and Karrman 1997; Hellstrom 2003; Larsen and Gujer 1996a, b; Larsen et al. 2001, 2004; Vinneras and Jonsson 2002). Separated urine is stored at on-site first, and then it is received treatment for recovering nutrients and removing micro-pollutants. A topic of urine storage was found in several reports for phosphate precipitation in which main discussion was developed on crystallization and ammonification (Lind et al. 2000; Udert et al. 2003a, d; Harada et al. 2006). Treatment processes of urine after storage have been well discussed in previous reports. Maurer et al. (2006) introduced several reports for application of ozonation for pharmaceuticals treatment in source-separated urine. Pronk et al. (2006a) reported the removal technology by using nano-filtration for source-separated urine. Application of electrodialysis was also evaluated for separation of micro-pollutants from urine in nutrient recovery process (Pronk et al. 2006b). Udert et al. (2003a, c, d) and Harada et al. (2006) described the precipitation kinetics for the struvite crystallization in stored urine. With combination of struvite crystallization, Ganrot et al. (2007) showed efficiency of some minerals (zeolite, etc.) on nitrogen recovery from urine.

In every study on urine processes, urine characteristics are basic information to evaluate performances. As well known, character of urine is affected by many factors: sexuality, the elapsed time after collection, temperature of urine, etc. Individual variations in character of fresh urine are relatively larger (Fittschen and Hahn 1998), and it was well understood that components of urine are unstable at room temperature although they were stable in refrigerator (Hanæus et al. 1996; Hotta and Funamizu 2008).

About organic matter in urine, Udert et al. (2003b) reported that approximately 85% of organic matter described by COD was easily biodegradable in aerobic condition. Hoglund et al. (1998, 2000) discussed organic matter of volatile solid and total solid in precipitate of stored urine. However, compared to interests in phosphate, nitrogen, and micro-pollutants, discussion about organic matter seemed to be not enough especially on the change in the characteristics of urine during storage process. Therefore, objectives of this study were (1) to prepare the list of urine components mainly focused on organic matter (2) to monitor the evolution of the urine components in storage process.

9.2 Collection of Urine and Storage Experiment

9.2.1 Collection and Storage Test of Fresh Urine

Fresh urine was collected from 189 volunteers in Hokkaido University within 4 h. Balance of gender was 145 males and 44 females. Large amount of the volunteers were younger people (early twenties). Every volunteer received a disposable cup for medical use at entrance of public toilet. Then fresh urine was obtained in the cup, and the urine was poured into a tank for storage. During the collection process, there was no control of temperature in the storage tank: According to thermometer, it was around 17 °C. The collected urine was immediately used for storage test (described below). After the collection, pharmaceuticals of carbamazepine, diclofenac, and Metoprolol were immediately added to fresh urine: Every concentration was prepared to 1.0 mg/L. Then 100 ml of the fresh urine was put into 100 ml of glass bottles (storage units). Little headspace of the storage units was negligible for oxygen supply because they were tightly sealed during storage process. Every storage unit was stored in incubator (30 °C). The mild condition of 30 °C was possible in usual storage unit of urine and was reasonable to observe typical reactions in stored urine (Hotta and Funamizu 2008).

9.2.2 Pretreatment

In 90 days storage test, sampling and analysis were conducted for several times. In every sampling process, urine pH was measured immediately after the storage units were opened. Then 1 ml of urine sample was obtained before any filtration process for analysis of DOC as described below. Further, urine was micro-filtrated (MF) by membrane filter with pore size of 0.45 μm. The filtrate was obtained for (1) further filtrated by UF membrane (MWCO 10 kDa, 1 kDa) and dialysis tube (MWCO 100 Da) to investigate the molecular weight distribution or (2) analysis of DOC, nitrogenous compounds, inorganic ions, organic carbon, and pharmaceuticals and (3) analysis of cyclic voltammetry (CV).

9.2.3 Determination of Amino Acid and Organic Acid in Urine

Determination of organic acid was conducted by using the ion exclusion chromatography system. Filtrate of urine was adequately diluted and then directly injected into the system. For amino acid analysis, HCl (6 N) was added into the urine and stored at −30 °C until analysis. Amino acid analysis was conducted by using amino acid analyzer of Hitachi (Model L-8500).

9.2.4 Molecular Weight Distribution of Organic Matter and Determination of Pharmaceuticals

After MF process, fractionation of organic matter in urine was done by using UF membrane and dialysis tube with molecular weight cutoff level was 10 kDa, 1 kDa, 100 Da, respectively. Then five fractions were obtained: they were (1) less than 100 Da, (2) 100–1 kDa, (3) 1 k–10 kDa, (4) 10 k–0.45 μm, and (5) more than 0.45 μm. The filtration process had been completed within 48 h at 5 °C, and the DOC of the each filtrate was determined by TOC analyzer (SHIMAZU: TOC-5000A). During the filtration process, transformation of organic matter seemed to be negligible because Hanæus et al. (1996) had shown that urea was quite stable for more than 40 days in a low temperature at 5 °C, and Udert et al. (2003c) concluded that urea hydrolysis triggered a series of degeneration of stored urine.

For the determination of pharmaceuticals, clean up and determination methods were followed by the OASIS technical notebook.

9.2.5 Analysis of Inorganic Species and Nitrogen Compounds

Phosphate ion was determined immediately after filtration process by colorimetric analysis of molybdenum blue method. Other species of inorganic ion were measured by using o-cresolphthalein complexone method for calcium and N-assay L Mg-H Nittobo (Nittobo Medical Co., Ltd.) for magnesium. Further, species of inorganic ion (Na^+, K^+, and Cl^-) were determined by the auto-analyzer 7070 with ion selective electrode (Hitachi Co., Ltd.).

Filtrate of membrane filter (0.45 μm) was analyzed for total nitrogen, urea, ammonia, creatinine, and uric acid. Colorimetric methods were applied in determination of urea (Hott and Funamizu 2008) and ammonia (Kempers and Zweers 1986). Total nitrogen was measured by semi-Kjeldahl method (Scheiner 1975). Further, two nitrogen species (creatinine and uric acid) were measured by enzymatic determination method using the CRE-L (Serotec Co., Ltd) for creatinine and the Determiner-L UA (Kyowa Medex Co., Ltd.) for uric acid.

9.2.6 Analysis of the Cyclic Voltammetry (CV)

Cyclic voltammetry (CV) was applied to evaluate the electrochemical character of components in urine. Applied potential of Pt working electrode was in the range of −0.5 to 1.5 V (vs Ag/AgCl) with 50 mV/s of scanning rate. Temperature of the system followed atmosphere around 15 °C. Dissolved oxygen in sample solution was eliminated by N_2 gas before the CV analysis.

9.3 Compounds in Very Fresh Human Urine

Table 9.1 lists the comparison of major components measured in this study. As seen, most comparable values obtained in this study were similar to previous reports (Udert et al. 2003a; Ciba-Geigy 1977). Furthermore, components in the urine collected in this study were shown: 40 species of amino acid (Table 9.2), 8 species of organic acid (Table 9.3), 4 species of nitrogen (Table 9.4), and the summary of obtained knowledge for organic matter (Table 9.5). From these tables, knowledge of the fresh urine from general public people was obtained as following: Approximately 84% of organic carbon in fresh urine was still unknown even though most of the measurable compounds were determined; nitrogen species in urine were mostly described by four nitrogen species: urea, ammonia, uric acid, and creatinine; major species of organic acid from fresh urine were the acetic acid, the lactic acid, and the propionic acid; in amino acids, the taurine and the histidine were slightly higher than other species. Table 9.6 lists the molecular weight distribution of organic matter. It was found that most of the organic matters except urea in fresh urine were distributed in the range of 100 Da–1 kDa.

Table 9.1 Comparison of components in fresh urine

		Urine [1]	Urine [2]	Urine [3]
Dilution	(–)	1	1	1
pH	(–)	6.2	7.2	6.5
DOC	mg-C/l	–	*5546*[a]	7300
COD	mg-C/l	–	3056	*4950*[a]
Total-N	mg-N/l	8830	–	6300
Urea	mg/l	–	12450	11800
Ammonia	mg-N/l	463	254	300
Phosphate	mg-P/l	800–2000[c]	367[b]	1100[b]
Na^+	mg/l	3450	2670	2940
Cl^-	mg/l	4970	3810	5180
Ca^{2+}	mg/l	233	129	120
Mg^{2+}	mg/l	119	77	60
K^+	mg/l	2737	2710	1760

[1]: Data from Ciba-Geigy(1977) cited from a report by Maurer et al. (2006)
[2]: Udert et al. (2003a), [3]: Results in this study
[a]Estimated by assumption of "DOC = C in Urea + C detected by COD"
[b]Including all system species of phosphate ion
[c]Total phosphorous

Table 9.2 Fate of amino acid from the fresh urine

	0 day	90 days
Taurine	106.2	114.0
Phosphoethanolamine	4.8	3.5
Aspartic acid	N.D.	6.5
Hydroxyproline	N.D.	N.D.
Threonine	15.2	0.9
Serine	31.3	N.D.
Asparagine	9.6	1.2
Glutamic acid	2.2	51.9
Glutamine	46.2	N.D.
Sarcosine	N.D.	29.2
α-Aminoadipic acid	6.3	4.1
Proline	N.D.	N.D.
Glycine	77.1	110.6
Alanine	21.2	51.7
Citrulline	2.5	3.6
α-Aminobutyric acid	0.9	1.0
Valine	2.8	2.7
Cystine	9.5	N.D.
Cystathionine	1.8	1.2
Methionine	1.2	0.8
Isoleucine	0.9	0.6
Leucine	2.5	0.9
Phenylalanine	6.2	4.5
γ-Amino-β-hydroxybutyric acid	N.D.	N.D.
β-Alanine	1.6	2.1
β-Amino-iso-butyric acid	41.6	33.8
γ-Aminobutyric acid	N.D.	N.D.
Monoethanolamine	16.9	14.7
Homocystine	N.D.	N.D.
Histidine	111.2	49.6
3-Methylhistidine	27.7	25.4
1-Methylhistidine	59.2	54.3
Carnosine	5.2	3.0
Anserine	7.6	6.4
Tryptophan	12.1	N.D.
Tryptophan	12.1	N.D.
Hydroxylysine	N.D.	2.7
Ornithine	1.6	2.0
Lysine	32.7	19.8
Arginine	2.7	1.1

Unit: mg/l-urine

Table 9.3 Fate of organic acid from fresh urine

	0 day	15 days	90 days
Malic acid	12.7	40.8	30.4
Lactic acid	190.3	567.0	66.0
Formic acid	39.7	63.5	8.8
Acetic acid	218.0	500.7	318.1
Propionic acid	177.2	98.8	145.6
Butyric acid	24.1	N.D.	N.D.
Valeric acid	23.9	23.1	3.7
Ascorbic acid	224.5	N.D.	N.D.
Unit: mg/l-urine			

Table 9.4 Fate of nitrogenous compounds in stored urine

Component	Unit	0 day	90 days
Total-N	g-N/l	6.3	6.3
Urea-N	g-N/l	5.4	0.3
Ammonia-N	g-N/l	0.3	5.9
Creatinine-N	g-N/l	0.2	0.0
Uric acid-N	g-N/l	0.3	N.D.

9.4 Fate of Urine Component During 90 Days Storage

9.4.1 Transformation of Organic Matter

In this study, we monitored the organic matter in urine during the storage process. Table 9.5 lists fate of organic matter in stored urine for 90 days. Main reduction of organic matter was degradation of urea. Figure 9.1 shows the evolution of DOC including urea and DOC except urea. As seen, DOC except urea was on the stable level throughout the storage process, whereas the DOC including urea constantly

Table 9.5 Fate of organic matter in stored urine for 90 days

	0 day		90 days	
	g-C/l	%	g-C/l	%
DOC	7.3	/	4.6	/
Urea	2.3	/	0.15	/
DOC without Urea	5.0	100	4.5	100
Organic acid	0.29	5.8	0.24	5.3
Amino acid	0.23	4.6	0.15	3.3
Uric acid	0.12	2.3	0.0	0.0
Creatinine	0.08	1.5	0.01	0.2
Sugar	0.0	0.0	0.0	0.0
Ascorbic acid	0.09	1.8	0.0	0.0
Other (unknown)	4.2	84	4.1	91

Table 9.6 Fate of organic matter in each fraction

Component	0 day		90 days	
	g-C/l	%	g-C/l	%
DOC	7.30	/	4.64	/
Carbon from urea	2.30	/	0.15	/
DOC except urea	5.00	100	4.49	100
More than 0.45 μm	0.41	8.3	0.25	5.5
10000 Da-0.45 μm	0.14	2.9	0.07	1.5
1000–10000 Da	0.68	13.5	0.34	7.5
100–1000 Da	3.7	73.9	0.07	1.6
Less than 100 Da	0.07	1.4	3.76	83.8

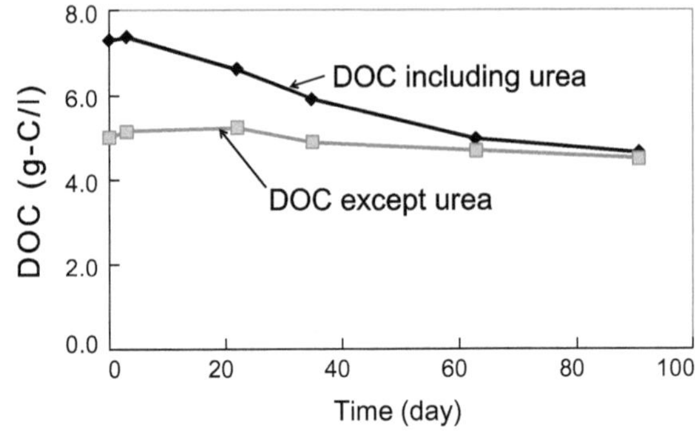

Fig. 9.1 Fate of dissolved organic matter

decreased. As listed in Table 9.6, 10% reduction was found in the organic carbon except urea. Udert et al. (2003b) reported more than 80% of dissolved COD of urine was biodegradable in aerobic condition, whereas they discussed possibility (Udert et al. 2006) for anaerobic biodegradation of organic matter in stored urine by citing previous reports (Hoglund et al. 1998, 2000). In this study, anaerobic process was indicated by temporal production of organic acids at day 15 (Table 9.3). Since, however, stored urine usually provide pH more than 7 with ammonia on high level, then it might be difficult to assume methane production as anaerobic reaction of organic matter (Hansen et al. 1998; Calli et al. 2005).

Figure 9.2 shows the fate of the DOC except urea in storage process of urine. Simple result of carbon transformation was found: 74% of organic matter were distributed in 100 Da–1 kDa fraction in fresh urine, whereas 84% of organic matter in stored urine presented in the fraction of less than 100 Da. Figure 9.3 shows the evolutions of the each fraction. As shown in Fig. 9.3a, the DOC from 100 Da to 1 kDa readily decreased in the first 3 days, whereas the DOC less than 100 Da except urea correspondingly increased. As listed in Table 9.3, it was found that the most of organic acid increased in the first 15 days during storage process. That also supported the transformation of organic matter into smaller size. Evolution of the DOC (more than 1 kDa) was shown in Fig. 9.3b. The DOC more than 0.45 μm

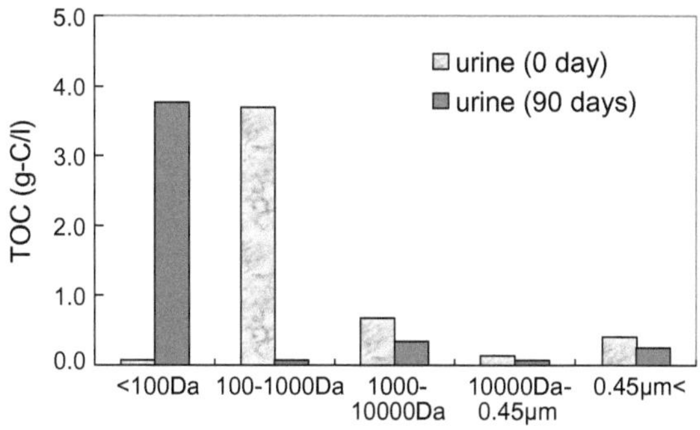

Fig. 9.2 Molecular distribution of organic matter

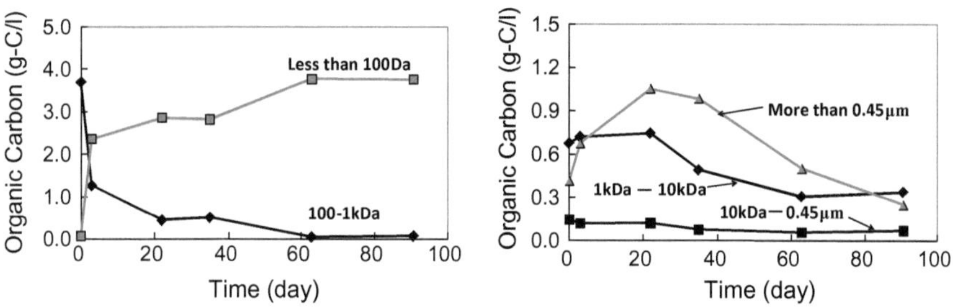

(a) Fraction less than 1kDa

(b) Fraction more than 10kDa

Fig. 9.3 Time courses of organic matter

showed unique evolution: It was increasing in the first 20 days, and then, it kept declining. The DOC between 10 kDa and 0.45 μm did not play an important role in transformation of organic matter in stored urine.

9.4.2 Electrochemical Character of Urine

As shown in Sect. 9.3 that 84% of organic carbon in fresh urine and 94% of organic carbon after 90 days storage are unknown. To characterize organic matter along with molecular weight distribution measurement, the cyclic voltammetry was applied in this study. The cyclic voltammograms give the information of oxidation potential constituents in urine.

Evolutions of cyclic voltammograms of the collected and stored urine were shown in Fig. 9.4. Every cycle in Fig. 9.4 started clockwise at 0 V. Positive currency indicated oxidation of some urine compounds on working electrode. As shown in Fig. 9.4, clear oxidation peaks of some compounds in very fresh urine were observed around 0.6 V compared with phosphate buffer as blank.

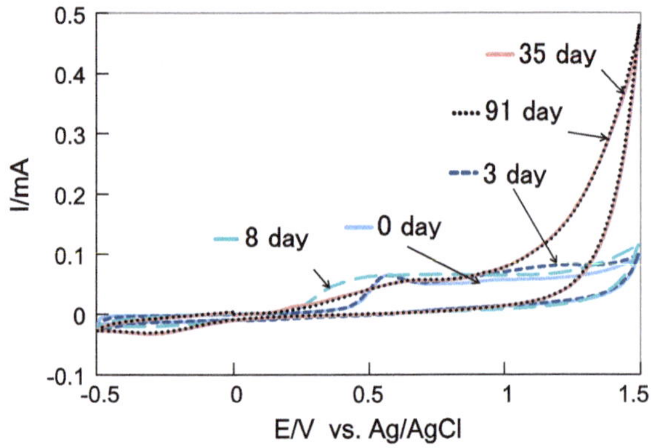

Fig. 9.4 Cyclic voltametry

Quite similar trends of the voltammogram were observed in day 3 although organic matter classified in 100–1000 Da significantly decreased within 3 days (Fig. 9.3a). Therefore, easily degradable organic matter in the range of 100–1 kDa seemed to be electrochemically inert. The voltammograms on day 8 remarkably differed from that of day 0 and day 3. Significant oxidation started after the applied potential went over 0.25 V. Therefore, we assumed that oxidation potential of organic matter in stored urine could change between 3 and 8 days because pH which strongly affects the oxidation potential of urine during 8 days was constant and pH was 6.5. Reactants on working electrode were different from before. It was previously found that urine compounds for easy oxidation such as ascorbic acid consumed first (Kakimoto and Funamizu 2006) and immediate reduction of ascorbic acid in Table 9.3 supported the result. At day 90, furthermore smooth profiles were obtained. From Fig. 9.3, transformations of organic matter also continued after day 8. Oxidation started after applied potential went over 0.3 V after day 8, whereas their peaks were slightly different from 0.06 mA (at day 8) to 0.05 mA (at day 90), respectively. After 8 days, some compounds that could be oxidized at the potential within 1.5 V remained throughout the storage process. The CV profiles showed different figures that indicated transformations of some compounds in stored urine. Large value of currency around 1.5 V on Fig. 9.4 was mainly caused by higher alkalinity (Osawa 2007).

9.4.3 *Fate of Pharmaceuticals*

As shown in Fig. 9.5, significant decrease of Tetracycline was found in stored urine at 30 °C. Kakimoto and Funamizu (2006) showed that the decomposition of Tetracycline profoundly related to levels of phosphate ion and ammonia level in composting process of human feces. From Sect. 9.4.5 shown later, concentrations of phosphate ion and ammonium ion in the stored urine were much higher than in usual wastewater. It was therefore assumed that the Tetracycline could decompose

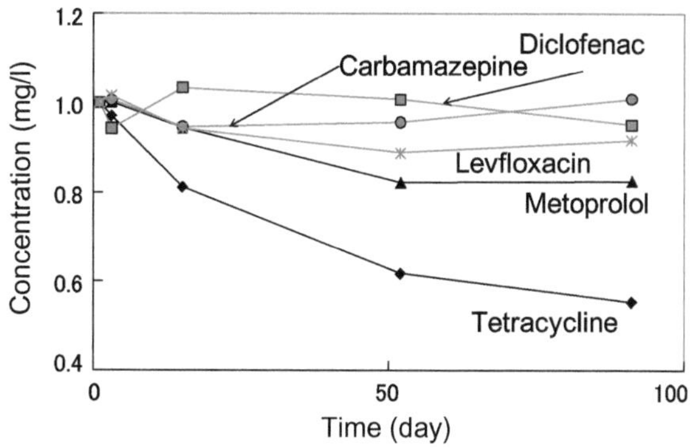

Fig. 9.5 Fate of pharmaceuticals

in storage process of urine due to high levels of phosphate, ammonium, and hydroxyl ion. Slight decrease of the Metoprolol was observed. Other pharmaceuticals used in this study did not decrease remarkably. It might be possible to assume that further three chemicals (Carbamazepine, Levofloxacin, and Diclofenac) could be stable in simple storage process of urine. Carbamazepine was very stable even in the electro-oxidation process (Osawa 2007). Several results from previous studies showed that Levofloxacin was hard to decompose in wastewater treatment plant (Yasojima et al. 2005) and composting process of human excreta (Kakimoto 2007).

Diclofenac could decompose in wastewater treatment plant mainly because of bacterial metabolism (Urase and Kikuta 2005). Biodegradability of the Diclofenac does increase in acidic condition that provides non-dissociation type. However, pH condition was relatively higher than pH 9 in stored urine. Thus, it might be hard for Diclofenac to decompose.

9.4.4 Nitrogen Transformation

Figure 9.6 is the observed results on urea and ammonium nitrogen, showing that ammonification occurred in stored urine at 30 °C. It is significantly dependent on reduction of urea. As shown in Fig. 9.7, decrease of the creatinine and the uric acid was also observed during storage process of urine. Until day 90, most of the nitrogen had decomposed to ammonia. From previous reports (Hotta and Funamizu 2008; Andrews et al. 1984), it is possible to assume that urea hydrolysis in stored urine is controlled by enzyme of urease. Further, enzymatic reactions for decomposition of creatinine and uric acid also were studied well (Suzuki and Matsugi 2005; Zhang et al. 2006). In this study, source-separated urine was collected in the unused disposable cup for medical use: Contamination in the toilet bowl was impossible in the collection process of fresh urine. Thus, it seemed that keeping the organic nitrogen spices as intact form was difficult even in source-separated fresh urine.

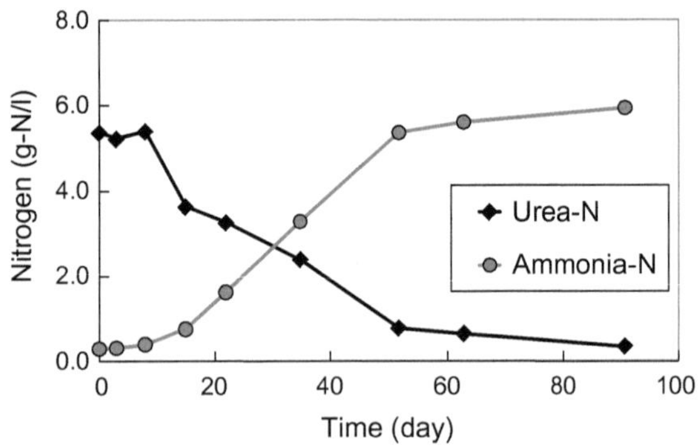

Fig. 9.6 Fate of urea and ammonia nitrogen

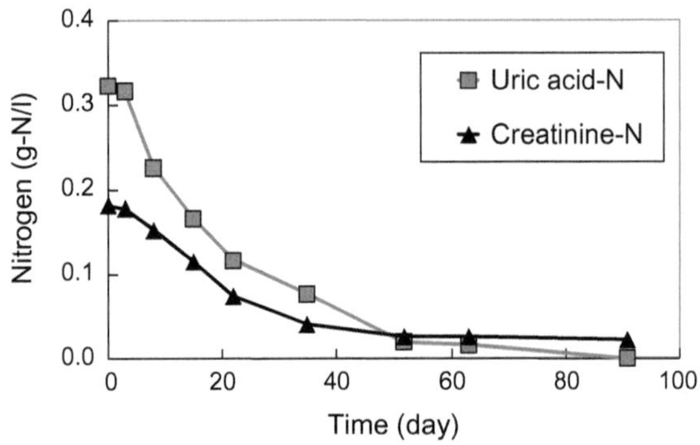

Fig. 9.7 Fate of uric acid and creatinine

9.4.5 *Evolution of Ion Species for* PO_4^{3-} *Precipitation*

As reported previously (Udert et al. 2003a; Wilsenach et al. 2006), reduction of PO_4^{3-} in stored urine is usually observed mainly because of precipitation of struvite and hydroxyapatite. Udert et al. (2003a) concluded that trigger of the precipitation was increase of pH caused by urea hydrolysis. As they had simulated the precipitation in stored urine, significant reduction of the main ions for PO_4^{3-} precipitation (Mg, Ca and PO_4^{3-}) was actually observed around pH 7.0 in urine.

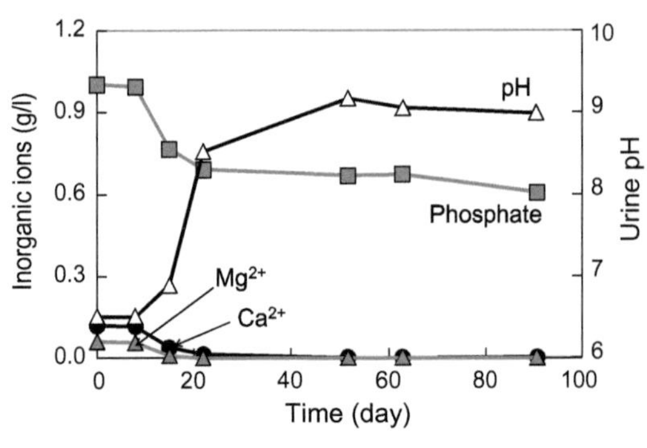

Fig. 9.8 Fate of pH and ion species related to PO_4^{3-} precipitation

Figure 9.8 shows that PO_4^{3-} was on the stable level after depletion of inorganic ions of Mg and Ca. Detected PO_4^{3-} at the end of the test was approximately 70% of initial amount. Phosphate precipitate of struvite (MAP) was equimolar compound of Mg and PO_4^{3-}, whereas another main precipitate of hydroxyapatite (HAP) was different balance on 10 (Ca ion) to 6 (PO_4^{3-}). The reduction of Mg and Ca ions (0.0043 mol/l-urine) almost corresponded with the reduction of phosphate (0.0042 mol/l-urine). This result led to the recognition that precipitates mainly composed by the MAP and the HAP. Udert et al. (2003c, d) have reported that their simulation result for precipitation of phosphate provided approximately 30% of precipitation in stored urine. Therefore, it was assumed that most phosphate reduction was caused by precipitation of the MAP and the HAP, and further, 70% still remained in the form of phosphate ion species.

9.5 Summary

Study on source-separated urine was conducted with special interest in organic carbon from fresh and stored urine. Conclusions obtained in this study were following:

(a) Lists of the components of very fresh urine from general public people were obtained: 40 species of amino acid, 7 species of organic acid, 4 species of nitrogen, and 7 species of inorganic ion. The list of distribution of molecular weight of organic carbon except urea was also prepared. However, more than 80% of organic carbon except urea was not determined in fresh urine.

(b) 90 days storage with tightly sealed at mild temperature provided smaller size of organic carbon. The majority of organic carbon in fresh urine was found in the distribution of 100–1000 Da, whereas it shifted into the distribution less than 100 Da after long storage. However, the main compounds less than 100 Da were still unknown. In the storage unit, more than 90% of organic carbon was not carbonated but they remained in dissolved type even after 90 days storage at mild temperature. Some species of organic acid temporally increased but finally decreased.

(c) Cyclic voltammetry (CV) indicated variation of characteristic in stored urine. It was found that oxidation of several compounds started over 0.4 V (vs Ag/AgCl electrode) in fresh urine, whereas onset of oxidation shifted around 0.3 V after 8 days storage. At day 90, more smooth shapes of CV profiles were obtained, whereas peak values shifted from 0.06 mA (at day 8) to 0.05 mA (at day 90).

(d) Declines of the Tetracycline and Metoprolol were found. However, pharmaceuticals of Carbamazepine, Levofloxacin, and Diclofenac were on stable level throughout the process.

(e) Most of the major organic nitrogen (urea, creatinine, and uric acid) decomposed to ammonia during the process. Source-separated urine free from contamination in toilet bowl also contained potential of ammonification.

Acknowledgements We thank all of volunteers with contribution to urine collection and all of staff in the Hokkaido University Hospital for their help in urine analysis of organic nitrogen, inorganic ion species. Mr. Yusuke Ino provided great helps with us especially in filtration process of urine. This work has been supported by CREST of the Japan Science and Technology Agency (JST).

References

Andrews KR, Blakeley LR, Zerner B (1984) Urea and urease. Adv Inorg Biochem 6:245–283

Calli B, Mertoglu B, Inanc B, Yenigun O (2005) Effects of high free ammonia concentrations on the performances of anaerobic bioreactors. Process Biochem 40(3–4):1285–1292. https://doi.org/10.1016/j.procbio.2004.05.008

Ciba-Geigy (1977) Wissenschaftliche Tabellen Geigy, Teilband Ko¨rperflu¨ ssigkeiten (Scientific Tables Geigy Volume: Body Fluids) 8th edn. Basel in German

Fittschen I, Hahn HH (1998) Characterization of the municipal wastewater part human urine and a preliminary comparison with liquid cattle excretion. Water Sci Technol 38(6):9–16

Ganrot Z, Dave G, Nilsson E (2007) Recovery of N and P from human urine by freezing, struvite precipitation and adsorption to zeolite and active carbon. Biores Technol 98(16):3112–3121. https://doi.org/10.1016/j.biortech.2006.10.038

Hanæus A, Hellstrom D, Johansson E (1996) Conversion of urea during storage of human urine. Vatten 96(4):263–270

Hansen KH, Angelidaki I, Ahring BK (1998) Anaerobic digestion of swine manure: inhibition by ammonia. Water Res 32(1):5–12. https://doi.org/10.1016/s0043-1354(97)00201-7

Harada H, Shimizu Y, Sato D, Miyagoshi Y, Matsui S, Matsuda T, Nagasaki T (2006) Development of a struvite-recovery plant from urine using a urine-diversion toilet. J Jpn Sci Water Environ 28(3):191–196

Hellstrom D, Karrman E (1997) Energy analysis and nutrient flows of various sewerage systems. Water Sci Technol 35(9):135–144. https://doi.org/10.1016/s0273-1223(97)00191-1

Hellstrom D (2003) Exergy analysis of nutrient recovery processes. Water Sci Technol 48(1):27–36

Hoglund C, Stenstrom TA, Jonsson H, Sundin A (1998) Evaluation of fecal contamination and microbial die-off in urine separating systems. Water Sci Technol 38(6):17–25. https://doi.org/10.1016/s0273-1223(98)00563-0

Hoglund C, Vinneras B, Stenstrom TA, Jonsson H (2000) Variation of chemical and microbial parameters in collection and storage tanks for source separated human urine. J Environ Sci Health A 35(8):1463–1475. https://doi.org/10.1080/10934520009377047

Hotta S, Funamizu N (2008) Evolution of ammonification potential in storage process of urine with fecal contamination. Biores Technol 1:13–17. https://doi.org/10.1016/j.biortech.2006.12.001

Kakimoto T, Funamizu N (2006) Effect of phosphate, ammonia and pH on the degradation of antibiotics in the composting toilet. Environ Eng Res 43:429–435

Kakimoto T (2007) Treatment of pharmaceuticals at source using composting process of human excrement. Ph. D Dissertation, Hokkaido University, Sapporo

Kempers AJ, Zweers A (1986) Ammonium determination in soil extracts by the salicylate method. Commun Soil Sci Plant Anal 17(7):715–723. https://doi.org/10.1080/00103628609367745

Larsen TA, Gujer W (1996a) Separate management of anthropogenic nutrient solutions (human urine). Water Sci Technol 34(3–4):87–94. https://doi.org/10.1016/0273-1223(96)00560-4

Larsen TA, Gujer W (1996b) The concept of sustainable urban water management. Water Sci Technol 35(9):3–10. https://doi.org/10.1016/s0273-1223(97)00179-0

Larsen TA, Peters I, Alder A, Eggen R, Maurer M, Muncke J (2001) Re-engineering the toilet for sustainable wastewater management. Environ Sci Technol 35(9):192A–197A. https://doi.org/10.1021/es012328d

Larsen TA, Lienert J, Joss A, Siegrist H (2004) How to avoid pharmaceuticals in the aquatic environment. J Biotechnol 113:295–304. https://doi.org/10.1016/j.jbiotec.2004.03.033

Lind B, Ban Z, Byden S (2000) Nutrient recovery from human urine by struvite crystallization with ammonia adsorption on zeolite and wollastonite. Biores Technol 73(2):169–174. https://doi.org/10.1016/s0960-8524(99)90157-8

Maurer M, Pronk W, Larsen TA (2006) Treatment processes for source-separated urine. Water Res 40(17):3151–3166. https://doi.org/10.1016/j.watres.2006.07.012

Osawa T (2007) Electro-oxidation of pharmaceuticals in urine. Master thesis, Hokkaido University, Sapporo (in Japanese)

Pronk W, Palmquist H, Biebow M, Boller M (2006a) Nanofiltration for the separation of pharmaceuticals from nutrients in source-separated urine. Water Res 40(7):1405–1412. https://doi.org/10.1016/j.watres.2006.01.038

Pronk W, Biebow M, Boller M (2006b) Treatment of source separated urine by a combination of bipolar electrodialysis and a gas transfer membrane. Water Sci Technol 53(3):139–146. https://doi.org/10.2166/2006.086

Scheiner D (1975) Determination of ammonia and kjeldahl nitrogen by idophenol method. Water Res 10(1):31–36. https://doi.org/10.1016/0043-1354(76)90154-8

Suzuki H, Matsugi Y (2005) Integrated microfluidic system for the simultaneous determination of ammonia, creatinine, and urea. Sens Actuators, B 108(1–2):700–707. https://doi.org/10.1016/j.snb.2004.12.032

Udert KM, Larsen TA, Biebow M, Gujer W (2003a) Urea hydrolysis and precipitation dynamics in a urine-collecting system. Water Res 37(11):2571–2582. https://doi.org/10.1016/s0043-1354(03)00065-4

Udert KM, Fux C, Munster M, Larsen TA, Siegrist H, Gujer W (2003b) Nitrification and autotrophic denitrification of source-separated urine. Water Sci Technol 48(1):119–130

Udert KM, Larsen TA, Gujer W (2003c) Biologically induced precipitation in urine-collecting systems. Water Sci Technol Water Supply 3:71–78

Udert KM, Larsen TA, Gujer W (2003d) Estimating the precipitation potential in urine-collecting systems. Water Res 37(11):2667–2766. https://doi.org/10.1016/s0043-1354(03)00071-x

Udert KM, Larsen TA, Gujer W (2006) Fate of major compounds in source-separated urine. Water Sci Technol 54(11–12):413–420. https://doi.org/10.2166/wst.2006.921

Urase T, Kikuta T (2005) Separate estimation of adsorption and degradation of pharmaceutical substances and estrogens in the activated sludge process. Water Res 39(7):1289–1300. https://doi.org/10.1016/j.watres.2005.01.015

Vinneras B, Jonsson H (2002) The performance and potential of faecal separation and urine diversion to recycle plant nutrients in household wastewater. Biores Technol 84(3):275–282. https://doi.org/10.1016/s0960-8524(02)00054-8

Wilsenach JA, Schuurbiers CAH, van Loosdrecht MCM (2006) Phosphate and potassium recovery from source separated urine through struvite precipitation. Water Res 41(2):458–466. https://doi.org/10.1016/j.watres.2006.10.014

Yasojima M, Kobayashi Y, Nakada H, Komori K, Suzuki Y, Tanakam H (2005) Behavior of human antibiotics in wastewater treatment plants. Environ Eng Res 42:357–368

Zhang Y, Wen G, Zhou Y, Shuang S, Dong C, Choi MMF (2006) Development and analytical application of an uric acid biosensor using an uricase-immobilized eggshell membrane. Biosens Bioelectron 22(8):1791–1797. https://doi.org/10.1016/j.bios.2006.08.038

Chapter 10
Volume Reduction of Urine

Muhammad M. Pahore, Ryusei Ito and Naoyuki Funamizu

Abstract Human urine contains nitrogen, phosphorus, and potassium, which can be applied as natural fertilizer for agriculture instead of commercially available fertilizer. However, due to low contents of nutrients in urine, its huge quantity is required to be transported to farmland to meet nutrient demand of crops, which increases transportation cost. To address transportation issue, a new onsite volume reduction system was tested at laboratory scale. Mathematical water transport model was proposed to evaluate performance of the system. Mass transfer coefficient and resistance of water flow through the vertical gauze sheet for the water transport model were obtained from laboratory-scale experiments using water as one of the model samples. The results agreed with simulated data thereby confirming the proposed model. The existing design procedures pertaining to an onsite volume reduction system were established incorporating the water transport model. The existing design procedure was then applied to dry climate of Southern Pakistan having an average air temperature of 30–40 °C and air humidity 20–40% for 80% volume reduction to reduce transportation cost. The findings revealed that the estimated size of the vertical gauze sheet is 2060–440 cm^2, which requires small area for setting up the system at individual household level.

Keywords Effective evaporation area · Capillarity · Mass transfer coefficient Saturated vapor pressure · Solubility · Water transport model

M. M. Pahore (✉) · R. Ito
Graduate School of Engineering, Hokkaido University,
kita13-nishi 8, Kita-ku, 060-8628 Sapporo, Hokkaido, Japan
e-mail: stranger.muhammad@gmail.com

N. Funamizu
Graduate School of Global Food Resources, Hokkaido University,
Kita-9, Nishi-9, Kita-ku, 060-8589 Sapporo, Japan

N. Funamizu
Research Institute of Humanity and Nature, Motoyama Kamigamo,
Kita-Ku, 603-8047 Kyoto, Japan

N. Funamizu (ed.), *Resource-Oriented Agro-sanitation Systems*,
https://doi.org/10.1007/978-4-431-56835-3_10

10.1 Introduction

The urine is rich in nutrient as compared to feces (Krichmann and Pettersson 1995) and is also free from microbial contamination. Therefore, it is highly recommended to be separated at source for proper treatment. The immediate benefits of urine separation would be an increased capacity and better effluent quality of existing wastewater treatment plants, with a lower overall resource consumption (Wilsenach et al. 2007). Likewise, another advantage of source separation of urine is saving energy cost in composting processes of feces in composting toilets. Urine source separation and nutrient recovery can significantly contribute for prevention of eutrophication of surface water (Wilsenach et al. 2007), protecting and maintaining quality of freshwater environment for betterment of human health and ensuring promotion of sustainable development.

Urine contains high concentration of nitrogen, phosphorus, and potassium as compared to other wastewater streams, but is less than 1.5% of the total volume of sewage (Jonsson et al. 1997). The heavy metal content in the urine is generally low as compared to commercially available phosphate fertilizers or manure (Krichmann and Pettersson 1995; Jonsson et al 1997; Maurer et al. 2006).

Because the composition of urine reflects the average requirements for plant growth (Heinonen-Tanski and Van Wijk-Sijbesma 2005), the use of urine as fertilizer in agriculture is the most obvious application (Maurer et al. 2006). Therefore, urine can be transported to farmland for its application as urine-based fertilizer instead of commercially available fertilizer.

Huge quantity of urine may be transported to farmland for meeting nutrient requirements of crops owing to low nutrient concentration in urine, as summarized in Table 10.1. Therefore, application of urine as fertilizer is problematic with regard to transportation (Lind et al. 2001; Maurer et al. 2006). To resolve the transportation issue, various steps have been suggested such as local storage and transport in sewer overnight (Larsen and Gujer 1996) and long-term local storage followed by truck transport in Sweden (Hanaeus et al. 1997). Likewise, some volume reduction techniques including reverse osmosis, electrodialysis, freezing and thawing, and thermal drying system have been tried (Maurer et al 2006). However, all these techniques require high energy and operation and maintenances cost, therefore, a new simple onsite volume reduction system (OVRS) in light of drying theory was conceived based on atmospheric energy.

Table 10.1 Concentration of nitrogen, phosphorus, and potassium in the fresh human urine*

Nutrient	Concentration (g L^{-1})
Nitrogen	6.3–8.8
Phosphorus	0.4–2.0
Potassium	1.7–2.7

*Based on the results in Table 9.1 in Chap. 9

The OVRS is comprised of a tank in which urine can be collected daily and a gauze sheet placed vertically hanging above the tank. The lower portion of the sheet will be in contact with urine. The diagram of proposed onsite volume reduction system is shown in Fig. 10.1. The water under capillary pressure is expected to move upward in the vertical sheet. It is important to know how easy does water move up in the sheet and which height it can reach in the sheet. Evaporation can take place through the sheet surface in the specified airflow, air temperature, and air humidity conditions. The vertical sheet defines the water balance during the experiments. This necessitates to understand the mechanism of the water supply from the tank to the vertical sheet, as well as the evaporation through the sheet surface.

In this chapter, the works by Pahore et al. (2010, 2011, 2012) on OVRS are demonstrated. In their works, a water transport model of a vertical sheet was developed and used for the evaluation of performance for a new OVRS proposed. At first, the mathematical model for water transport was proposed. The experiments were designed based on the parameters stated in the proposed water transport model. The objectives of these experiments were focused on obtaining data of the mass transfer coefficient and the water penetration or resistance of the water flow through a vertical gauze sheet. The experiment results and the simulated data were compared for the confirmation of the proposed model. In this paper, a comprehensive design procedure was formulated for the estimation of a vertical gauze

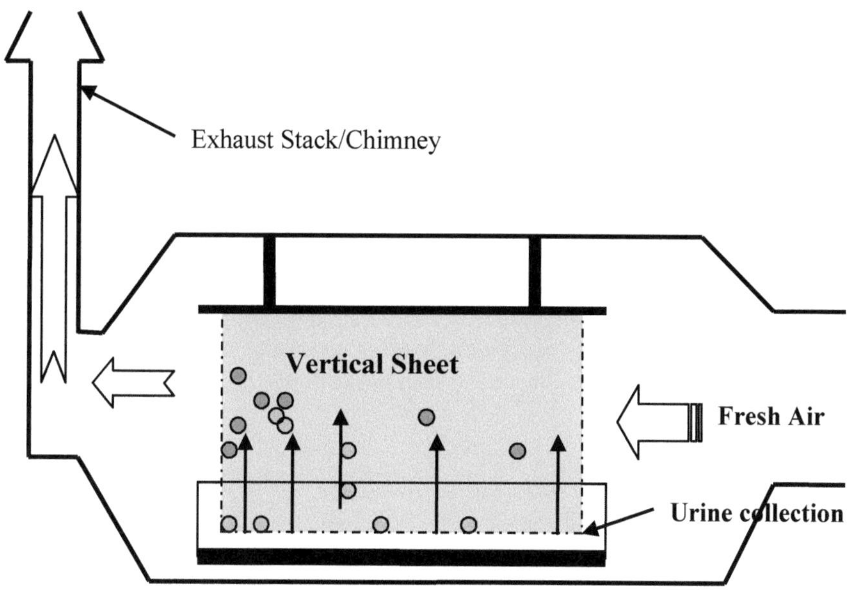

Fig. 10.1 Diagram illustrating concept of onsite volume reduction system. Reprinted from Environmental Technology 31(4):399–408 (2010) with permission from the copyright holders, Taylor & Francis Ltd., www.tandfonline.com

sheet size for the OVRS, suitable to the dry climate of Southern Pakistan. Subsequently, the same procedure can also be adopted for estimating the size of the vertical gauze sheet of the OVRS, suitable to other climate conditions based on the air temperature and the air humidity data.

10.2 The Water Transport Model of the Vertical Gauze Sheet

10.2.1 The Model for Describing Water in the Vertical Gauze Sheet

The key component of the OVRS is a vertical sheet whose bottom end is in contact with urine collected in a tank. In the drying process, water penetrates into the sheet through the capillaries along the sheet fibres. Then, the water on the sheet surface transfers into the atmosphere. For the penetration process, the water moves up toward the top of the sheet by the capillary force against gravity and drag forces. The concept of the OVRS is based on the following mass balance over the vertical sheet.

$$\text{Accumulation of water in the gauze sheet} = q_{\text{supply}} - q_{\text{evaporation}} \tag{10.1}$$

The supplying rate of the water q_{supply} (g/hr) is estimated from the Lucas–Washburn equation (Lucas 1918; Washburn 1921);

$$q_{\text{supply}} = WT\frac{\sigma}{\mu}\frac{H_{\max} - H}{H} \tag{10.2}$$

where W and T are, respectively, width and thickness of the vertical sheet (cm), H and $H_{\max}$ are, respectively, current and maximum water level in the sheet (cm), σ is the penetration factor of water into the sheet (g cm^{-1} h^{-2}), and μ is the viscosity of the water (g cm^{-1} s^{-1}). σ includes the effect of surface tension and the capillary diameter. According to the evaporation process, the water molecules vaporize on the surface of the sheet, then, they are transported into the air through the boundary layer by diffusion.

The drying theory expresses the evaporation rate $q_{\text{evaporation}}$ (g h^{-1}) of water from wet surface in the following mass transfer equation (McCabe and Smith 1967);

$$q_{\text{evaporation}} = M_{\text{air}}\,Ky\,(Xi - X)\,A \tag{10.3}$$

where M_{air} is the molecular weight of the air (g $kmol^{-1}$), Ky is the mass transfer coefficient (kmol cm^{-2} h^{-1}), Xi and X are, respectively, humidity at the wet surface and in the air (g-water/g-dry air), and A is the effective drying area (cm^2). In this paper, saturated humidity at air temperature was applied for Xi.

Now, therefore, substituting Eqs. (10.2) and (10.3) into (10.1) gives the following model equation,

$$WT\left(\frac{dH}{dt}\right) = WT\left(\frac{\sigma}{\mu}\right)\left(\frac{H_{max} - H}{H}\right) - M_{air}Ky(Xi - X)A \qquad (10.4)$$

Regarding the estimation of the mass transfer coefficient, the mass transportation theory gives the following relationship among three non-dimensional numbers: Sherwood, *Sh*, Reynolds, *Re,* and Schmidt, *Sc*, numbers (McCabe and Smith 1967);

$$Sh = \alpha\, Re^{\beta} Sc^{\gamma} \qquad (10.5)$$

where α, β, and γ are constant values.

Here, these non-dimensional numbers are calculated by following equations,

$$Sh = \frac{Ky\, R\, T\, L}{PT\, D_{AB}} \qquad (10.6)$$

$$Re = \frac{\rho\, L\bar{u}}{\mu_{air}} \qquad (10.7)$$

$$Sc = \frac{\mu_{air}}{\rho\, D_{AB}} \qquad (10.8)$$

where L is the characteristic length (cm), D_{AB} is the diffusivity of water in the air (cm^2 s^{-1}), ρ is the density of air (g cm^{-3}), μ_{air} is the viscosity of the air (g cm^{-1} s^{-1}), T is the air temperature (in K), R is the gas constant (Pa cm^3 $kmol^{-1}$ K^{-1}), $\bar{u}$ is the mean fluid velocity (m sec^{-1}), and P_T is the total system pressure (k Pa.).

The diffusivity of water in air, $D_{AB,}$ is calculated from the following equation (Brokaw 1960, 1965; Perry and Green 1984);

$$D_{AB} = \frac{0.001858\, T^{3/2} M_{AB}^{3/2}}{(P\sigma_{AB}^{1/2}\Omega_D)} \qquad (10.9)$$

where P is the pressure (Pa s), M_{AB} is the mean molecular weight of air and water (g $kmol^{-1}$), σ_{AB} is the binary pair characteristic length (in Å), Ω_D is the collision function for diffusion (dimensionless).

10.2.2 Experimental Setup and Material

Figure 10.2 illustrates the experimental setup for the laboratory-scale OVRS. The vertical sheet was placed over the urine collection tank, while its bottom end was in permanent contact with the urine. The wind tunnel was enclosed and it comprised of a vertical gauze sheet, a urine collection tank, and a power controllable fan placed behind the vertical sheet to ensure turbulent flow. The tunnel was connected by an air recirculation tube to a chamber with constant air temperature and humidity.

The drying rate was estimated measuring weight of the tank by an electric balance located under the tank. A gauze vertical sheet was chosen because it has large pore size. At first, synthetic urine (Wilsenach et al. 2007) was used in evaporation rate experiments to confirm that OVRS can be applied for volume reduction of urine. Subsequently, de-ionized water was used instead of urine in order to understand the development of the water transport model based on the water penetration through the sheet and the evaporation.

The water level in the vertical gauze sheet, H was periodically measured along with the maximum water level in the sheet, H_{max} during the 24 h long experiments. The observed results were compared with the simulated data from Eq. (10.2) by Runge–Kutta method (Nyhoff and Leesthma 1985) to estimate the penetration factor, σ. The measurement was performed at 100% air humidity, 25 °C air temperature and 0 m sec^{-1} wind velocity without allowing evaporation.

The evaporation rate ($q_{evaporation}$) was estimated by measuring decrease of the tank weight in three experiment setups, RUN-1, RUN-2, and RUN-3, with a

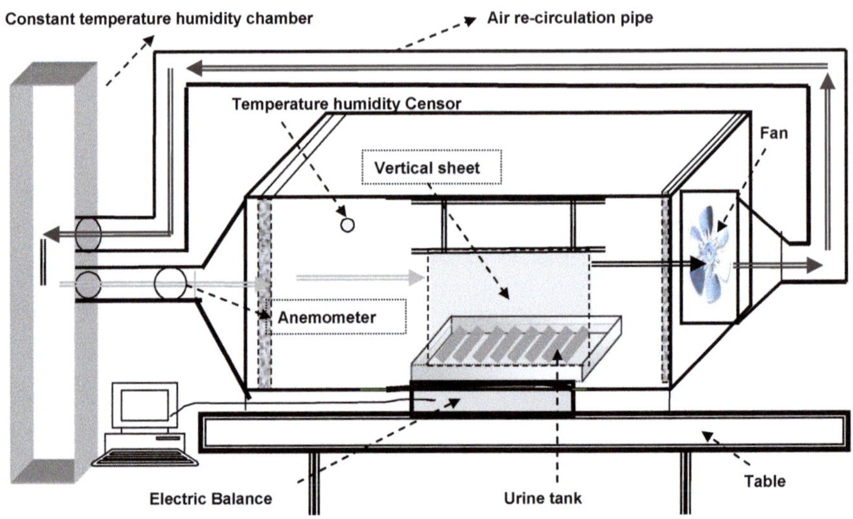

Fig. 10.2 Laboratory-scale experimental setup of onsite volume reduction system. Reprinted from Environmental Technology 31(4):399–408 (2010) with permission from the copyright holders, Taylor & Francis Ltd., www.tandfonline.com

Table 10.2 Experimental conditions for the evaporation rate experiments of the onsite volume reduction system

Run no.	Air humidity (%)					Air temperature (°C)					Wind velocity (m sec^{-1})				
RUN-1	40					40					2	3	4	5	6
RUN-2	50	60	70	80	90	25					5				
RUN-3	50					20	25	30	35	40	5				

combination of constant and variable conditions as summarized in Table 10.2. The water level in the sheet was also measured. The mass transfer coefficient, *Ky*, was estimated from the Eq. (10.3).

10.3 Experimental Results: Water Evaporation Rate and Confirmation of the Model

10.3.1 Estimation of the Evaporation Rate Using Synthetic Urine

Figure 10.3 shows the changes in the evaporation rate for one of the conditions set in RUN-1 during the 48 h long experiment with synthetic urine.

The evaporation rate in the initial stage was observed to be high owing to the clean virgin sheet. The steady state condition was established after about a couple of

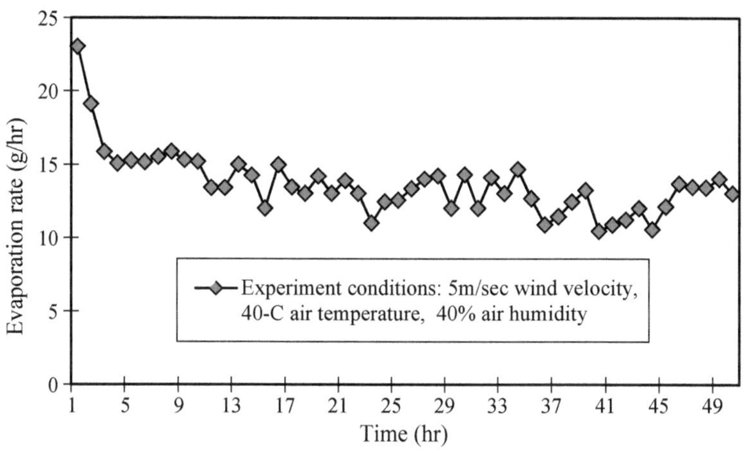

Fig. 10.3 Time course of evaporation rate for one of the conditions set for RUN-1 and operated for 48 h long experiment using synthetic urine. Reprinted from Environmental Technology 31 (4):399–408 (2010) with permission from the copyright holders, Taylor & Francis Ltd., www.tandfonline.com

hours, when the evaporation rate did not decrease anymore. Therefore, achieving steady state condition confirmed that the evaporation rate was not significantly decreased by an increase of the salt concentration in the vertical sheet during the experiment.

10.3.2 Estimation of the Water Supply Rate Using de-Ionized Water

Figure 10.4 shows the water level rise in the vertical gauze sheet during the 24 h long experiment with de-ionized water. The water level increased with time. The rising rate of water level in the initial stage was observed to be high owing to the empty sheet during the first couple of hours. The approximation by Eq. (10.2) agreed with the observed data, while the values of σ and H_{max} were found to be 1.0×10^{-5} g cm^{-1} s^{-2} and 31 cm, respectively.

10.3.3 Estimation of the Evaporation Rate Using de-Ionized Water and Mass Transfer Coefficient

Figure 10.5 shows the evaporation rate for RUN-1 during 9 h of experiment with de-ionized water. The evaporation rate in the initial stage was observed to be high in all experiments. The steady state condition was established after the water supply to the sheet through the capillaries was balanced by the evaporation.

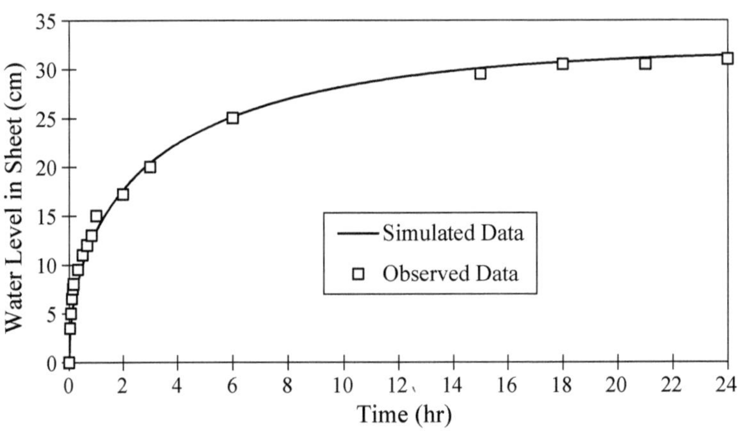

Fig. 10.4 Time course of water level in the gauze sheet using observed and simulated data. Reprinted from Environmental Technology 31(4):399–408 (2010) with permission from the copyright holders, Taylor & Francis Ltd., www.tandfonline.com

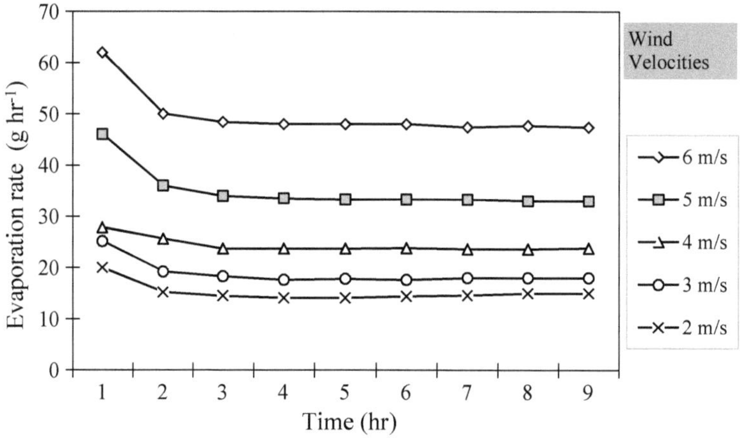

Fig. 10.5 Time course of the evaporation rate for experiments of RUN-1. Reprinted from Environmental Technology 31(4):399–408 (2010) with permission from the copyright holders, Taylor & Francis Ltd., www.tandfonline.com

The evaporation rate increased significantly with the increase of the wind velocity, because the high wind velocity makes strong turbulent flow resulting in thin boundary layer. Under this condition, a high concentration gradient is formed for fast molecular transportation.

The mass transfer coefficient can be characterized in a non-dimensional form by Sherwood Number, which is a function of Reynolds Number and Schmidt Number (Perry and Green 1984). The observed experimental data was summarized in the non-dimensional form. The change in *Sc* was observed to be very small due to the fact that the air temperature range of the performed experiments was narrow. Thus, the value of *Sc* was considered as constant. Therefore, the experimental data was summarized in the non-dimensional form of *Sh* and *Re*. *Sh* was obtained from the mass transfer coefficient with Eq. (10.6). The values of α and β were evaluated in approximation to obtain data and calculated by Eq. (10.6) and presented in Fig. 10.6. The values of α and β came out to be 0.08 and 0.266, respectively. The following equation is formulated from the experimental data.

$$Sh = 0.08\,Re^{0.266} \tag{10.10}$$

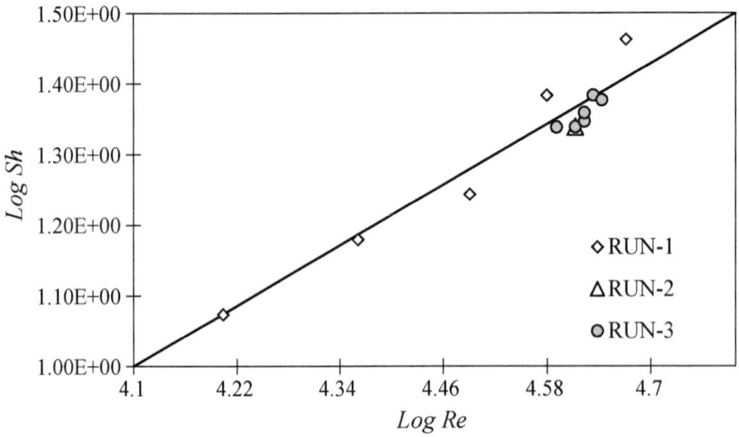

Fig. 10.6 Summary of the mass transfer coefficient in non-dimensional form of *Re* and *Sh* using data of RUN-1, RUN-2, and RUN-3. Reprinted from Environmental Technology 31(4):399–408 (2010) with permission from the copyright holders, Taylor & Francis Ltd., www.tandfonline.com

10.3.4 Confirmation of the Water Transport Model of the Vertical Sheet

Estimated water level, *H*, at steady state was compared with the experimental results obtained at various experimental conditions. The result is illustrated in Fig. 10.7. Every point is plotted in equivalent line, while the accuracy of the water transport model is authenticated and validated.

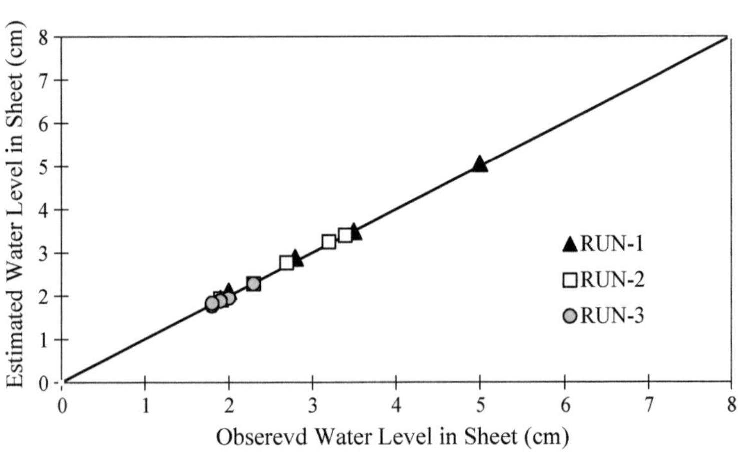

Fig. 10.7 Comparison of the observed and the estimated water level in the sheet using the evaporation rate experiment data. Reprinted from Environmental Technology 31(4):399–408 (2010) with permission from the copyright holders, Taylor & Francis Ltd., www.tandfonline.com

10.4 Preliminary Design Procedure of the Vertical Sheet for OVRS

10.4.1 Design Procedure

A step-by-step preliminary procedure to design the vertical sheet based on different climate conditions and daily water loads is explained in the following sections.

- Step 1: Setting the volume reduction target.
  It is essential to set targets for urine volume reduction through evaporation from a vertical sheet used in the OVRS and applied at household level on daily basis. In order to provide financial incentives to farmers to use urine instead of commercially available fertilizer, higher volume reduction rates, like 80% volume reduction per day, may be considered.
- Step 2: Available data and design parameters.
  The readily available data from the climate and the established data from the experiments can be used for estimation of the size of the sheet required for the volume reduction set in the step 1. Unknown data, given data, and climate-based data are explained in Table 10.3.
- Step 3: Estimation of the mass transfer coefficient.
  The OVRS is based on atmospheric energy, so it is necessary to take into account the air temperature, the air humidity, and the wind velocity of the region for which the system has to be designed. For a particular climate, the mass transfer coefficient can be estimated from Eq. (10.10).
- Step 4: Estimation of height for the vertical sheet.
  The height for the vertical sheet can be estimated from the following simplified form of the Eq. (10.4).

$$H = -\left(\frac{\sigma}{\mu}\right) + \sqrt{\frac{\frac{M_{\mathrm{air}} Ky(Xi-X)}{T} - 4\left(\frac{\sigma}{\mu}\right)}{2\left(\frac{\sigma}{\mu}\right)}} \tag{10.11}$$

Table 10.3 The given parameters and the unknowns for the design of the vertical gauze sheets of onsite volume reduction system

Given				Unknown
Target	Constant	From climate	From experiment	
Quantity of urine to be treated per day	Molecular weight of air	Wind velocity air temperature air humidity viscosity of air	Inter-relationship between *Re*, *Sh* and *Sc*, σ, H_{max} type of material for the vertical sheet	Size of the vertical sheet (W and H)

– Step 5: Estimation of width for the vertical sheet.
  The width for the vertical sheet can be calculated from the following form of Eq. (10.2):

$$W = \frac{q_{\text{evaporation}}}{24 M_{\text{air}} K y (Xi - X) H} \tag{10.12}$$

where

$q_{\text{evaporation}}$ is quantity of urine to be treated per day

– Step 6: Application of more than one sheet.
  For practical application purpose, it is recommended to split the sheet width into suitable numbers in order to achieve adequately smaller system requiring small area for installation. In our expected conditions, airflow should be turbulent resulting in forming thin viscous sublayer, so that we could treat each sheet independently. However, effect of the distance between the sheets on the drying may need to be investigated through a separate study.

10.4.2 Estimation of Size for a Vertical Sheet for OVRS Adopted to Dry Climate Conditions of Southern Pakistan

After confirmation, the water transport model was applied to estimate the size of a vertical gauze sheet for OVRS of urine in Southern Pakistan as an example.

Because Southern Pakistan has dry and hot climate with poor piping system, so urine drying with natural energy like sun drying of laundry is expected to work well in this area. Hence, it is anticipated that the proposed OVRS may be effectively used in dry climate conditions as compared to the highly humid or extremely cold climate. A higher volume reduction rate, for example, 80% volume reduction, may be preferred to provide financial incentives to the farmers to transport urine from household to their farmland and apply it as fertilizer instead of commercially available fertilizer. The size of the vertical sheet required for 80% volume reduction of urine for dry climate of Southern Pakistan was estimated to assess how much area of land would be required to install the OVRS at household level. The area of land is particularly important for developing countries due to the fact that human settlements are closely concentrated due to limitation of availability of utilities. In Southern Pakistan, each household is comprised of about 10 family members. Considering 80% residence time of each person at house, each household is

Table 10.4 Estimated size of the vertical gauze sheet for the onsite volume reduction system to be adopted for dry climate conditions of Southern Pakistan

Given data	Design parameters			
Wind velocity (m sec^{-1})	Required dimensions of vertical drying sheet			
	H (cm)	*W* (cm)	*w* = *W*/4 (cm)	Size (cm^2)
2	5	411	103	2060
3	3.5	395	98	1385
4	2.9	342	85	993
5	2.3	276	69	636
6	1.9	231	57	440

expected to generate about 10 L of urine per day. Therefore, target quantity of urine to be evaporated per day becomes 8 L, which is 80% of 10 L. The size of vertical sheet estimated for dry climate of Southern Pakistan (air temperature 40 °C, air humidity 40%, and wind velocity 2–6 m sec^{-1}) is illustrated in Table 10.4.

10.4.3 Other Issues Which May Influence OVRS

Three issues are concerned with urine treatment by OVRS, including high salt concentration, the effect of organic matter and the pH increase caused by hydrolysis of urea to ammonia nitrogen. The salts in urine may be expected to move up in the vertical gauze sheet along with the water affecting the evaporation rate during the urine concentration process. However, the gauze is expected to have enough capacity to retain a certain quantity of salts, beyond which the salts may fall back to the tank or the tank cover due to gravitational force, as a result of increase in weight of salt particles and high frequency of vibrations caused by the wind velocity. The salts accumulated in small capillaries of the gauze may also be moved up under capillary pressure clearing the pathway in the small threads of the cloth, and in this way maintaining the flow of water in the sheet. Thus, the water flow and the evaporation through the surface of the sheet probably will not be affected, or the drying process significantly inhibited. Similarly, as an alternative approach, the possibility of changing out the vertical sheet from time to time may be explored, if such a need arises. In the experiment, using synthetic urine significant inhibition of the process was not found. However, 48 h of testing may be itself a limiting factor for more accurate calculations.

The concentration of the organic substances in urine is high about 10,000 gCOD m^{-3}, which includes organic acids, creatinine, amino acids, carbon hydrates, and micropollutants (Ciba-Geigy 1977). The concentration of organic substances in the urine sludge is substantially higher than in the urine solution (Udert 2003). The synthetic urine used in the present experiment contained creatinine which is one of the important organic matters. However, effect of all the organic substances on drying process could not be covered in this study. Therefore, a separate study may be carried out to assess the implications of organic matter on drying process using actual human urine. Likewise, effect of the pH increase in the urine during this drying method may also need to be studied separately.

10.5 Conclusion

In this study, a mathematical water transport model of a vertical gauze sheet was developed to be able to evaluate the performance of onsite volume reduction system (OVRS), which is a new volume reduction method based on drying process for nutrient recovery from source-separated urine. At first, the mathematical model was conceived considering the water penetration into the vertical gauze sheet and the water evaporation into the atmosphere. The experiments were designed based on the parameters stated in the water transport model of the vertical gauze sheet. The values of mass transfer coefficient and the penetration factor were obtained from the experiments. The findings of this study are explained as under:

- The experimental results agreed with the simulated data thereby confirming and validating the accuracy of the proposed water transport model of the vertical gauze sheet. It is clear that the proposed water transport model is reliable enough to be applied for estimation of the size of a vertical gauze sheet required for various climate conditions.
- The water transport model was applied to the conditions of Southern Pakistan, which has hot and dry climate with air temperatures around 30–40 °C and an air humidity about 20–40%. Based on the mass transfer coefficient and the penetration factor, the estimated size of a vertical gauze sheet was found to be in the range of 440–2060 cm^2, optimal for 80% volume reduction of 10 L urine per day. This size is quite adequate in densely populated areas, requiring small places to install and run the system at household level. It can be concluded that the proposed OVRS can be effectively used for volume reduction of urine in areas with dry climate.

References

Brokaw RS (1960) Predicting transport properties of dilute gases. J Chem Phys 32:1005–1006

Brokaw RS (1965) Predicting transport properties of dilute gases. J Chem Phys 42:1140–1146

Ciba-Geigy (1977) Wissenschaftliche Tabellen Geigy, Teilband Korperflussigkeiten (Scientific tables Geigy, volume body fluid) 8th edn. Basel in German

Hanaeus J, Hellstrom D, Johansson E (1997) A study of urine separation in an ecological village in northern Sweden. Water Sci Technol 35:153–160

Heinonen-Tanski H, Van Wijk-Sijbesma C (2005) Human excreta for plant production. Biores Technol 96:403–411. https://doi.org/10.1016/j.biortech.2003.10.036

Jonsson H, Stenstorm TA, Sundin A (1997) Source-separated urine-nutrient and heavy metal content, water saving and fecal contamination. Water Sci Technol 35:145–152. https://doi.org/10.1016/S0273-1223(97)00192-3

Krichmann H, Pettersson S (1995) Human urine—chemical composition and fertilizer use efficiency. Fertilizer Res 40:149–154. https://doi.org/10.1007/BF00750100

Larsen TA, Gujer W (1996) Separate management of anthropogenic nutrient solution (human urine). Water Sci Technol 34:87–94. https://doi.org/10.1016/0273-1223(96)00560-4

Lind BB, Ban Z, Byden S (2001) Volume reduction and concentration of nutrients in human urine. Ecol Eng 16:561–566. https://doi.org/10.1016/s0925-8574(00)00107-5

Lucas VR (1918) Ueber das Zeitgesetz des kapillaren Aufstiegs von Flüssigkeiten. Colloid Polym Sci 23:15–22. https://doi.org/10.1007/bf01461107

Maurer M, Pronk W, Larsen TA (2006) Treatment processes for source separated urine. Water Res 40:3151–3166. https://doi.org/10.1016/j.watres.2006.07.012

McCabe WL, Smith JC (1967) Unit operations of chemical engineering, 2nd edn. McGraw Hill Book Company and Kogakusha Co. Limited, Pennsylvania

Nyhoff J, Leesthma S (1985) FORTRAN 77 and numerical methods for engineers and scientists. Prentice Hall Englewood Cliffs, New Jersey

Pahore MM, Itoh R, Funamizu N (2010) Rational design of an onsite volume reduction system for source-separated urine. Environ Technol 31(4):399–408. https://doi.org/10.1080/09593330903505654

Pahore MM, Ito R, Funamizu N (2011) Performance evaluation of onsite volume reduction system with synthetic urine using water transport model. Environ Technol 32(9):953–970. https://doi.org/10.1080/09593330.2010.521954

Pahore MM, Ushijima K, Ito R, Funamizu N (2012) Fate of nitrogen during volume reduction of human urine using an onsite volume reduction system. Environ Technol 33(2):229–235. https://doi.org/10.1080/09593330.2011.560192

Perry RH, Green DW (1984) Perry's Chemical Engineers Handbook, 7th edn. McGraw Hill Book Company, Pennsylvania

Udert KM (2003) The fate of nitrogen and phosphorus in source-separated urine. Schriftenreihe des Instituts fur Hydromechanik und Wasserwirtschaft 16 Institute for Hydrodynamics and Water Resource Management Swiss Federal Institute of Technology, Zurich

Washburn EW (1921) The dynamics of capillary flow. Phys Rev 17:273–283

Wilsenach JA, Schuurbiers CAH, Van Loosdrecht MCM (2007) Phosphate and potassium recovery from source separated urine through struvite precipitation. Water Res 41:458–466. https://doi.org/10.1016/j.watres.2006.10.014

Chapter 11
Recovery of Nitrogen and Phosphorus from Urine

Ryusei Ito and Naoyuki Funamizu

Abstract Human excreta, especially urine, are rich in nitrogen and phosphorous that can be utilized for agricultural purposes. The recovery of nitrogen from urine has potential to reduce energy consumption for producing nitrogen fertilizer, and the recovery of phosphorus from excreta will reduce the pressure on phosphorus reserves. The direct formation of slow-released fertilizer—methylene urea—from urine was being proposed in this study. The experiments were tried to prove the formation of methylene urea from human urine, and the NMR, IR analyses, and element analysis showed the product is methylene urea. It is experimentally showed that phosphorous in urine can be recovered on the shell surface by immersing shell particles in urine. Scallop (*Mizuhopecten yessoensis*) shell was used in the experiments. SEM/EDS, powder XRD analysis, and composition measurements showed that the precipitate from the experiment with the Ca/P ratio = 1 was DCPD, and amorphous of hydroxyapatite was developed in the experiment with the initial Ca/P = 10. Elemental analysis of the cross section of shell particle by SEM/EDS showed that the reaction started from particle surface and then propagated to inside the particles.

Keywords Phosphorous recovery by shell particles · Slow-released nitrogen fertilizer · Urine

R. Ito (✉)
Graduate School of Engineering, Hokkaido University,
Kita13-Nishi 8, Kita-Ku, Sapporo City, Hokkaido 060-8628, Japan
e-mail: ryuusei@eng.hokudai.ac.jp

N. Funamizu
Graduate School of Global Food Resources, Hokkaido University,
Kita-9, Nishi-9, Kita-Ku, Sapporo 060-8589, Japan

N. Funamizu
Research Institute of Humanity and Nature, Motoyama Kamigamo,
Kita-Ku, Kyoto 603-8047, Japan

N. Funamizu (ed.), *Resource-Oriented Agro-sanitation Systems*,
https://doi.org/10.1007/978-4-431-56835-3_11

11.1 Introduction

Nitrogen and phosphorous fertilizers are important elements for agriculture. The FAO (2015) showed that total fertilizer nutrient ($N + P_2O_5 + K_2O$) consumption is estimated at 183 million tons in 2013. The production of ammonia, which is the main resource of nitrogen fertilizers, requires 34.5 GJ/ton-N of energy by the steam reforming method (Kongshaug 1998). This fact shows the energy consumption of ammonia in the world corresponds to a few percent of the annual net electricity consumptions in the world. It is reported that 850 g of nitrogen is evacuated from a person per year (Almeida et al. 1999), while a human being excretes about 6 Mton of nitrogen per year. The human urine contains 86% of nitrogen in the excreta (Malisie et al. 2007). Most of the nitrogen exists as urea formation which easily decomposes into ammonium formation by biological activity and the hydrolysis reaction resulting from the contamination of feces (Hotta and Funamizu 2008). Therefore, the recovery of nitrogen from urine has potential to reduce energy consumption for producing nitrogen fertilizer.

The main source of phosphorous fertilizer is phosphorous rock, whose production rate is expected to reach a maximum in 2030 (Cordell et al. 2009). Mihelcic et al. (2011) estimated that 3.4 million tons of phosphorus was excreted from humans worldwide in 2009, and 1.7 million tons of which was in urine. MacDonald et al. (2011) reported that 9.6 million tons of phosphorus is excreted from livestock globally. Thus, the excreta are an essential source of phosphorus, and the recovery of phosphorus from excreta will reduce the pressure on phosphorus reserves.

11.2 Production of Slow-Released Nitrogen Fertilizer from Urine (Ito et al. 2013)

11.2.1 Slow-Released Nitrogen Fertilizer

The conventional nitrogen fertilizers are very soluble in water, resulting in low efficiency by the flushing out with rain water. These phenomena cause the redundant consumption and nitrification of surface water. Therefore, the controlled release fertilizers are developing to avoid this problem. Methylene urea, isobutylidenediurea, and cyclodiurea have low solubility and low degradation rate, and they are known as a slow-released nitrogen fertilizer.

The methylene urea formation from urea solution and formaldehyde is well known, and this reaction is used for industrial production. Urine is rich in urea, but urine also contains salts and organic matter with high concentrations. Therefore, it should be examined that these salts and organic matter may affect the methylene urea formation reaction or not. This section shows the experimental results on making methylene urea directly from human urine.

11.2.2 Production of Methylene Urea from Synthetic Urine

The synthetic urine having the composition (Wilsenach et al. 2007) in Table 11.1 was used for the experiments. Figure 11.1 shows the time course of molecular distribution of nitrogen component during methylene urea formation reaction at pH 2. At the beginning, main nitrogen component in the synthetic urine was urea along with relatively small amount of creatinine and ammonium chloride. The urea concentration decreased to form larger molecular size components, but the total amount of nitrogen did not change. The components with the range of 10 k–100 kDa appeared after one hour and then disappeared to show polymerization process of urea derivatives.

Table 11.1 Components of synthetic urine

Chemicals	Concentration (mmol/L)
$MgCl_2{\cdot}6H_2O$	3.2
NaCl	78.7
Na_2SO_4	16.2
$Na_3(C_6H_5O_7){\cdot}2H_2O$	2.6
KCl	21.5
$C_4H_7N_3O$	9.7
$CaCl_2$ $2H_2O$	4.4
KH_2PO_4	30.9
NH_4Cl	18.7
$(NH_2)_2CO$	417
$Na_2(COO)_2$	0.15
$C_6H_8O_6$	0.57

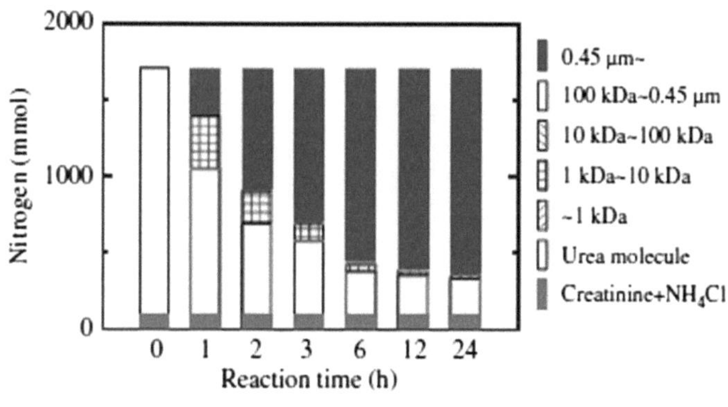

Fig. 11.1 Molecular distribution of urea derivatives at pH 2 (Reprinted from Ito et al. (2013) with permission from the copyright holders, Taylor & Francis Ltd, www.tandfonline.com)

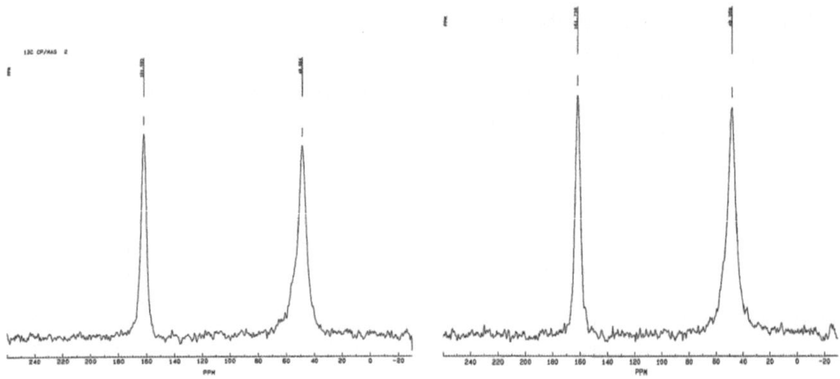

(a) Prepcitates from Synthetic urine **(b) Precipitates from Urea solution**

Fig. 11.2 NMR spectrum of precipitates from synthetic urine (Reprinted from Ito et al. (2013) with permission from the copyright holders, Taylor & Francis Ltd, www.tandfonline.com)

The precipitates were analyzed by means of C13-NMR, IR spectrometry, and element analysis. Figure 11.2a shows the NMR chart of the precipitates. There are two peaks which represent the methylene group (–CH_2—50 ppm) and the ketone group (–C=O—160 ppm). This chart is similar to the precipitates from the urea solution as illustrated in Fig. 11.2b. The element analysis of the precipitates showed that the weight ratio of each element was C: 28.52%, H: 5.57%, N: 35.33% for precipitate from synthetic urine and C: 29.78%, H: 5.59%, N: 36.82% for precipitate from urea solution, respectively. Since these values are almost equal, it seems that the same substances are produced from synthetic urine and urea solution.

11.2.3 Application to Real Urine

The NMR spectrum of precipitate from real urine in Fig. 11.3 shows two big peaks and three small peaks around 60 ppm. The two big peaks seem to be in the same functional groups (–CH_2—50 ppm and –CO—160 ppm) in Fig. 11.2. The small peaks may suggest the existence of unknown functional groups or unknown derivatives of the –CH_2—group. The element analysis showed that the ratio of elements was C: 29.61%, H: 5.85%, and N: 30.35%, and this means that the nitrogen content was lower than precipitates by synthetic urine or urea solution. From the above results, it might be concluded that the main constituent of the precipitation from real urine was methylene urea and the precipitates may contain trace unknown functional groups, which is deficient in nitrogen and requires more analysis for other elements or functional groups in the precipitation.

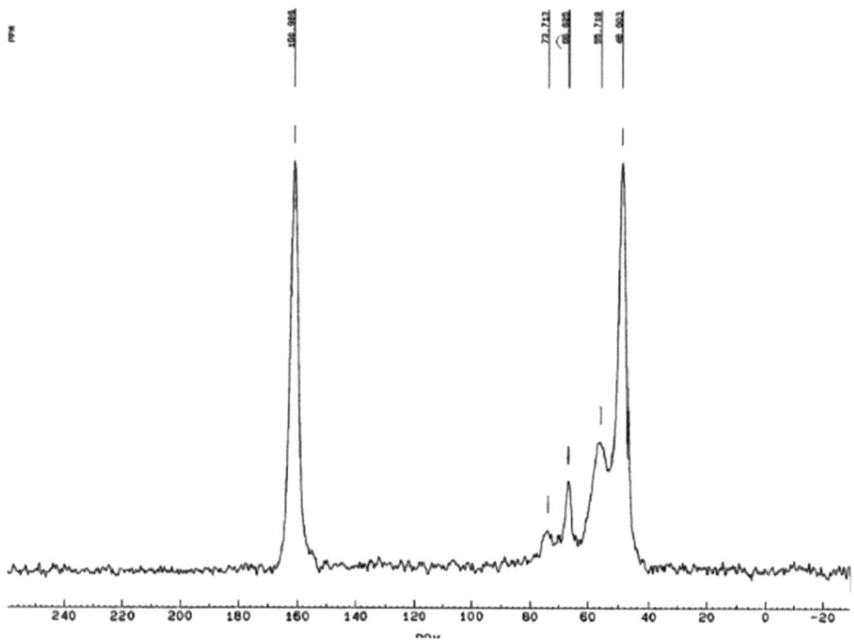

Fig. 11.3 NMR spectrum of precipitates from urine (Reprinted from Ito et al. (2013) with permission from the copyright holders, Taylor & Francis Ltd, www.tandfonline.com)

11.3 Phosphorous Recovery from Urine by Heterogeneous Crystallization (Ito and Funamizu 2016)

*11.3.1 Heterogeneous Crystallization of Phosphorous by Using Scallop (*Mizuhopecten yessoensis*) Shell*

Heterogeneous crystallization occurs on the surfaces of the crystals of calcium salts, resulting in easy liquid–solid separation (Schmelzer 2005). The heterogeneous crystallization process is used for effective phosphate recovery from wastewater (Song et al. 2006; Kato and Matsuoka 1991). Yoshimura and Shiomi (2013) reported that phosphate can be recovered from 50 mmol/L potassium hydrogen phosphate solution by shell particles, and dicalcium phosphate dihydrate (DCPD, $CaHPO_4{\cdot}2H_2O$) was precipitated on the shell particles in 1000 mmol/L potassium hydrogen phosphate solution. In order to develop a phosphorous recovery system from urine by using scallop (*Mizuhopecten yessoensis*) shell, the following heterogeneous crystallization experiments were conducted.

11.3.2 Experiment

Figure 11.4 shows the experimental setup. Crushed and sieved scallop (*Mizuhopecten yessoensis*) shell particles with 0.71–1.19 mm size were packed in the mesh box. Three mesh boxes containing shell particles were vertically connected by a wire, and they were held under a plastic plate of 100 mm in diameter. The plate was placed on the 500 ml conical beaker with all the mesh boxes. The beaker was filled with 500 ml of synthetic urine, and the precipitation process was started by dipping the 27 mesh boxes under the plate in the beaker. The composition of the synthetic urine is summarized in Table 11.1. The conical beaker was maintained at 20 °C, and the solution in the beaker was stirred with a magnetic stirrer at rotational speed of 500 rpm. In order to change the initial molar ratio of calcium concentration to phosphate concentration (Ca/P) from 0.2 to 10.0, the total amount of shell particles was changed in the experiments.

11.3.3 Results-1 Observation of Particle Surface by SEM/EDS

Figure 11.5 shows SEM images of the shell particle surfaces before the experiment, the final products from the experiments with initial Ca/P = 1 and 10, respectively. Pillar-shaped crystals were found on the virgin shell particles as in Fig. 11.5a. After the experiments, square pillar-shaped crystals (Fig. 11.5b, Ca/P = 1.0) and spherical crystals (Fig. 11.5c, Ca/P = 10) were, respectively, grown on the surface. It is reported that among the compounds of calcium phosphate, DCPD forms square pillar crystals, octacalcium phosphate (OCP; $Ca_4(HPO_4)(PO_4)_2{\cdot}2.5H_2O$) forms platy crystals and amorphous HAP forms spherical crystals and amorphous (Amjad 1998). The elements were mapped onto cross-sectional SEM images of the final products for the initial Ca/P ratio of 0.3, 1, and 5, as shown in Fig. 11.6.

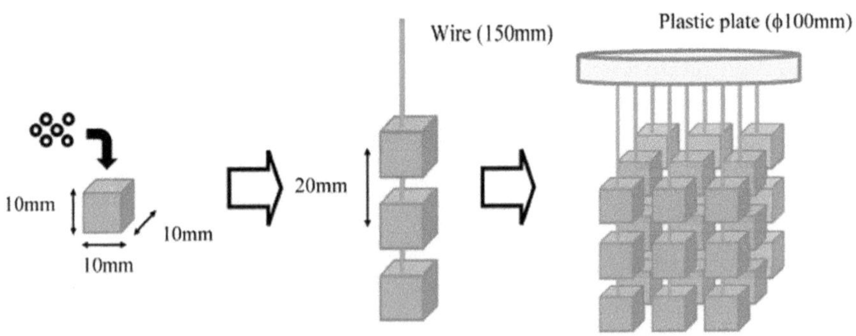

Fig. 11.4 Illustration of experimental setup (Reprinted from Ito et al. (2016) with permission from the copyright holders, JSWE)

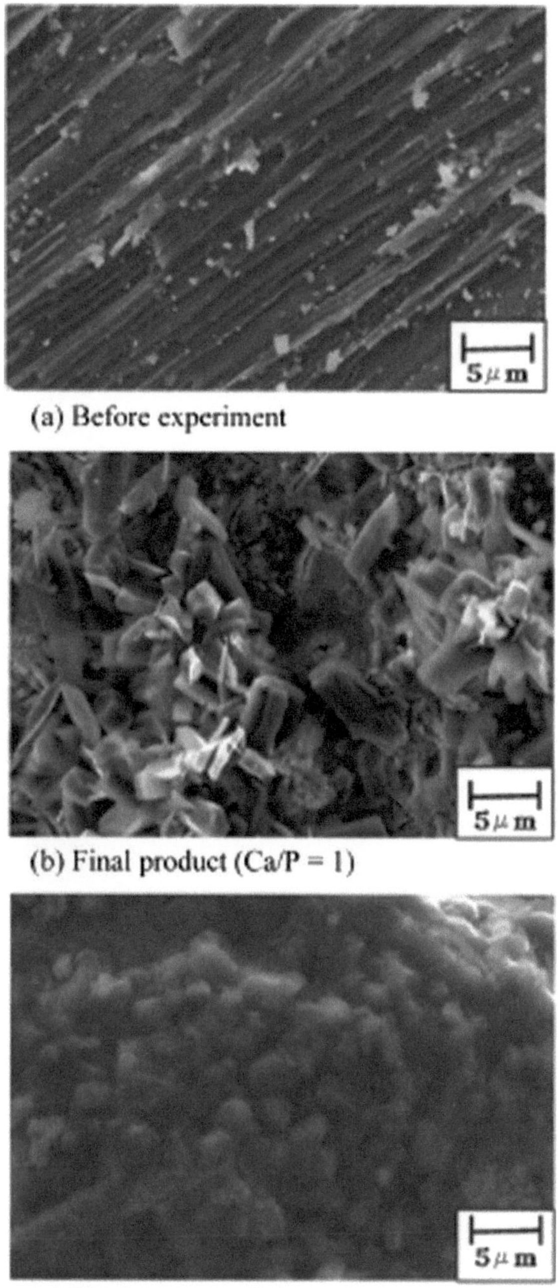

Fig. 11.5 SEM images of shell surfaces before and after the experiment (Reprinted from Ito et al. (2016) with permission from the copyright holders, JSWE)

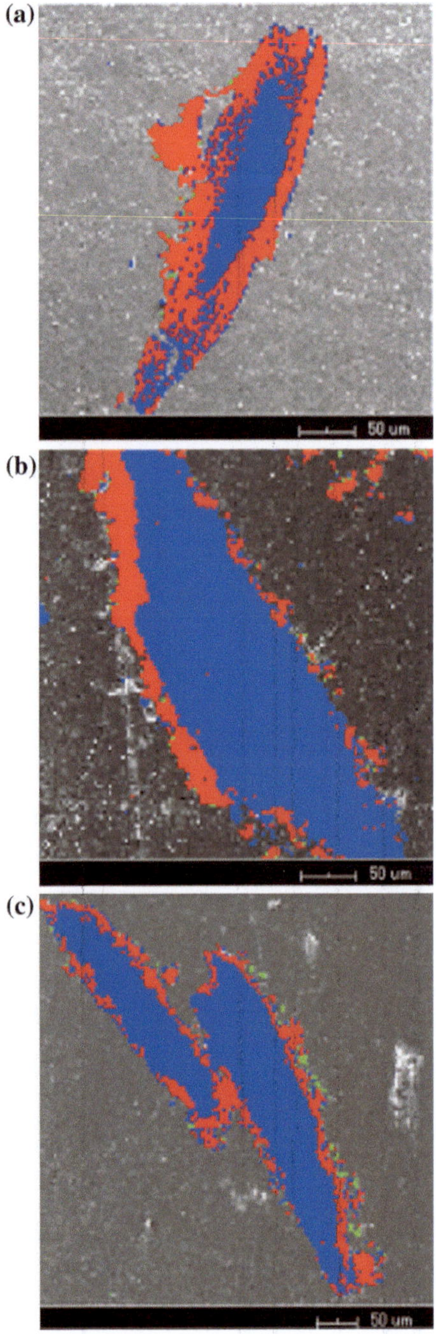

Fig. 11.6 Element mapping of calcium and phosphorous on the cross section of the final products of the experiments. Experimental conditions: **a** Initial Ca/P ratio = 0.3; **b** Initial Ca/P ratio = 1; **c** Initial Ca/P ratio = 5. Blue color indicates the existence of calcium atoms but no phosphorous atoms. Green color shows the existence of phosphorous atoms but no calcium atoms. Red color indicates the existence of both calcium and phosphorous atoms (Reprinted from Ito et al. (2016) with permission from the copyright holders, JSWE)

Blue indicates the existence of calcium atoms which represents shell particle, and red indicates the existence of calcium and phosphorous atoms which represents calcium phosphate. Calcium phosphate was found at the surface of the shell particles for all conditions. Thus, a forming of calcium phosphate starts from the particle surface and then penetrates inside the particles. The depth of permeation of phosphate was high at the low initial Ca/P ratio.

11.3.4 Results-2 Crystal Structure of Precipitate on the Shell Surface

The precipitate obtained after the experiments was analyzed by powder XRD to identify the final products. Figure 11.7a is the XRD profile of the final product from the experiment with initial Ca/P = 1, showing very clear peaks. This profile was compared with the XRD profile of DCPD as shown in Fig. 11.7d, and these two profiles are very similar. The element analysis of the final product showed that the ratio of Ca and P in the final product was Ca/P = 1.25. On the other hand, the ratio of Ca and P of DCPD is Ca/P = 1.0. It is reasonable to consider that the contamination of calcium carbonate in the shell particles occurs in the element analysis of the final product. Therefore, it is concluded that the final product from the experiment with initial Ca/P = 1 might be DCPD.

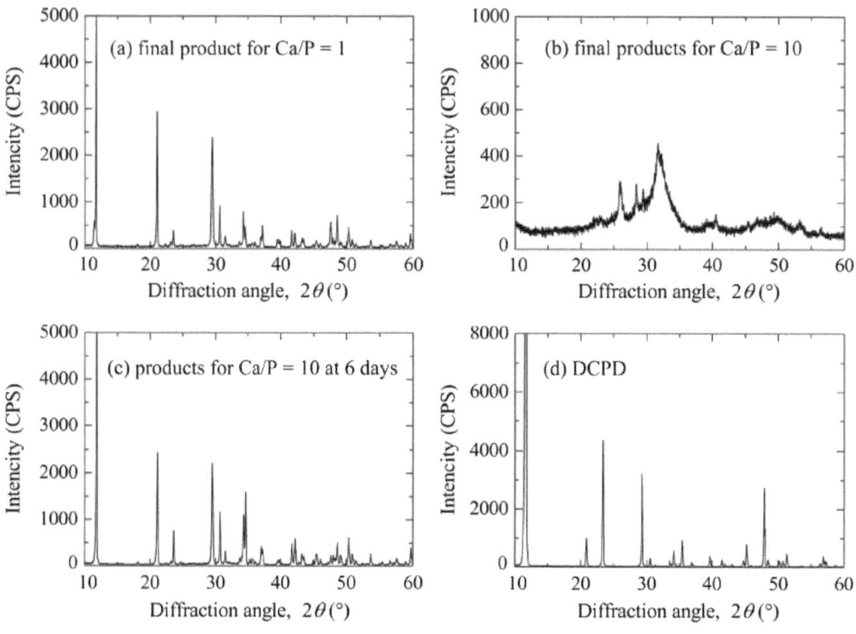

Fig. 11.7 Powder XRD profiles of precipitate

The products obtained from the experiment with the initial Ca/P = 10 had the profile as shown in Fig. 11.7b. This profile does not have clear peaks, and this profile is similar to the profile of the poorly crystallized apatite reported by Amjad (1998). And the element analysis of the final product showed that the ratio of Ca and P was 1.59, and this value is very similar to the Ca and P ratio of hydroxyapatite, 1.67. Therefore, it might be said that the final product from the experiment with the initial Ca/P = 10 is amorphous of hydroxyapatite. In order to examine that changes of crystal structure during the reaction process, XRD of the intermediate precipitates at 6 days was also measured, and the profile shown in Fig. 11.7c was obtained. This profile is similar to the profile of DCPD. And the element analysis of the intermediates showed the Ca and P ratio was 1.17. Therefore, in the experiment with initial Ca/P ratio = 10, the crystal structure changed from DCPD to amorphous of hydroxyapatite.

11.4 Summary

Current studies on nitrogen and phosphorous from source-separated urine were summarized in this chapter. The slow-released nitrogen fertilizer of methylene urea was produced from urine solutions by author's method (mixing with urine and formaldehyde at pH = 2 condition). The NMR, IR analyses, and element analysis showed the product is methylene urea.

Phosphorous in urine is able to be recovered by shell particles. Calcium phosphate is precipitated onto the surface of shell particles. SEM/EDS, powder XRD analysis, and composition measurements showed that the precipitate from the experiment with the Ca/P ratio = 1 was DCPD, and amorphous of hydroxyapatite was developed in the experiment with the initial Ca/P = 10. Elemental analysis of the cross section of shell particle by SEM/EDS showed that the reaction started from particle surface and then proceeded to inside the particles.

References

Almeida M, Butler D, Friedler E (1999) At-source domestic wastewater quality. Urban Water 1:49–55. https://doi.org/10.1016/s1462-0758(99)00008-4

Amjad Z (1998) Calcium phosphates in biological and industrial systems. Kluwer Academic Publishers, Boston

Cordell D, Drangert JO, White S (2009) The story of phosphorus: global food security and food for thought. Glob Environ Change 19(2):292–305. https://doi.org/10.1016/j.gloenvcha.2008

Food and Agriculture Organization of the United Nations (2015) World fertilizer trends and outlook to 2018. p 8

Hotta S, Funamizu N (2008) Evolution of ammonification potential in storage process of urine with fecal contamination. Bioresour Technol 99:13–17. https://doi.org/10.1016/j.biortech.2006.12.001

Ito R, Takahashi E, Funamizu N (2013) Production of slow-released nitrogen fertilizer from urine. Environ Technol 34(20):2813–2820. https://doi.org/10.1080/09593330.2013.790069

Ito R, Funamizu N (2016) Phosphate recovery from synthetic urine with shell of Mizuhopecten yessoensis. J Water Environ Technol 14(6):437–446. https://doi.org/10.2965/jwet.16-007

Kato Y, Matsuoka I (1991) A study on phosphorus removal from wastewater by using calcium salt: the characteristics of calcium phosphate precipitates. Shigen-to-Sozai 107(9):627–632 (in Japanese)

Kongshaug G (1998) Energy consumption and greenhouse gas emissions in fertilizer production. IFA Technical Conference; Sept 28–Oct 1, Marrakech, Morocco

MacDonald GK, Bennett EM, Potter PA, Ramankutty N (2011) Agronomic phosphorus imbalances across the world's croplands. Proc Natl Acad Sci USA 108(7):3086–3091

Malisie AF, Prihandrijanti M, Otterpohl R (2007) The potential of nutrient reuse from a source-separated domestic wastewater system in Indonesia—case study: ecological sanitation pilot plant in Surabaya. Water Sci Technol 56:141–148. https://doi.org/10.2166/wst.2007.566

Mihelcic JR, Fry LM, Shaw R (2011) Global potential of phosphorus recovery from human urine and feces. Chemosphere 84(6):832–839. https://doi.org/10.1016/j.chemosphere.2011.02.046

Schmelzer EJ (2005) Nucleation theory and applications. Wiley-VCH Verlag GmbH & Co, KGaA, Weinheim, FRG, Germany

Song Y, Weidler PG, Berg U, Nüesch R, Donnert D (2006) Calcite-seeded crystallization of calcium phosphate for phosphorus recovery. Chemosphere 63(2):236–243. https://doi.org/10.1016/j.chemosphere.2005.08.021

Wilsenach JA, Schuurbiers CAH, van Loosdrecht MCM (2007) Phosphate and potassium recovery from source separated urine through struvite precipitation. Water Research 41: 458–466. https://doi.org/10.1016/j.watres.2006

Yoshimura K, Shiomi H (2013) Phosphorus removal by crystallization for scallop shell adsorbent. J Soc Mater Sci Jpn 62(6):373–376 (in Japanese)

Part IV
Technologies for Gray Water

Chapter 12
Slanted Soil System

Ken Ushijima, Ryusei Ito, Nowaki Hijikata, Ynoussa Maïga and Naoyuki Funamizu

Abstract Slanted soil system is low cost, simple system suitable for greywater treatment. Performance of the slanted soil system for greywater was evaluated in light of required water quality for irrigation reuse. Removal rate of COD and BDOC, those may cause clogging of irrigation facility, was 52–83% and 88–89%, respectively. LAS, one of the major surfactants, was removed more than 80%. Phytotoxic test showed that more than 2.82 m treatment is required. Only fine soil (1–4 mm) performed 5 $\log_{10}$ and 3 $\log_{10}$ reductions of *Escherichia coli* and MS2 phage, while coarse soil could not remove those pathogens. Clogging was observed in fine soil after 3–5 weeks operation; however, combination of coarse soil chamber and fine soil chamber could extend it to 8 weeks, without decrease of treatment performance. Reductions of total COD and LAS were described by first-order reaction model, and reaction coefficient k was described by equation of per area discharged rate.

Keywords Decentralized system · Water reuse · Irrigation · Arid zone

K. Ushijima (✉)
Northern Regional Building Research Institute, Hokkaido Research Organization (HRO), 3-1-20, Midorigaoka Higashi 1, Asahikawa, Asahikawa 078-8801, Japan
e-mail: ushijima-ken@hro.or.jp

R. Ito · N. Hijikata
Graduate School of Engineering, Hokkaido University, Kita13-Nishi 8, Kita-Ku, Sapporo City, Hokkaido 060-8628, Japan

Y. Maïga
International Institute for Water and Environmental Engineering (2iE), Ouagadougou, Burkina Faso

N. Funamizu
Graduate School of Global Food Resources, Hokkaido University, Kita-9, Nishi-9, Kita-Ku, Sapporo 060-8589, Japan

N. Funamizu
Research Institute of Humanity and Nature. Motoyama Kamigamo, Kita-Ku, Kyoto 603-8047, Japan

N. Funamizu (ed.), *Resource-Oriented Agro-sanitation Systems*,
https://doi.org/10.1007/978-4-431-56835-3_12

12.1 Introduction

The slanted soil system is one of the simplest treatment technologies, and suitable for the concept of Resource-Oriented Sanitation System (see Chap. 1). It consists of several chambers containing soil (Fig. 12.1). These chambers are able to be stacked vertically; therefore, its footprint is very small (e.g. approximately 1.0 × 0.5 m). This system accepts direct discharge of greywater (Kondo et al. 2011); therefore, no pump and no septic tank are required. Itayama et al. (2006) and Kondo et al. (2011) performed continuous monitoring of slanted soil system which is used in real household in Japan as treatment facility to avoid water pollution of river. Itayama et al. (2006) reported averaged removal ratio of the COD, the suspended solid (SS), the total nitrogen (TN) and the total phosphorus (TP) as 85, 78, 78 and 86%.

Regarding water reuse point of view, Li et al. (2009) referred this result and concluded that the treated water is not suitable for reuse because it remains high in organic load and suspended solids, which can limit the chemical disinfection. This is however only the discussion for potable reuse. In the Resource-Oriented Sanitation concept, greywater reuse for irrigation becomes important especially in arid or semi-arid zone (see Chap. 1). Onsite separated greywater has generally lower concentration of pollutant load (Gajurel et al. 2003) than other wastewater and its treatment would be easier than mixed wastewater. For example, WHO wastewater reuse guideline (WHO 2006) stated that 2–7 $\log_{10}$ pathogen reduction by treatment is required for irrigation use; however, separated greywater contains 2 $\log_{10}$ or more lower bacterial pathogen than mixture wastewater which WHO

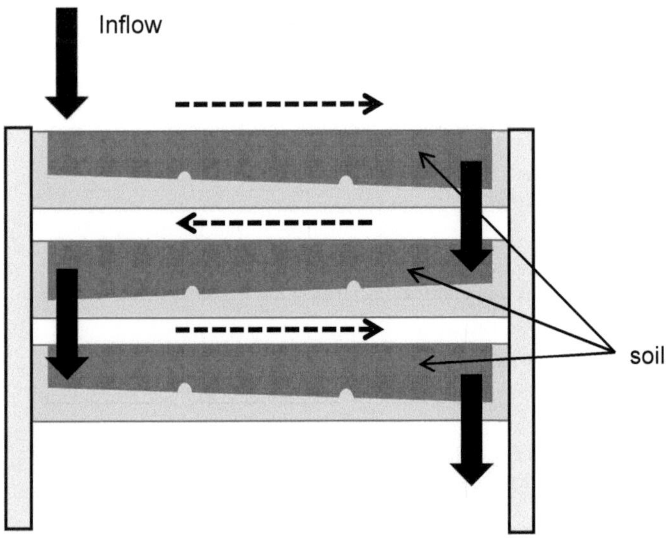

Fig. 12.1 Structure of slanted soil system. Reprinted from Ecological Engineering, 50, Ushijima et al. (2013b) with permission from Elsevier

assumed (Ottoson and Stenström 2003). Therefore, we are able to set 5 $\log_{10}$ as a target level of bacterial pathogen reduction for separated greywater treatment. If reuse purpose was focused on irrigation only, advanced treatment technology for nitrogen and phosphorus removal would not be necessary. The more important point is to be low cost and simple, because most of those who need greywater reuse are low-income people. Thus, in this chapter, we evaluated the performance of the slanted soil system as a treatment facility for irrigation reuse in arid or semi-arid area, and proposed its design criteria.

12.2 Materials and Methods

12.2.1 Experimental Apparatus and Operation

Pilot-scale slanted soil systems were installed in laboratory; four soil chambers were connected as shown in Fig. 12.2; three lines are parallel; the size of each chamber was 0.10 m of depth, 0.145 m (Cases 1–6) or 0.10 m (Cases 7–9) of width and 0.94 m of length; chamber bed has gradient of 1/20. We performed nine cases

Fig. 12.2 Picture of slanted soil system for this experiment. Reprinted from Ecological Engineering, 50, Ushijima et al. (2013b) with permission from Elsevier

of experiments, in variety of soil type, soil particle size and amount of discharged wastewater (Table 12.1). In all experiments, greywater was discharged to first chamber three times in a day because actual discharge pattern of greywater shows 2 or 3 peaks in a day (Funamizu et al. 2002; Kondo et al. 2011). In this study, 5 h intervals were set between three discharges; morning, noon and evening. Volume and components of each discharge were listed in Table 12.1. The volume ratio of kitchen sink wastewater, laundry wastewater and shower wastewater in daily total was fixed at 3:2:5, which was determined with referring observed data in developing countries (Otaki 2007; Ushijima 2008) for Cases 1–6. Regarding Cases 7–9, kitchen wastewater was not included because it was especially focused on real case of Burkina Faso, in which kitchen wastewater reuse for livestock was observed (Ushijima et al. 2013a).

Each greywater was prepared synthetically. Kitchen sink wastewater (KSWW) was prepared following the method which was described in detail by Huelgas (2009). The components listed in Table 12.2 were well mixed in 5 L of water solution. After 30 min settling, 4 L of supernatant was diluted with water until 94 L. Incubated *E. coli* (NBRC13168) and MS2 phage (NBRC 102619) were added to this solution as an indicator of bacterial and virus pathogen, respectively. Reported concentrations of *E. coli* in real greywater were $10^{1.2}$–$10^{5.4}$ mL^{-1} (Ottoson and Stenström 2003); therefore, we set the concentration of *E. coli* as 1.0×10^5 CFU mL^{-1}. Regarding MS2 phage, we set the concentration of 1.0×10^3 PFU mL^{-1}, which is same level as wastewater not separated with toilet wastewater (WHO 2006). Laundry wastewater was prepared by washing new laboratory eplon or T-shirts in automatic laundry machine with laundry detergents. Total water use in the laundry machine was 60 L. Shower wastewater was prepared by mixing shampoo, rinse, liquid soap and toothpaste into tap water (Cases 1–6), or just simply mixed liquid soap to tap water with assuming specific case in Burkina Faso (Cases 7–9). Table 12.3 represents characteristics of synthetic greywater.

Two types of soil were used in this study. Kanuma soil, which was used in Case 1, consists of alumina and hydrated silica (Itayama et al. 2006). Performance of Kanuma soil had already been examined by previous studies (Kondo et al. 2011; Itayama et al. 2006); however, it is very unique soil and available only in Japan. As alternative soil, which is available in the world, crushed baked mud brick was applied in Cases 2–9. It was prepared by crushing baked mud brick and sieving. In case that soil was clogged while experimental period, it was replaced by new soil (Cases 4 and 5), or was washed and refilled (Cases 7 and 8). After replacement, experiment was continued under same condition as before replacement.

12.2.2 Sampling and Analysis

Effluent of each chamber was sampled every 1 or 2 weeks. Total COD (T-COD), suspended solid (SS), *E. coli*, and MS2 phage were measured for some of those

Table 12.1 Variation of greywater discharge, experimental period and soil type

		Greywater discharge	Discharge rate ($m^3/m^2/d$)	Experimental period	Soil type
Case 1	Morning	KSWW 6L + LWW 12L	4.1	6 weeks	Kanuma soil (4–11 mm)
	Noon	KSWW 6L			
	Evening	KSWW 6L + SWW 30L			
Case 2	Morning	KSWW 6L + LWW 12L	4.1	6 weeks	Crushed baked mud brick (4–11 mm)
	Noon	KSWW 6L			
	Evening	KSWW 6L + SWW 30L			
Case 3	Morning	KSWW 6L + LWW 12L	4.1	5 weeks[a]	Crushed baked mud brick (1–4 mm)
	Noon	KSWW 6L			
	Evening	KSWW 6L + SWW 30L			
Case 4	Morning	KSWW 3L + LWW 6L	2.1	8 weeks[b]	Crushed baked mud brick (1–4 mm)
	Noon	KSWW 3L			
	Evening	KSWW 3L + SWW 15L			
Case 5	Morning	KSWW 3L + LWW 6L	2.1	8 weeks[c]	Crushed baked mud brick (1st chamber 1–4 mm, 2nd–4th chamber 4–11 mm)
	Noon	KSWW 3L			
	Evening	KSWW 3L + SWW 15L			
Case 6	Morning	KSWW 1.5L + LWW 3L	1.0	8 weeks[b]	Crushed baked mud brick (1–4 mm)
	Noon	KSWW 1.5L			
	Evening	KSWW 1.5L + SWW 7.5L			

(continued)

Table 12.1 (continued)

		Greywater discharge	Discharge rate ($m^3/m^2/d$)	Experimental period	Soil type
Case 7	Morning	SWW 12L + LWW 4L	5.2	8 weeks[d]	Crushed baked mud brick (1st chamber 4–9.5 mm, 2nd–4th chamber 1–4 mm)
	Noon	SWW 12L			
	Evening	SWW 24 L			
Case 8	Morning	SWW 12L + LWW 4L	5.2	8 weeks[d]	Crushed baked mud brick (1st chamber 2–6.5 mm, 2nd–4th chamber 1–4 mm)
	Noon	SWW 12L			
	Evening	SWW 24 L			
Case 9	Morning	SWW 12L + LWW 4L	5.2	8 weeks	Crushed baked mud brick (1–9.5 mm)
	Noon	SWW 12L			
	Evening	SWW 24 L			

KSWW Kitchen sink wastewater, *LWW* Laundry wastewater, *SWW* Shower wastewater
[a]Experiment was ended at 5th week because overflow due to clogging was observed
[b]The soil in 1st chamber was replaced by new soil at 3rd week, when overflow was observed
[c]Experiment was ended at 8th week because overflow due to clogging was observed
[d]The soil in 2nd chamber was washed at 3rd or 4th week, when overflow was observed

Table 12.2 Components of synthetic kitchen sink wastewater

Components	Amount (g)
Kitchen detergents	5.12
Soybean flower	34.6
Fish powder	34.6
Dextrine	6.9
Powder cream for coffee	21.6
Beef extract	5.2
Carrot	34.6
Cabbage	34.6
Banana peel	25.9
Apple	25.9
Grapefruit peel	25.9
Dried horse mackerel	25.9
Cooked rice	25.9
Tea leaves	7.6

Table 12.3 Characteristics of each greywater

	KSWW	LWW	SWW
T-COD (mg/L)	600–1000	500–600	200–400
SS (mg/L)	150–250	70–100	40–70
pH	7.0	10.1	7.3
EC (mS/cm)	0.69	1.15	0.28
E. coli (CFU/mL)	10^5	–	–
MS2 phage (PFU/mL)	10^3	–	–
LAS (mg/L)	–	50	–

samples. Additional to these items, dissolved COD (D-COD), biodegradable dissolved organic carbon (BDOC) and liner alkylbenzene sulphonate (LAS) were also measured.

COD and SS were measured in standard method of APHA (1985). *E. coli* in the sample was incubated on “Compact Dry EC (NISSUI PHARMACEUTICAL CO. LTD)” 24 h under 37 °C, and blue-coloured *E. coli* colonies were counted. MS2 phage was measured in PFU method with the double-layer method (Adams 1959). *E. coli* (NBRC 13965) is used as bacterial host. Bacteriophage was propagated for 22–24 h at 37 °C in *E. coli*. LAS was measured by liquid chromatography–mass spectrometry (LC-MS). Targeted LAS was C10–C14. Eluents were methanol and water, containing ammonium acetate. Solid-phase extraction as the pretreatment for LAS measurement was done using BOND ELUT PPL (500 mg) cartridge. BDOC was measured by incubating filtered samples in BOD bottles at 20 °C for 7 days and 14 days and measured dissolved organic carbon with TOC analyzer (TOC-5000A, Shimazu corp.). Decrease from initial TOC to 7 or 14 days was regarded as $BDOC_7$ and $BDOC_{14}$ in this study.

Phytotoxicity of each sample was evaluated for Cases 7–9 by germination test with komatsuna (*Brassica rapa* var. *peruviridis*) in petri-dish based on a modified method of Zucconi et al. (1981). Seeds of Komatsuna were stocked in 4 °C with dark for several days, and 50 seeds were sown on two-ply filter paper (No. 5C, ADVANTEC) in petri-dish. Then, 8 mL of sample water was gently applied in the petri-dish. The petri-dish was kept in dark incubator at 23 °C. After one day from sowing, the germination number was count. After 5 days from sowing, root lengths were measured. Same procedure was repeated three times for each sample.

12.3 Results

12.3.1 SS Removal and Clogging

SS removal by treatment length was almost linear in logarithmic graph (Fig. 12.3). Total removal rates through four chambers were 60–94%, and these were higher in

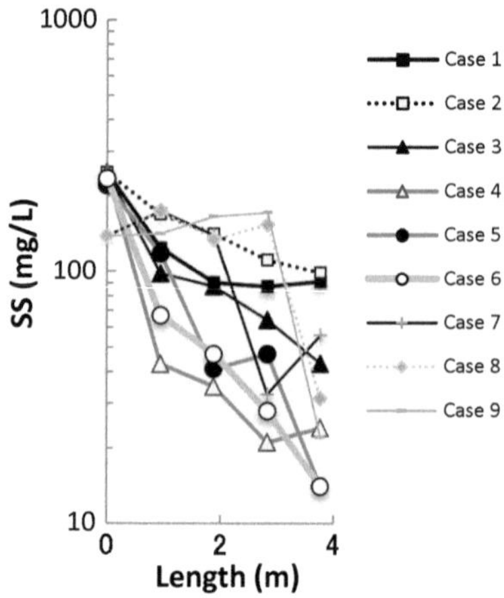

Fig. 12.3 Concentration of suspended solid (SS) at influent and effluent of each chamber on 4th week (Cases 1, 2, 3 and 5), 3rd week (Cases 7, 8 and 9), and 2nd week (Cases 4 and 6). Reprinted from Ecological Engineering, 50, Ushijima et al. (2013b) with permission from Elsevier

finer particle and lower discharge. Discharged water was overflowed from the chamber due to clogging in Cases 3, 4, 5, 6, 7 and 8 at the 5th, 3rd, 8th, 3rd, 3rd and 4th week, respectively. As general tendency, time periods until clogging (T) were longer in coarser soil and higher discharge. In Case 5, combination of fine and coarse soil, T was longer than those using fine soil only (Cases 3, 4 and 6). In the Case 9, which has wider size range also did not clogged in 8 weeks experimental period. However, according to those results, no specific relationship was observed between T and estimated cumulative amount of SS removal.

12.3.2 Removal of Organic Materials and LAS

T-COD at the outlet of 4th chamber on 4th week was 154–394 mg L^{-1}, while the results of Cases 4 and 6 were at 2nd week because these were clogged at 3rd week. The removal rates were 52–83%, higher in finer particle and lower discharge. In Cases 4, 5 and 6, those measured D-COD and BDOC, more than 90% of particle COD (P-COD), derived by deducting D-COD from T-COD, were removed by this system (Fig. 12.4). Although the removal rates of D-COD were lower (58–68%), those of both $BDOC_7$ and $BDOC_{14}$ were removed 87–91% (Fig. 12.4). This is probably because biologically degradable organic matters in dissolved part were selectively removed and non-degradable or less degradable matters remained.

LAS at the outlet of 4th chamber of on 3rd week of the Cases 4 and 6 or 4th week of the Case 5 was 2.3–3.3 mg L^{-1} and removal rates were 93–96%. LAS at the outlet of 4th chamber of 3rd week of the Cases 7, 8 and 9 was 8.4–13.6 mg L^{-1} and removal rates were 81–88%.

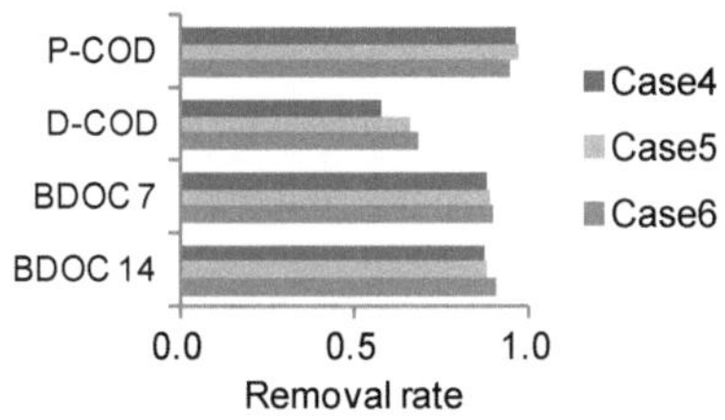

Fig. 12.4 Removal rate of particle COD (P-COD), dissolved COD (D-COD), and biodegradable dissolved organic carbon upon 7 days incubation ($BDOC_7$) and 14 days incubation ($BDOC_{14}$). Reprinted from Ecological Engineering, 50, Ushijima et al. (2013b) with permission from Elsevier

12.3.3 Pathogens Removal

Figures 12.5 and 12.6 show *E. coli* and MS2 phage concentrations in each case, respectively. Most of *E. coli* and MS2 phage were not removed in all chambers containing coarse soil; Cases 1, 2 and 1st chamber of Case 5. In contrast to those coarse soil, 5 $\log_{10}$ reduction of *E. coli* and 3 log10 reduction of MS2 phage were performed through one or two chambers until 3rd or 4th week, in Cases 3, 4, 5 and 6, those used fine soil. After 4th or 5th week, breakthrough of *E. coli* and MS2 was observed in Cases 3 and 5. No breakthrough was observed in Cases 4 and 6, probably because experimental periods were only 3 weeks.

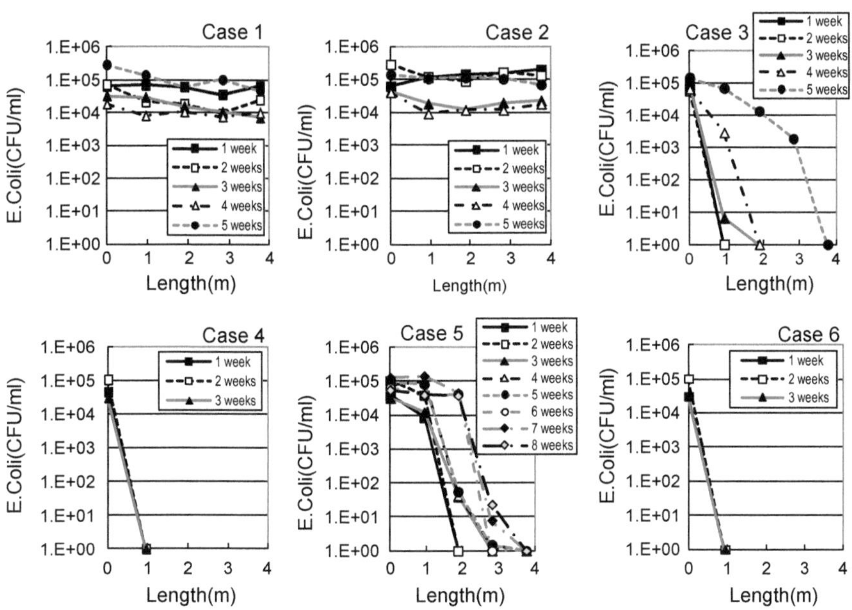

Fig. 12.5 *E. coli* concentration at influent and effluent from each chamber. Reprinted from Ecological Engineering, 50, Ushijima et al. (2013b) with permission from Elsevier

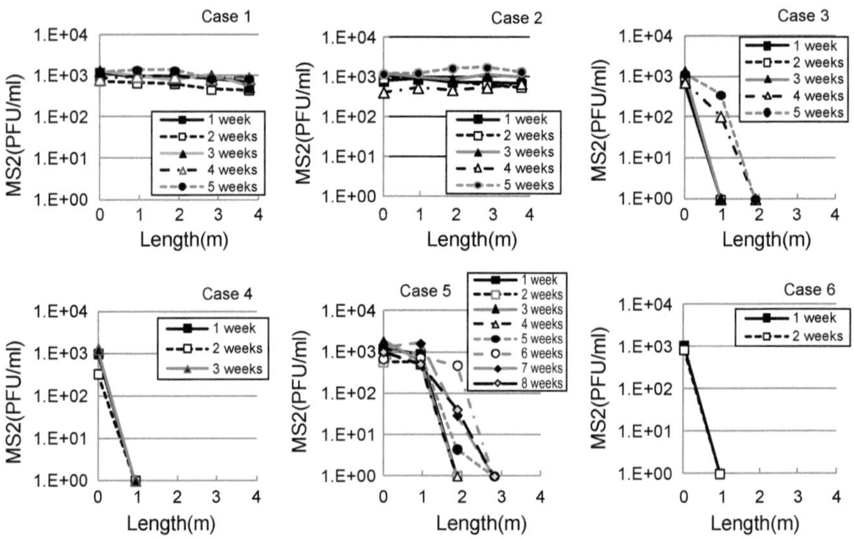

Fig. 12.6 MS2 phage concentration at influent and effluent from each chamber. Reprinted from Ecological Engineering, 50, Ushijima et al. (2013b) with permission from Elsevier

12.3.4 Phytotoxicity Test

At the 7th week, the germination tests using the effluents from each chamber and raw greywater were performed. The results show no definite difference between all samples. In this Fig. 12.7, 0.94, 1.88, 2.88, 3.74 m show the effluent from 1st, 2nd, 3rd and 4th chamber, respectively. On the other hand, root lengths of the effluents from 1st (0.94 m) and 2nd (1.88 m) chamber showed significantly lower value than the control, but those from 3rd (2.88 m) and 4th (3.74 m) chamber did not show any statistically significant difference from control (Fig. 12.8). Among three soil size variations of Cases 7, 8 and 9, clear difference was not observed in both roots length and germination test.

12.4 Discussions

12.4.1 Treatment Performance

Fine soil presented better removal performance of pathogen, SS and T-COD. Particularly, sufficient pathogen removals were performed only by fine soil chamber, while most part of pathogen passed through coarse soil chamber. Disadvantage of fine soil was shorter time periods until clog; however, result of Case 5 represented that combination of coarse and fine soil can extend this time

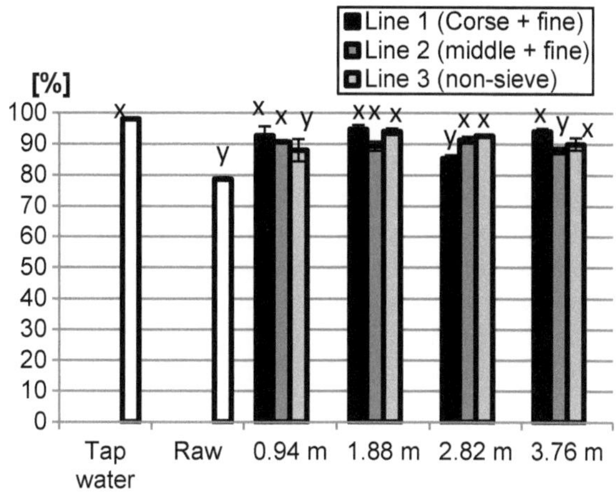

Fig. 12.7 Germination ratio by sample of 7th week. Reprinted from Proceedings of the IWA International Conference Cities of the Future Xi'an. Xi'an, China (2011)

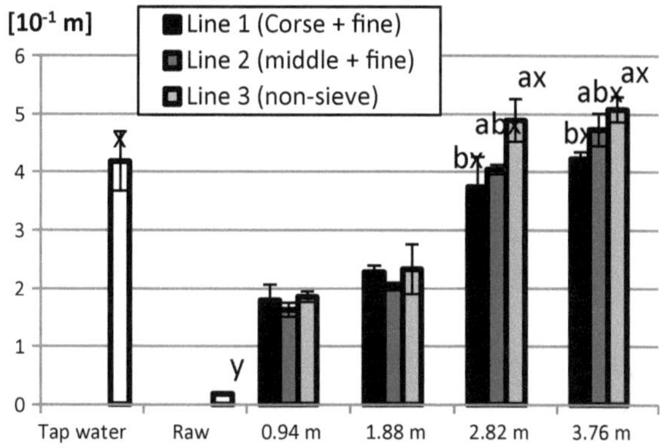

Fig. 12.8 Root lengths by sample of 7th week. Reprinted from Proceedings of the IWA International Conference Cities of the Future Xi'an. Xi'an, China (2011)

period to more than double. Furthermore, wider ranged soil particle in Case 9 showed also longer time period until clog (more than 8 weeks); however, pathogen removal has not yet checked. More discussion based on pathogen analysis for wider ranged soil particle is necessary. Clogging was not observed in these previous studies (e.g. Itayama et al. 2006; Kondo et al. 2011); however, most of previous studies used Kanuma soil, which is generally 10 mm or larger particle size (Itayama et al. 2006), and they did not measure pathogens. Tada et al. (2007) used combination of several local soils; pumiceous soil (4–20 mm) for 1st and 2nd chambers, river sand (0.5–4 mm) for 3rd chamber, and limestone soil (0.25–2 mm) for 4th chamber, and its time period until clogging was half year. Two of coarse soil chambers (2 m), and much lower q (0.06–0.36 $m^3m^{-2}d^{-1}$, which is 1/10–1/50 of our experiment) probably caused this longer period until clogging than our results.

Originally, slanted soil chamber system was developed to reduce environmental impact, and therefore, focus of the previous studies was removal rate of organic matter and nutrients (Itayama et al. 2006; Kondo et al. 2011; Tada et al. 2007).

In the light of required water quality for irrigation reuse, removal of nutrient is not important but removal of organic matters is important because they cause clogging of irrigation facilities such as drip irrigation system, and therefore, particle COD (P-COD) and BDOC would be especially important. Figure 12.4 implies that the slanted soil system performed high removal rate in both P-COD (94–97%) and BDOC (88–89%), while removal rate of D-COD (58–68%) was comparatively low. Although BDOC levels in treated water (11–12 mg L^{-1}) were much lower than untreated greywater, these exceed 0.15–0.25 mg L^{-1} those were reported as maximum level to avoid bacterial regrowth (Servais et al. 1995; Niquette et al. 2001). Of course, the volume of biofilm and increasing speed of biofilm volume is supposed to become smaller than untreated greywater; however, farther practical study is needed for quantitative evaluation.

LAS removal rates were 81–96%, and final concentrations were 2.3–13.6 mg L^{-1}. Hijikata et al. (2011) proposed 8 mg L^{-1} as target level for irrigation use. The Cases 4, 5 and 6 showed lower concentration; however, the Cases of 7, 8 and 9 showed larger concentration. Furthermore, this study could not cover byproduct of LAS degradation, which was found in treated greywater by another system (Huelgus 2009). However, as long as seeing the results of phytotoxicity assessment for the Cases 7, 8 and 9, 2.82–3.76 m treatment seems sufficient, under discharge rate of 1.0–5.2 m^3/m^2/day.

12.4.2 Reduction Characteristics of COD and LAS

Reduction curves of T-COD (Fig. 12.9) and LAS (Fig. 12.10) against chamber length seem linear in logarithmic graph (Ushijima et al. 2013c). When we assume uniform flow velocity inside of each system, these reductions can be regarded as first-order kinetic reaction, which is described as

$$C_{\mathrm{COD}} = C_{0\,\mathrm{COD}} \cdot \exp(-\,k_{\mathrm{COD}}\,L) \tag{12.1}$$

$$C_{\mathrm{LAS}} = C_{0\,\mathrm{LAS}} \cdot \exp(-k_{\mathrm{LAS}}\,L) \tag{12.2}$$

where C_{COD} is COD concentration, $C_{0\mathrm{COD}}$ is initial concentration, k_{COD} is the reaction coefficient with m^{-1} unit, and L is length of soil, C_{LAS} is LAS concentration, $C_{0\mathrm{LAS}}$ is initial concentration, k_{LAS} is the reaction coefficient with m^{-1} unit.

Because of the assumption above, the values of k_{COD} and k_{LAS} values would change by different flow velocity. Therefore, correlations between these two reaction coefficients and possible control factors of flow velocity; elapsed time, soil particle size and q defined by Q/V where Q is daily discharge ($m^3\ d^{-1}$), V is volume of chamber were examined, and finally, the following relationships were obtained:

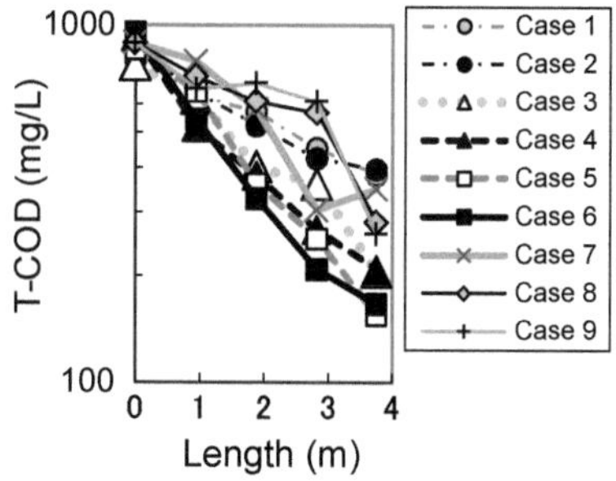

Fig. 12.9 Concentration of total COD (T-COD) at influent and effluent from each chamber on 4th week (Cases 1, 2, 3 and 5), 3rd week (Cases 7, 8 and 9), and 2nd week (Cases 4 and 6). Reprinted from Proceedings of 7ème édition Journées scientifiques de 2iE, 1–5 April 2013, Ouagadougou, Burkina Faso (2013)

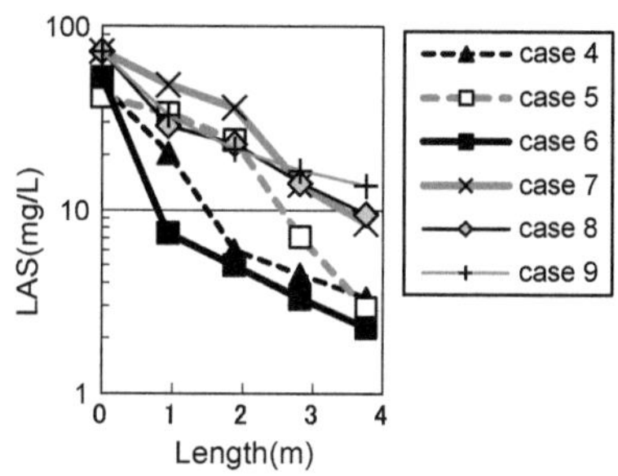

Fig. 12.10 Concentration of liner alkylbenzene sulphonate (LAS) at influent and effluent from each chamber on 4th week (Case 5), 3rd week (Cases 7, 8 and 9), and 2nd week (Cases 4 and 6). Reprinted from Proceedings of 7ème édition Journées scientifiques de 2iE, 1–5 April 2013, Ouagadougou, Burkina Faso (2013)

$$k_{\mathrm{COD}} = -0.05\,q + 0.5 \tag{12.3}$$

$$k_{\mathrm{LAS}} = -0.05\,q + 0.8 \tag{12.4}$$

12.4.3 Design Criteria

First of all, fine soil chamber is required in order to remove pathogen. According to the results of Cases 3 and 5, those observed breakthrough, removal capacity levels of *E. coli* and MS2 phage in one chamber (0.145 m width, 0.1 m height and 0.94 m length) were estimated 10^{10} CFU and 10^{8} PFU, respectively. When the capacity became full, it has to be replaced. Required replacement interval (*INT*) would be explained by

$$INT = 6.9\,q^{-1}[\mathrm{d}]. \tag{12.5}$$

On the other hand, outlet concentrations of LAS are able to simulate by Eqs. (12.2) and (12.4). In case applying target level of 8 mg L^{-1} proposed by Hijikata et al. (2011), adequate q which satisfies $C_{LAS} < 8$ has to be set. Regarding organic matters, farther study is required in order to simulate outlet concentration, not only of T-COD but also of P-COD and BDOC. The q is determined by balance between chamber size and discharged volume of greywater, and therefore, possible chamber size can be designed according to expected discharged volume.

Regarding clogging issue, combination of coarse and fine soil is preferable. According to result of Case 5, first 1 or 2 chambers were recommended to use coarse soil, and it extends time periods until clogging. Under the condition of Case 5, time period until clogging was longer than D; however, quantitative evaluation on clogging under different condition is difficult so far.

In the case of rural area of Burkina Faso belong to Sahel Africa, water usage was about 20 L d^{-1} $capita^{-1}$(Ushijima et al. 2012). Although family size varies in wide range, core family consists of parents and several children. Therefore, we assumed that one slanted soil system is suitable for one core family, and set the discharge rate as 100 L d^{-1}, which corresponds to the maximum greywater volume from 5 persons per day.

Regarding soil depth, no former research tested variety of depth. In terms of practical design, depth increase is not preferable because it needs higher position of sink. Therefore, this study applied 0.10 m as other study set. Gradient also set as 1/20 same as other studies, by same reason as depth setting. Regarding soil width and length, this study used 0.10–0.145 m width and total 3.76 m soil chamber; however, most of other study used about 0.45 m width and various length soil chambers (e.g. Itayama et al. 2006; Kondo et al. 2011).

LAS concentrations in slanted soil system of various soil width and length were simulated based on the condition above (Fig. 12.11) (Ushijima et al. 2013c).

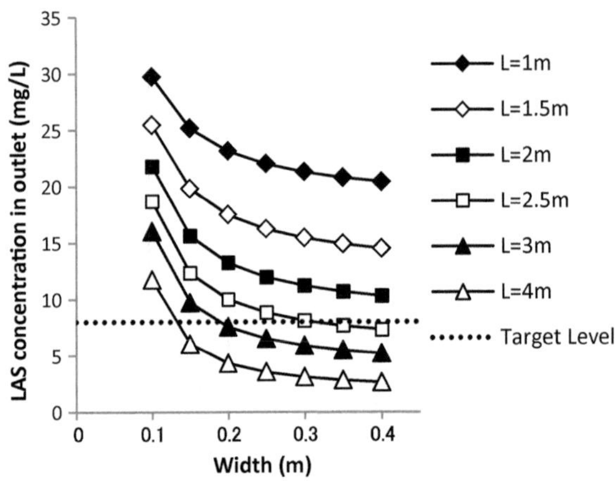

Fig. 12.11 Simulated LAS concentration by width and length. Reprinted from Proceedings of 7ème édition Journées scientifiques de 2iE, 1–5 April 2013, Ouagadougou, Burkina Faso (2013)

Initial concentration was estimated by LAS concentration in famous laundry detergent in Burkina Faso and its instruction. Effect on LAS removal by width increase was drastically increased 0.1–0.2 m width but not so much after 0.20 m. This tendency was common in all tested length; therefore, 0.20 seems reasonable width. Effect of lengths was more drastic than width. Length of 2.0–3.0 m was considered as minimum length to make it lower than 8 mg L^{-1} which was proposed as target level for irrigation use (Hijikata et al. 2011).

12.5 Conclusions

Performance of the slanted soil system for greywater was evaluated in the light of required water quality for irrigation reuse. The slanted soil system performed high removal rate in both P-COD (94–97%) and BDOC (88–89%), while removal rate of D-COD (58–68%) was comparatively low. LAS removal rates were more than 80% and some of final concentrations (2.3–13.6 mg L^{-1}) exceeds 8 mg L^{-1} which was proposed as target level for irrigation use (Hijikata et al. 2011); however, phytotoxicity test showed that 2.82–3.76 m length of treatment can provide sufficient removal of phytotoxicity. First-order reaction equation of T-COD and LAS was determined and $k_{\text{T-COD}}$ and k_{LAS} were expressed by the equation of q. Fine soil (1–4 mm) presented better removal performance of pathogen, SS and T-COD than coarse soil (4–11 mm). Particularly, sufficient pathogen removals were performed only by fine soil chamber, while most part of pathogen passed through coarse soil chamber. Disadvantage of fine soil was recognized in shorter time periods until clog; however, result of Case 5 represented that combination of coarse and fine soil can extend this time period.

References

Adams MH (1959) Bacteriophages. Interscience Publishers, New York

Apha (1985) Standard Method for the Environment of water and wastewater, 16th edn. American Public Health Association, Washington, D.C

Funamizu N, Mizukubo T, Lopez Zavala MA, Takakuwa T (2002) Fractioning gray-water in the differentiable onsite-wastewater treatment system. In: Proceedings of XXVIII Congreso Interamericano de Ingenieria Sanitariay Ambiental, Cancun, Mexico

Gajurel DR, Li Z, Otterpohl R (2003) Investigation of the effectiveness of source control sanitation concepts including pre-treatment with Rottebehaelter. Wat Sci Tech 41:111–118

Hijikata N, Nakaoka E, Sawadogo B, Ushijiam K., Funamizu N (2011) Phytotoxicity assay of several gray waters for reuse as agricultural irrigation. In: Proceedings of the IWA international conference Cities of the Future Xi'an. Xi'an, China (CD-ROM)

Huelgas AP (2009) Onsite Treatment of Higher-load Graywater by Membrane Bioreactor. Dissertation for the Doctoral Degree Hokkaido University Sapporo

Itayama T, Kiji M, Suetugu A, Tanaka N, Saito T, Iwami N, Mizuochi M, Inamori Y (2006) On site experiments of the slanted soil treatment systems for domestic gray water. Wat Sci Tech 53(9):193–201. https://doi.org/10.2166/wst.2006.290

Kondo T, Ebie Y, Kiji M, Sugiura N, Xu K, Inamori Y (2011) Decentralized domestic gray water treatment by using slanted soil chamber system. Sustain Environ Res 21:81–87

Li F, Wichmann K, Otterpohl R (2009) Review of the technological approaches for grey water treatment and reuses. Sci Total Environ 407(11):3439–3449. https://doi.org/10.1016/j.scitotenv.2009.02.004

Niquette P, Servais P, Savoir P (2001) Bacterial dynamics in the drinking water distribution system of brussels. Water Res 35:675–682. https://doi.org/10.1016/S0043-1354(00)00303-1

Otaki M (2007) Water use in Thailand. J Human Env Eng 9:124–127 (in Japanese)

Ottoson J, Stenström TA (2003) Faecal contamination of greywater and associated microbial risks. Water Res 37(3):645–655. https://doi.org/10.1016/S0043-1354(02)00352-4

Servais P, Laurent P, Randon G (1995) Comparison of the bacterial dynamics in various French distribution systems. Aqua 44:10–17

Tada C, Itayama T, Iwami N, Kuwabara T, Tanaka N, Ebie Y, Inamori Y, Ushijima K, Moritani A (2007) Characteristics of kitchen wastewater treatment by slanted soil treatment system using various soils. In: Proceedings of the 5th international symposium on sustainable sanitation Tokyo Japan, pp 101–107

Ushijima K, Irie M, Sintawardani N, Triastuti J, Ishikawa T (2008) Field research on domestic wastewater from a slum in Bandung City. J Hydrosci Hydraul Eng 26:129–140

Ushijima K, Hijikata N, Ito R, Funamizu N (2012) Effect estimation of Dry-Toilet application for rural farmer family in Burkina Faso. J Arid Land Study 22(1):99–102

Ushijima K, Tanaka E, Hijikata N, Ito R, Funamizu N (2013a) Grey water treatment by slanted soil system considering practical condition in Sahel Africa. J Jap Soc Civ Eng, Ser G 69(7): III-83-III-89

Ushijima K, Ito K, Ito R, Funamizu N (2013b) Greywater treatment by slanted soil system. Ecol Eng 50:62–68. https://doi.org/10.1016/j.ecoleng.2012.04.028

Ushijima K, Tanaka E, Hijikata N, Ito R, Funamizu N, Maïga Y, Moyenga D, Maïga AH (2013c) Design of grey water treatment system for agricultural reuse. In: Proceedings of 7ème édition Journées scientifiques de 2iE, 1st–5th April 2013, Ouagadougou, Burkina Faso

World Health Organization (2006) WHO guidelines for the safe use of wastewater, excreta and greywater wastewater use in agriculture, vol 2. WHO publication, Geneva

Zucconi F, Forte M, Monac A, De Bertoldi M (1981) Biological evaluation of compost maturity. Biocycle 22:27–29

Chapter 13
Membrane System for Gray Water

Aileen Huelgas-Orbecido and Naoyuki Funamizu

Abstract Higher-load gray water (HLGW) consists of kitchen-sink and washing machine discharges. This type of wastewater needs to be treated before ultimately discharging this to bodies of water. Treatment technologies should be at least secondary treatment technology, i.e., involving biological treatment processes, in order to treat organic matter in the wastewater. Membrane bioreactor is a promising technology that can be used to treat HLGW. This technology requires small footprint and could produce an effluent that is of high quality that may be suitable for recycling. Membrane bioreactor was used to treat HLGW, i.e., kitchen-sink wastewater only and a mixture of kitchen-sink and washing machine wastewaters. The permeate of the system (ultrafilter hollow-fiber submerged MBR) treating the mixture has a higher chemical oxygen demand (COD) compared to that of kitchen-sink wastewater only. In both cases, the % COD removal is >90.7%. Furthermore, >99% of the linear alkylbenzene sulfonate (LAS) was removed.

Keywords Membrane bioreactors · Higher-load gray water · Kitchen-sink wastewater · Washing machine wastewater · Linear alkylbenzene sulfonate

A. Huelgas-Orbecido (✉)
De La Salle University, 2401 Taft Avenue, Manila, Philippines
e-mail: aileen.orbecido@dlsu.edu.ph

N. Funamizu
Graduate School of Global Food Resources, Hokkaido University,
Kita-9, Nishi-9, Kita-Ku, Sapporo 060-8589, Japan

N. Funamizu
Research Institute of Humanity and Nature, Motoyama Kamigamo,
Kita-Ku, Kyoto 603-8047, Japan

N. Funamizu (ed.), *Resource-Oriented Agro-sanitation Systems*,
https://doi.org/10.1007/978-4-431-56835-3_13

13.1 Introduction

In On-site Wastewater Differentiable Treatment System (OWDTS), the gray water is further divided into higher-load gray water (HLGW), consisting of discharges from kitchen sink (KSWW) and washing machine (WMWW), and lower load gray water (LLGW), consisting of discharges from shower, bath, and washbasin (Lopez Zavala et al. 2002). The latter has been the focus of researches that employs technologies that vary from simple filtration system to a complicated process trains (Jefferson et al. 1999). These technologies include rotating biological contactors (RBC) and fluidized bed reactor (Nolde 1999); biological aerated filters (BAF) followed by membrane filtration (Hills et al. 2001); and membrane technology (Jefferson et al. 2001).

Membrane bioreactors (MBRs) became a popular treatment facility for domestic and industrial wastewaters. A MBR has been investigated to treat HLGW. A submerged MBR was preferred over a cross-flow MBR since it is reported to be less energy intensive. Furthermore, a constant transmembrane pressure (TMP) was employed to reduce energy consumption. This is achieved by setting a difference between the reactors' liquid level and the effluent tube.

13.2 Higher-Load Gray Water Characteristics

HLGW contains high concentration of primary pollutants which is supported by a report by Friedler et al. (2005) that states that washing machine, kitchen sink, and dishwasher are major contributors of pollutants. Almeida et al. (1999) reported that KSWW contributes to about 23.2% of chemical oxygen demand (COD), 38% of nitrate–nitrogen (NO_3–N), 9.4% of phosphate–phosphorus (PO_4–P), and 13% by volume of the total daily wastewater discharge from a household. On the other hand, WMWW accounts to about 22.3% COD, 7.6% NO_3–N, 4.3% PO_4–P, and 16.2% by volume of the total daily wastewater discharge.

Table 13.1 shows the characteristics of HLGW (Huelgas et al. 2009; Huelgas and Funamizu 2010). The reported COD values are much higher compared to domestic raw wastewater with a typical range of 250–800 mg/L (Tchobanoglous et al. 2003) and to total gray water discharge with a typical range of 495–682 mg/L (Palmquist and Hanaeus 2005). Moreover, the COD:N:P ratio indicates nutrient imbalance as compared to 100:5:1 ratio for typical domestic wastewater discharge. The same characteristics were observed by Jefferson et al. (2004) with regard to other gray water sources wherein the COD:N:P ratio for bath, shower, and hand basins was at 100:2.25:0.06, 100:2.91:0.05, and 100:1.77:0.06, respectively. The nutrient deficiency is due to the fact that urine and feces are the main sources of these nutrients and these are considered as blackwater, which is treated separately in OWDTS (Huelgas and Funamizu 2010).

Table 13.1 HLGW characteristics

Parameters	HLGW (KSWW + WMWW)	KSWW	Semisynthetic HLGW (KSWW + WMWW)
	Huelgas et al. (2009)		Huelgas and Funamizu (2010)
COD (mg/L)	540–1200 (890)[a]	770–2050 (1300)[a]	675
Total nitrogen (TN) (mg/L)	10–35	21.9–43.5	
NH_4–N (mg/L)	0.9–2.6	0.3–2.7	0.17
NO_3–N (mg/L)	0.2	0.9–5.3	0
Total phosphorus (TP) (mg/L)	0.6–1.3	2.9–14.5	2.37
PO_4–P (mg/L)			1.31
LAS (mg/L)	10–23		30.8
COD:N:P	100:2.53:0.11	100:2.3:0.5	100:3.75:0.35

[a]Values in parenthesis show the average value

Linear alkylbenzene sulfonates (LAS) is a surfactant mostly used in the formulation of laundry detergent. The LAS concentrations shown in the table are higher than the concentrations reported for municipal wastewater treatment plants. The typical range is 1–15 mg/L (Temmink and Klapwijk 2004). The main source of detergents is laundry wastewater which is concentrated in HLGW, thus higher values were obtained compared to domestic wastewater discharges.

13.3 Membrane Bioreactor

Membrane bioreactor (MBR) process was introduced in the late 1960s. However, it was only further developed from 1990 onwards, especially in Canada, Japan, and UK.

The MBR process involves a suspended growth activated sludge process that utilizes microporous membranes for solid–liquid separation in place of secondary sedimentation tank. These membranes can be immersed inside an activated sludge reactor or can be a separate external module (Fig. 13.1). A former configuration is commonly known as submerged MBR and is widely used compared to an external MBR. The membranes are subjected to a vacuum in order to draw water through the membrane while retaining the solids in the reactor. In submerged MBR, the aeration inside the reactor serves as oxygen source for biological reaction and also used for scouring the membrane surface from attached biofilms. The liquid surrounding the membranes is rapidly agitated, usually by coarse bubble aeration, which could minimize fouling of the membranes with the high solid concentration in the activated sludge reactor.

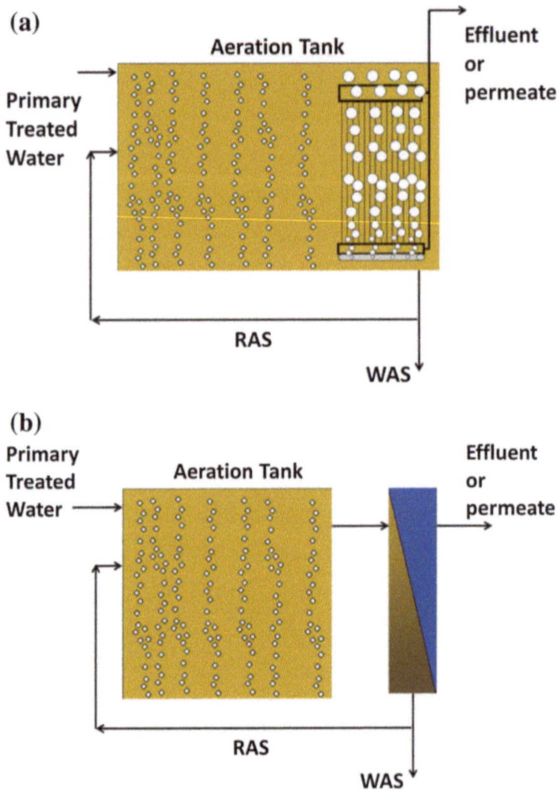

Fig. 13.1 Two types of MBR system. **a** Submerged MBR and **b** external MBR

Most membrane systems use either microfiltration (MF) or ultrafiltration (UF) membranes with pore size ranging from 0.04 to 0.4 μm (Asano et al. 2007). MBRs have the potential to produce very high-quality effluent because MF or UF membranes provide a physical barrier to protozoa, bacteria, and some viruses.

The MBR process is significantly more compact than conventional biological systems, with a potentially lower sludge production and comparable effluent quality to conventional processes. As solids' separation does not require floc formation and settling, the biomass can operate at very high levels of MLSS, generally in the order of 10,000–15,000 mg/L. This high concentration results in a smaller reactor size and footprint and a long sludge age with a corresponding reduction in biological sludge production and very good nutrient removal. The membrane separation provides the advantage of good particulate removal and effective disinfection of the effluent down to the particle size of a membrane pore.

Reports showed >91% COD removal and >95% BOD_5 removal for both external and submerged MBR treating domestic wastewater (Cicek 2003). Table 13.2 enumerates the advantages and disadvantages of the MBR system.

Table 13.2 Advantages and disadvantages of MBR system

Advantages	Disadvantages
Low suspended solids' concentration and removal of large particles in the permeate enables effective disinfection to be accomplished	Use more electricity: High pressure systems can be energy intensive
Can be designed with long sludge age, hence low sludge production	Technical complexity limited availability of experienced operating personnel
Secondary clarifiers and tertiary filtration processes are eliminated, thereby reducing plant footprint	High capital expenses (CAPEX) and operational expenses (OPEX)
No reliance upon achieving good sludge settleability, hence quite amenable to remote operation	Requires replacement of membranes every five years which is relatively expensive
Produces a MF/UF-quality effluent suitable for reuse applications	All MBR systems make use of propriety equipment and designs; there is no standard configuration
	Pilot testing is often required for full-scale design

13.4 Nature of Conventional Sanitation Concept and Required New Policy

A hollow-fiber submerged MBR was used to treat the higher-load gray water (Huelgas et al. 2009). It is tested for treatment of KSWW only and for treatment of the mixture of KSWW and WMWW (HLGW). Figure 13.2 shows the configuration of a laboratory-scale MBR. The hollow-fiber membrane is an ultrafilter with molecular weight cutoff (MWCO) of 100 kDa. The 0.5 m water level difference corresponds to a TMP of 5 kPa. The dissolved oxygen was monitored and maintained above 4 mg/L inside and the temperature was maintained at 20 °C.

13.4.1 KSWW Only

The reactors used to treat KSWW were operated for 82 days (Huelgas et al. 2009). The hydraulic retention time (HRT) for all three reactors are at 4.5, 7, and 12 h which correspond to organic loading rates (OLR) of 6.9 kg COD/m^3-d, 4.5 kg COD/m^3-d, and 2.6 kg COD/m^3-d, respectively. Table 13.3 summarizes the operating conditions for three reactors. Furthermore, the corresponding food-to-microorganism (F/M) ratios are 0.58 kg COD/kg MLSS-d, 0.38 kg COD/kg MLSS-d, and 0.22 kg COD/kg MLSS-d. These F/M ratios are relatively higher compared to some literature values which indicated a typical F/M ratio of <0.2 kg COD/kg MLSS-d and even approaches zero for long sludge retention time (SRT) and high sludge concentration (Stephenson et al. 2000).

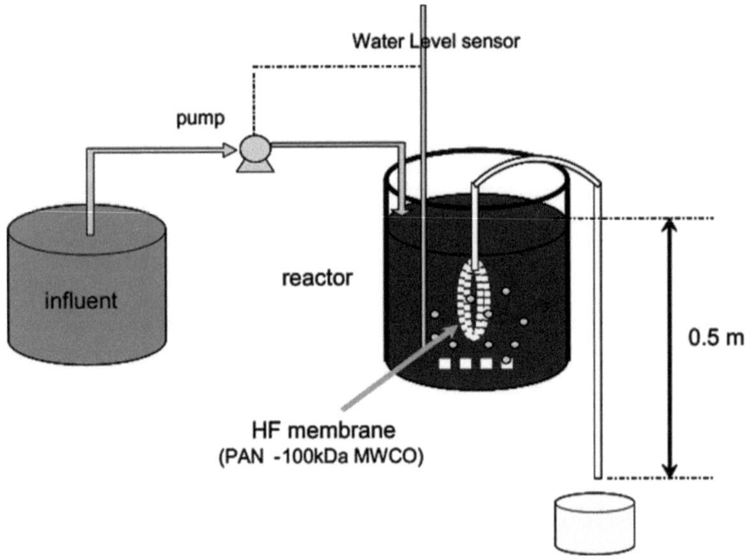

Fig. 13.2 Configuration of a hollow-fiber submerged MBR. Reprinted from Environmental Technology 30(1):111–117 (2009) with permission from the copyright holders, Taylor & Francis Ltd. www.tandfonline.com

Table 13.3 Operating conditions at different KSWW loading rates

Operating parameters	Reactor 1	Reactor 2	Reactor 3
HRT (h)	4.5	7	12
OLR (kg COD/m^3-d)	6.9	4.5	2.6
MLSS (g/L)	11–13	11–13	11–13
F/M (kg COD/kg MLSS-d)	0.58	0.38	0.22

Table 13.4 shows the permeate quality of reactors treating KSWW only. The reactor with shorter HRT and higher OLR has a higher COD in the permeate with an average of 53 mg/L while the least value was observed at longer HRT and lower OLR. Similar results were reported by Stephenson et al. (2000) treating domestic

Table 13.4 Permeate quality of the reactors treating KSWW only

Parameters (mg/L)	Influent	Permeate		
		Reactor 1	Reactor 2	Reactor 3
COD	770–2050 (1300)[a]	35–73 (53)[a]	11–53 (29)[a]	10–26 (17)[a]
NH_4–N	0.3–2.7	–	–	–
NO_3–N	0.9–5.3	–	–	–
PO_4–P	–	–	0.1	0.8

[a]Values in parenthesis show the average value

wastewater. It was reported that for a system with loading range between 1.2 and 3.2 kg COD/m^3-d, the corresponding effluent COD concentration ranges from 16 to 40 mg/L. The % COD removal rates are 95.9, 97.8, and 98.7% for reactors 1, 2, and 3 respectively.

13.4.2 KSWW + WMWW Mixture

The reactors used to treat HLGW were operated for 72 days (Huelgas et al. 2009). Table 13.5 shows the operating conditions for all three reactions. The HRT are 4, 8, and 12 h which correspond to OLR of 5.3 kg COD/m^3-d, 2.7 kg COD/m^3-d, and 1.8 kg COD/m^3-d, respectively. On the other hand, the corresponding F/M ratios are 0.53 kg COD/kg MLSS-d, 0.27 kg COD/kg MLSS-d, and 0.18 kg COD/kg MLSS-d. The OLR for the HLGW mixture is lower compared to the corresponding OLR for the KSWW at the same HRT. This can be explained by lower pollutant characteristics of the HLGW as presented in Table 13.1.

Table 13.6 shows the permeate quality for the reactors treating HLGW. The COD concentrations are higher compared to the reactors treating KSWW only (refer to Table 13.4). This implies that KSWW is more biodegradable compared to HLGW. There is the need to further investigate the factors that contribute to higher COD in the permeate. The % COD removals are 90.7 and 93.3% for reactors 2 and 3, respectively.

It was reported (Huelgas et al. 2009) further that the reactor operated at HRT of 4 h was terminated due to problems in foaming. Thus, no permeate quality was presented in Table 13.6.

Table 13.5 Operating conditions at different HLGW loading rates

Operating parameters	Reactor 1	Reactor 2	Reactor 3
HRT (h)	4	8	12
OLR (kg COD/m^3-d)	5.3	2.7	1.8
MLSS (g/L)	9–11	9–11	9–11
F/M (kg COD/kg MLSS-d)	0.53	0.27	0.18

Table 13.6 Permeate quality of the reactors treating HLGW

Parameters (mg/L)	Influent	Permeate		
		Reactor 1	Reactor 2	Reactor 3
COD	540–1200 (890)[a]	–	48–123 (82.5)[a]	28–124 (59.4)[a]
NH_4–N	0.9–2.6	–	0.1	0.55
NO_3–N	0.2	–	0–0.01	0–0.41
PO_4–P	–	–	–	–

[a]Values in parenthesis show the average value

13.5 Flat-Plate Submerged MBR

A flat-plate submerged MBR was also used to treat the higher-load gray water (Huelgas and Funamizu 2010). It is tested for treatment of the mixture of KSWW and WMWW (HLGW). Figure 13.3 shows the configuration of a laboratory-scale MBR. The flat-plate membrane is a microfilter with pore size of 0.4 μm. The 30 cm water level difference corresponds to a TMP of 3 kPa. The dissolved oxygen was monitored and maintained above 4 mg/L at the system was at 20 °C. The reactor was operated intermittently at 10-min ON–2-min OFF permeation.

Table 13.7 shows the operating parameters of a laboratory-scale MBR. It has an average HRT, OLR, and F/M ratio of 13.6 h, 1.21 kg COD/m^3-d, and 0.07 kg COD/kg MLSS-d, respectively.

Table 13.8 shows the influent and permeate characteristics. An average of 96% removal was achieved with a mean permeate concentration of 26 mg/L. On the other hand, >99% removal was observed for LAS giving an average permeate concentration of 0.025 mg/L.

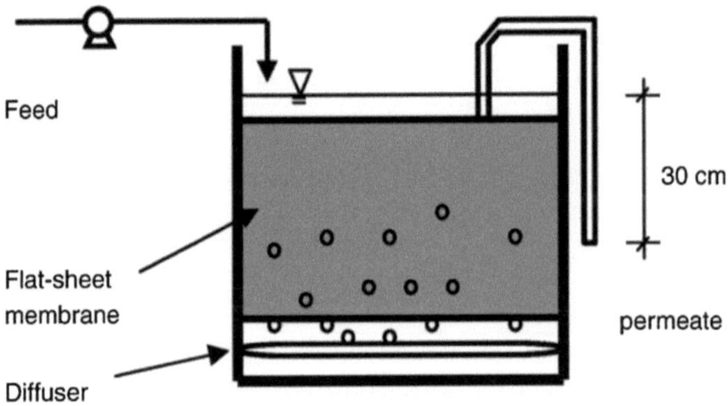

Fig. 13.3 Configuration of a flat-plate submerged MBR Reprinted from Desalination, 250/1, Huelgas and Funamizu (2010) with permission from Elsevier

Table 13.7 Operating parameters of a flat-plate submerged MBR

Operating parameters	Average	Minimum	Maximum
HRT (h)	13.6	10	16
TMP (kPa)	3	3	3
Flux (m^3/m^2-d)	0.22	0.18	0.28
MLSS	16	10	25
OLR (kg COD/m^3-d)	1.21	0.97	1.62
F/M (kg COD/kg MLSS-d)	0.07	0.04	0.12

Table 13.8 Characteristics of the influent and permeate

Parameters (mg/L)	Influent	Permeate
COD	675	26.28
TKN	25.31	12.06
NH_4–N	0.17	0.16
NO_3–N	0	9.85
TP	2.37	–
PO_4–P	1.31	0.78
LAS	30.8	0.025

References

Almeida MC, Butler D, Friedler E (1999) At-source domestic wastewater quality. Urban Water 1:49–55. https://doi.org/10.1016/s1462-0758(99)00008-4

Asano T, Burton F, Leverenz H, Tsuchihashi R, Tchobanoglous G (2007) Water reuse: issues, technologies, and applications. McGraw-Hill

Cicek N (2003) A review of membrane bioreactors and their potential application in the treatment of agricultural wastewater. Biosystems Engineering, University of Manitoba, Winnipeg, Manitoba, Canada

Friedler E, Kovalio R, Galil NI (2005) On-site greywater treatment and reuse in multi-storey buildings. Water Sci Technol 51:187–194

Hills S, Smith A, Hardy P, Birks R (2001) Water recycling at the Millennium dome. Water Sci Technol 43:287–294

Huelgas A, Nakajima M, Nagata H, Funamizu N (2009) Comparison between treatment of kitchen sink wastewater and mixture of kitchen sink and washing machine wastewaters. Environ Technol 30(1):111–117. https://doi.org/10.1080/09593330802445549

Huelgas A, Funamizu N (2010) Flat-plate submerged membrane bioreactor for the treatment of higher-load greywater. Desalination 250(1):162–166. https://doi.org/10.1016/j.desal.2009.05.007

Jefferson B, Laine A, Parsons S, Stephenson T, Judd S (1999) Technologies for domestic wastewater recycling. Urban Water 1:285–292. https://doi.org/10.1016/s1462-0758(00)00030-3

Jefferson B, Judd S, Diaper C (2001) Treatment methods for grey water decentralized sanitation and reuse concepts, systems and implementation. IWA, London, pp 344–353

Jefferson B, Palmer A, Jeffrey P, Stuetz R, Judd S (2004) Grey water characterization and its impact on the selection and operation of technologies for urban reuse. Water Sci Technol 50:157–164

Lopez Zavala MA, Funamizu N, Takakuwa T (2002) Onsite wastewater differentiable treatment system: Modeling approach. Water Sci Technol 46:317–324

Nolde E (1999) Greywater reuse systems for toilet flushing in multi-storey buildings—over ten years experience in Berlin. Urban Water 1:275–284. https://doi.org/10.1016/s1462-0758(00)00023-6

Palmquist H, Hanæus J (2005) Hazardous substances in separately collected grey- and blackwater from ordinary Swedish households. Sci Total Environ 348:151–163. https://doi.org/10.1016/j.scitotenv.2004.12.052

Stephenson T, Judd S, Jefferson B, Brindle K (2000) Membrane bioreactors for wastewater treatment. IWA Publishing, London

Tchobanoglous G, Burton FL, Stensel HD (2003) Wastewater engineering treatment and reuse. 4th ed McGraw-Hill, New York, pp 186, 857

Temmink H, Klapwijk B (2004) Fate of linear alkylben-zene sulfonate (LAS) in activated sludge plants. Water Res 38:903–912. https://doi.org/10.1016/j.watres.2003.10.050

Chapter 14
Treatment of Greywater by Geotextile Filter and Intermittent Sand Filtration

Sebastian Ignacio Charchalac Ochoa, Ken Ushijima, Nowaki Hijikata and Naoyuki Funamizu

Abstract Intermittent sand filtration (ISF) is an efficient system for treatment of greywater; however, the high quality of effluent and the simple setup contrast with the high failure rate due to clogging of surface layers. The efficacy of several polypropylene nonwoven geotextiles (apparent opening size from 0.10 to 0.18 mm) used as primary treatment filters to remove suspended particles from domestic greywater and the effects of this pretreatment in the performance of fine and small media size (0.3 and 0.6 mm) ISFs was examined. Results showed geotextile achieved SS removal rates from 25 to 85% and COD from 3 to 30%; although the portion larger than 75 μm was removed at higher rates (55–90%), particles smaller than the nominal pore size of the filter were also captured. Geotextile used as pretreatment resulted in improvement of lifetime of the ISFs over an experimental run of 60 days. The vertical profile of volatile organic matter in the ISFs was

This chapter is reproduced from Journal of Water Reuse and Desalination, volume 5, issue number 1, pages 39–49 (Charchalac Ochoa et al. 2015), with permission from the copyright holders, IWA Publishing.

S. I. Charchalac Ochoa (✉)
Division of Engineering Sciences, Western Regional Center (CUNOC),
University of San Carlos of Guatemala (USAC), Quetzaltenango, Guatemala
e-mail: charchalac.sebastian@usac.edu.gt

K. Ushijima
Northern Regional Building Research Institute, Hokkaido Research
Organization (HRO), Midorigaoka Asahikawa, Memuro 078-8801, Japan

N. Hijikata
Graduate School of Engineering, Hokkaido University, Kita13-Nishi 8,
Kita-Ku, Sapporo City, Hokkaido 060-8628, Japan

N. Funamizu
Graduate School of Global Food Resources, Hokkaido University, Kita-9,
Nishi-9, Kita-Ku, Sapporo City 060-8589, Japan

N. Funamizu
Research Institute of Humanity and Nature, Motoyama Kamigamo,
Kita-Ku, Kyoto 603-8047, Japan

N. Funamizu (ed.), *Resource-Oriented Agro-sanitation Systems*,
https://doi.org/10.1007/978-4-431-56835-3_14

evaluated at the end of the experiment, and it showed a clear reduction in the accumulation of organic material on the top layer of the ISFs, effectively avoiding its early failure by accumulation of solids.

Keywords Greywater · Intermittent sand filter · Geotextile · Water reuse

14.1 Introduction

Decentralized reclamation and reuse of domestic wastewater is considered as one of the alternatives to alleviate the water availability for agricultural irrigation and reduce the health risks associated with lack of improved sanitation in rural communities of developing countries, which heavily depend on agricultural production (Morel and Diener 2006). Reclamation of domestic greywater is of particular interest for water reuse in domestic gardens or small-scale irrigation systems since it generally contains lower concentration of organic matters and pathogens compared to mixed wastewater (Eriksson et al. 2002; Gajurel et al. 2003; Friedler and Hadari 2006).

The Onsite Wastewater Differentiable Treatment System (OWDTS) is an ecological sanitation alternative based on the "don't mix" and "don't collect" principles that proposes source separation and treatment of reduced volume blackwater (excreta and urine), higher load greywater (HLGW), and lower load greywater (LLGW) (López-Zavala et al. 2002). Typically, HLGW includes greywater from kitchen sink, laundry (totally or partially), and dishwasher, while LLGW is formed by effluents from shower, bath, and handbasin. Treatment of blackwater through dry bio-toilet was proposed for production of compost to be used as fertilizer. The separation of greywater into two types is based on the assumption that greywater from different sources varies in volume and concentration of pollutants (Almeida et al. 1999; Funamizu et al. 2002; Friedler 2004), thus reducing the treatment required for certain effluents. LLGW was intended to be treated by natural soil infiltration systems and HLGW to be treated by onsite conventional systems and discharged afterward. In this sense, the treatment of HLGW by membrane bioreactors for urban non-potable reuse (Huelgas and Funamizu 2010) and by slanted soil system for reuse in irrigation (Ushijima et al. 2013) has been explored. However, the complexity of these systems, added to the frequent maintenance, requirement of qualified labor, and energy requirements (in the case of the MBR) may still be a barrier for wide implementation in low- and medium-income developing countries.

On the other hand, intermittent sand filtration (ISF) is one of the simplest and most efficient methods to treat wastewater; this technology combines physical and biological aerobic processes, enhanced by the aeration given by intermittent load of influent, as in the case of domestic greywater discharge (Reed et al. 1995). High-quality effluents in terms of organic matter and suspended solids (SS) can be achieved with shallow depths (Darby et al. 1996; Rodgers et al. 2010), depending on the physical configuration of the system and the hydraulic loading rate (HLR).

However, the weak point of this system is the relatively high failure rate due to clogging (Leverenz et al. 2009).

The reasons for the porosity reduction that causes clogging have been widely discussed, including simple accumulation of solids or fibers (Langergraber et al. 2003; Spychala and Blazejewski 2003; Zhao et al. 2009), biofilm growth (Rodgers et al. 2004), bacterial production of slimes (Spychala and Blazejewski 2003), precipitation of chemicals (e.g., $CaCO_3$), mechanical compaction of media (Rolland et al. 2009), and others. Nevertheless, clogging is not caused by only one reason, but rather due to a complex combination of all these processes (Leverenz et al. 2009; Hua et al. 2010). Compared to the accumulation of solids, biofilm growth is an inherent process to bioreactors since they rely on biological degradation and is therefore controllable to certain extent by avoiding overloading and allowing enough resting time between loads (Darby et al. 1996). High loads and the particle size distribution of SS have been suggested to influence the clogging rate of ISFs (Winter and Goetz 2003); these loads can be significantly affected by lack of or inefficient primary treatment, which has been stressed as the main reason of failure in practical use of ISFs and other similar systems (Morel and Diener 2006).

We propose the use geotextile fabric filters as pretreatment for domestic greywater prior to treatment by simple ISF systems as a measure of protection to overload of SS. Geotextile fabrics and similar materials have been used as primary treatment for removal of SS and lint from domestic greywater (Christova Boal et al. 1996), municipal wastewater (Kotha 2001), combined runoff and sewage (Marino 2006), surface waters (Mulligan et al. 2009) and rainwater (Silva Vieira et al. 2013); as media for a recirculating bioreactor for treatment of domestic wastewater (Roy et al. 1998); and as biofilm attachment filters for municipal wastewater treatment (Korkut 2003). These materials, particularly nonwoven needle-punched geotextiles, have been shown to be efficient in trapping SS and other large materials (Marino et al. 2006), given their complex fiber structure compared to simple strainers. Additionally, they may have some capacity in removal of pathogens (Keraita et al. 2008), especially helminthes. The main mechanisms for capture of particles by geotextiles are sieving and cake filtration (Korkut 2003); consequently, capture of particles will be more influenced by their size rather than density or the flow rate of the influent; this could be an advantage if compared to traditional grease traps, given the typical unsteady flow of domestic greywater sources. However, geotextiles and ISFs could be especially sensitive to oils and fats; therefore, kitchen sink greywater is excluded from treatment in this study; thus, the target influent of this study is a mixture of LLGW plus laundry greywater and is referred in this paper only as greywater.

The aim of this paper is to evaluate the ability of different geotextiles to remove SS from greywater and extend the lifetime of simple ISFs by controlling the accumulation of particles. Two features are discussed: first, the lifetime and efficiency of geotextile to remove SS and chemical oxygen demand (COD) from greywater and second the effects of using this pretreatment on the permeability and accumulated materials in the surface of ISFs fed with greywater.

14.2 Materials and Methods

14.2.1 Evaluation of Geotextiles

Six geotextile materials (provided by Toyobo Co. and Mexichem S.A.B. de C.V.) were selected to evaluate their capability as single-layer direct filters to remove SS and COD from greywater (Table 14.1). All the materials were polypropylene nonwoven needle-punched geotextiles with apparent opening size (AOS) ranging from 0.10 to 0.18 mm and thickness from 1.4 to 2.3 mm. Samples of each material were cut from a roll (2 × 20 m or 1 × 20 m) of geotextile provided as is in general distribution.

To evaluate the reduction in permeability of the geotextile, a device designed according to ASTM D4491-99a (American Society for Testing and Materials 2004) requirements was constructed with PVC pipes and accessories. The procedure was as follows: A sample of geotextile was cut in circular shape and installed in the device sample holder. The effective diameter was 5 cm. Then, permittivity was measured three consecutive times using water and the arithmetic average was calculated. After that, a batch of 50 cm of greywater was loaded to the geotextile sample and the filtrate collected. Permittivity was measured again with water. The process of loading batches and measuring permittivity was repeated until permittivity was under ten percent of original, and it was considered as clogged.

The quality assessment of influent greywater and collected filtrates was done in terms of SS and COD. The material with the best performance, i.e., higher removal of SS and COD, was subsequently examined for a more detailed analysis regarding the particle size distribution of SS in the effluent. Fractions of SS by size were determined by filtering enough volume (5–6 L) of the filtrate through US Standard steel mesh sieves number 100, 200, 400 (nominal pore size of 150, 75, and 32 μm, respectively) and glass fiber filter ADVANTEC GB-140 (pore size 0.40 μm) and measuring the individual suspended solids concentration of each filtrate. The load and sampling of greywater was done in similar method as the preliminary experiment in batches of 50 cm.

Table 14.1 Characteristics of geotextile fabrics

Type	Manufacturer	Model	AOS (mm)	Permeability (cm s^{-1})	Thickness (mm)	Weight (g m^{-2})
NP	Toyobo	4061N	0.14	0.150	1.00	83
NP	Toyobo	4101N	0.12	0.200	1.40	106
NP	Toyobo	4161N	0.10	0.250	2.00	171
TBNP	Mexichem	NT 1800	0.18	0.040	1.70	144
TBNP	Mexichem	NT 2000	0.18	0.042	1.90	174
TBNP	Mexichem	NT 3000	0.15	0.042	2.10	235

AOS Apparent opening size (pore); *NP* Needle-punched; *TBNP* Thermal-bonded needle-punched

14.2.2 Evaluation of Intermittent Sand Filters

Four ISFs were constructed with acrylic Lucite plastic tubes with 0.05 m inner diameter and 1.32 m length (Fig. 14.1). Media depth was 1 m, leaving 0.16 m between the sand surface and the geotextile filter to avoid mix-up and interference between the clogging either in sand surface or in the geotextile. Two filters (each constructed with and without geotextile) contained sand with an effective size (d_{10}, i.e., size of screen opening through which 10% by weight of sand passes) of either 0.30 mm (fine) or 0.60 mm (medium). The sand was obtained from a company that supplies the material typically used in large sand filters of local wastewater treatment plants; it was washed, dried, sieved (mesh nominal opening sizes: 0.125, 0.3, 0.425, 0.60, 0.85, 1.20, 2.0 mm) and then graded. The uniformity coefficients (UC), calculated as $d_{60}/d_{10,}$ were 1.4 and 1.66 for fine and medium sand, respectively. The inclusion of only fine- and medium-size media was also intended to observe more clearly the effects of deposited solids, since the porosity is supposed to be smaller than in coarser media and therefore more susceptible to clogging.

Sampling ports of 2.5 cm diameter were installed at depths 0.1, 0.2, 0.4, and 0.6 m from the media surface to determine the COD and LAS removal efficiency in the ISFs. These ports were used only in the last two weeks to avoid any disturbance to the ISFs and when the microbial populations had likely already stabilized. The ISFs were operated for 60 days. The conditions were set such as to be able to

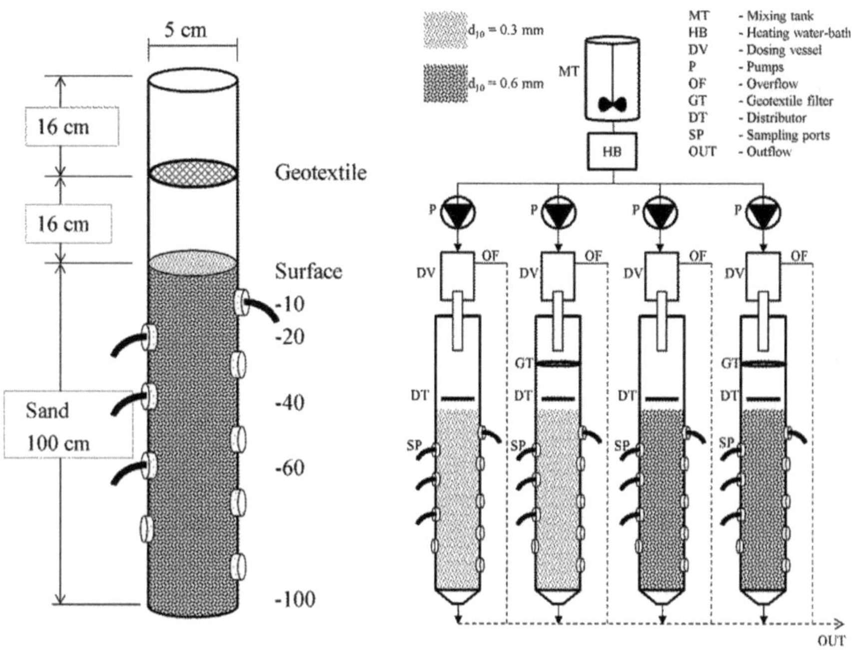

Fig. 14.1 Individual ISF and experimental setup

observe clear differences due to overloading of SS. For the first six weeks, effluent was collected only at the bottom of each ISF. Since the greywater volume of each discharge was small, only one port was opened in each sampling, starting from depth 0.6 and going upwards; this took a week for evaluation at all depths.

The geotextile filter for pretreatment was installed in two of the ISFs through a stainless steel ring equipped with rubber seals to ensure the flow of greywater would go only through the material; the inner diameter of the steel ring was 0.05 m. The geotextile filter was replaced every time it became clogged (i.e., incomplete infiltration after 24 h), which occurred every 7–10 days.

A greywater hydraulic loading rate (HLR) of 16 cm d^{-1} was applied based on the recommendation of Darby et al. (1996). SS and COD loading rates were 16 and 57 g m^{-2} d^{-1}, respectively. The total daily volume was split into three equal single-time discharges during daytime (9 AM, 12 PM, and 6 PM), corresponding to a theoretical morning, noon, and evening time peak water use (Funamizu et al. 2002).

Effects of geotextile filter on ISF performance were measured by two factors: permeability of surface (in terms of infiltration time of a single given greywater dose) and effluent quality. To use a conventional permeability test was deemed impractical since this could cause disturbances to the normal accumulation of SS and the natural development of ISFs biofilm under greywater loading. Therefore, the permeability of media was measured by the total time in seconds that it took for a single morning discharge of 5.66 cm to completely infiltrate (i.e., no water line observed above sand) the surface. Additionally, postexperiment weight loss on ignition (LOI) of sand layers at different depths (three samples were taken from each depth: 0.025, 0.075, 0.15, 0.25, 0.35, 0.5, 0.7, and 0.9 m, respectively) was performed to determine the difference in organic material accumulated as an indicator of physical clogging (Rodgers et al. 2004). Samples for LOI were first evaluated for moisture content and subsequently burned in a muffle furnace at 550 °C for two hours. Difference between initial dry weight and burnt weight in percentage was calculated as LOI.

14.2.3 Greywater Characteristics

Composition of greywater was a mixture of shower, laundry, and handbasin sources (65:30:5 in volume) with the characteristics summarized in Table 14.2. Shower and washbasin greywater was collected from a domestic household to include the organic waste originated from human body. Constituents included common brand shampoo (5 g 30 L^{-1}), conditioner (5 g 30 L^{-1}), body soap (5 g 30 L^{-1}), toothpaste (3 g 30 L^{-1}), and hand soap (3 g 30 L^{-1}). Prior to preparation of greywater, each constituent was weighed in separate plastic containers with precision of 0.01 g. Then shower was taken by one person using the contents of the plastic containers. The shower drain was connected to semitransparent container tanks

Table 14.2 Influent characteristics

	Average	Min.	Max.
TCOD (mg L^{-1})	357	210	505
DCOD (mg L^{-1})	187	65	335
SS (mg L^{-1})	100	75	124
DS (mg L^{-1})	240	196	272
TN (mg L^{-1})	7.5	4.1	13.7
TP (mg L^{-1})	7.5	5.2	10.0
LAS (mg L^{-1})	25	–	–

previously marked to the desired volume (20 L each), so the volume of accumulated greywater was visible and was collected with relative accuracy.

Laundry greywater was prepared by washing four pieces of used clothes in a washing machine using 12.5 g of detergent (Unilever OMO) per 40 L of water; this detergent was chosen since linear alkyl benzene sulfonates (LASs) are the active surfactant. LAS has been estimated to be the most widely used synthetic anionic surfactants and may have a negative impact on environment (Mungray and Kumar 2009). LAS can exist in high concentrations in greywater, and thus, it is a useful indicator compound for greywater treatment systems that aim for reuse of effluent.

After preparation, both types of greywater were transported, mixed in the volume ratio indicated, and stored in refrigerator at 3–4 °C under constant stirring to maintain a homogeneous particle suspension. Fresh greywater was prepared every five days. Typical temperature of greywater is in the range of 18–38 °C (Eriksson et al. 2002); therefore, the influent was passed through a water bath that heated it to a temperature of 20 °C before being supplied to the ISFs. In general, the quality of the experimental greywater (SS and COD) was in the low to middle range for raw greywater (Almeida et al. 1999; Morel and Diener 2006), but higher than usual influents from septic tanks or pond effluents loaded to ISFs (Darby et al. 1996).

14.2.4 Sample Analysis

Permittivity of geotextiles was evaluated using the ASTM D4491-99a Falling-Head Method; normalized values were used for a better comparison. Influent and effluent samples were analyzed for suspended and dissolved solids (SS and DS), total and dissolved COD (TCOD and DCOD), total phosphorus (TP), total nitrogen (TN), and linear alkyl benzene sulfonates (LASs). Standard analytical methods were used for SS and DS (American Public Health Association et al. 2012). HACH kits were used for COD, TN, and TP (Method 8000, Method 10071, and Method 8190, respectively) with the HACH DR2800 Spectrophotometer.

Concentration of LAS (C10–C14) was determined using liquid chromatography–mass spectrometry (LC-MS) based on Ushijima et al. (2013). Column used was Wakopak® WS-Aqua (4.5 × 250 mm), and mobile phase was 0.2 mmol ammonium acetate and 100% acetonitrile, ratio 63:37 at 0.3 mL min^{-1} in isocratic mode. Samples for solid phase extraction were spiked with recovery standard C8 surrogate and then diluted to a concentration in the range of the limit of detection (0.01–0.25 mg L^{-1}). Solid phase extraction was performed according to Managaki et al. (2005) as follows: The pH was adjusted to 3 ± 0.05 with 10% HCl solution and loaded to a BOND ELUTE® PPL cartridge previously conditioned with 10 mL of methanol and 10 mL of pure water adjusted to pH 3 ± 0.05. LAS was eluted from the cartridge with 10 ml of methanol and subsequently filtered with a 0.2 μm PTFE Dismic filter, and aliquots were loaded to the LC-MS vials. Standard LAS solution (C10–14) and recovery standard C8 reagent were obtained from WAKO Industries (>99% purity). Recovery rates were 92, 90, 87, 75, and 64%, respectively.

14.3 Results

14.3.1 Performance of Geotextiles

As shown on the range in Table 14.2, the SS and COD concentration in greywater fluctuated significantly each time despite following the same recipe for greywater in each preparation; this is probably due to the alternate type of clothes used for laundry greywater as well as the variability of the load of human waste that depends on personal habits and activities.

The permeability of geotextiles, indirectly estimated in terms of permittivity, followed a steep reduction over the course of loading greywater (Fig. 14.2). From the six models evaluated, only model 4061 N showed a slower reduction of permeability, just 20.99% loss after 300 mL cm^{-2} of greywater were loaded. At the same point, all the other models were already below 10% of original permittivity. The materials captured by geotextile included hair, cloth fibers, and small clumps of particles or biofilm.

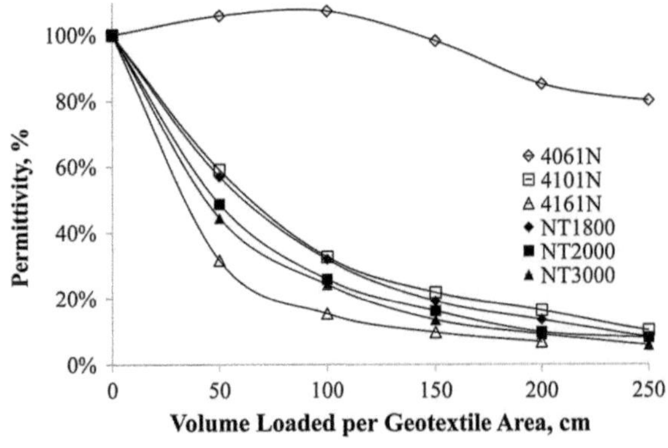

Fig. 14.2 Permittivity reduction in geotextiles

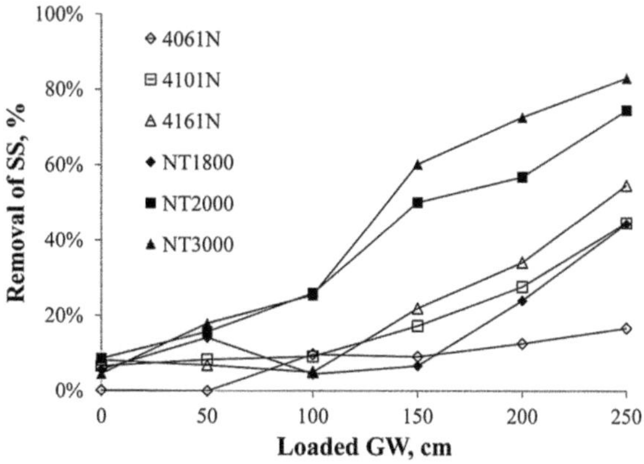

Fig. 14.3 SS removal by geotextiles

Differences in removal rate of SS were negligible in the initial conditions, but generally, after loading 100 cm it was possible to observe difference in the behavior of the materials (Fig. 14.3). Overall, only the geotextile models with density above 170 g m^{-2} (4161 N, NT2000 and NT3000) achieved removal rates above 50% of the SS. Table 14.3 shows the average SS and TCOD removal by geotextiles, where the maximum value of the range refers to the last measured removal before permittivity value was below 10% of original.

For the filter experiment, geotextile NT3000 was chosen due to its ability to capture more SS compared to other models. Since the quality of greywater is variable, several samples of NT3000 were individually assessed about the specific size of the particles removed; this is of importance since the size of deposited particles has been cited as one possible factor in ISF clogging. Figure 14.4 shows fractioned removal of SS by geotextile NT3000 for one sample batch test when greywater had a SS concentration of 143 mg L^{-1}, from which 77 mg L^{-1} were solids larger than 0.075 mm. SS removal had a high initial value and sustained increase in quality, from 79 to 85%, the remaining SS in filtrate being mostly solids smaller than 0.032 mm. Although the pore size of NT3000 is 0.15 mm, the material was able to remove SS in smaller fractions. Permeability was significantly reduced after only 150 cm.

Table 14.3 SS and COD removal by geotextiles

Geotextile	SS		COD	
	Removal (%)	Range (%)	Removal (%)	Range (%)
4061 N	8.1	0.0–16.8	3.0	0.0–5.4
4101 N	19	6.7–44.7	10.3	2.0–20.5
4161 N	21.8	5.0–54.7	11.3	2.9–26.0
NT1800	16.6	4.5–44.5	6.7	1.0–15.4
NT2000	38.6	8.6–74.5	20.3	5.2–32.0
NT3000	44	4.6–83.1	19.6	7.5–34.2

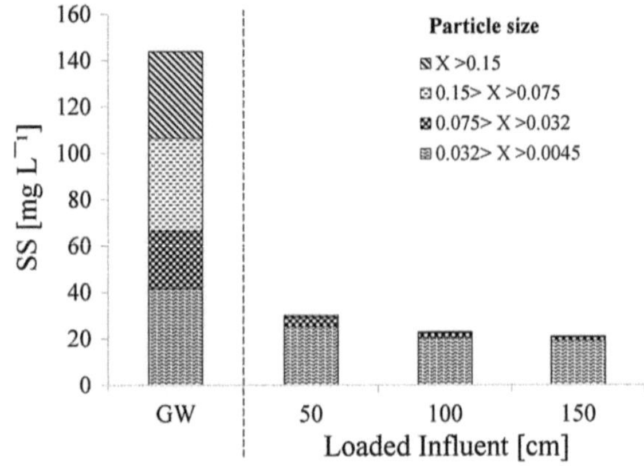

Fig. 14.4 Particle size distribution in GW filtered by NT3000

14.3.2 *Performance of ISFs and Geotextile Filter*

The performance of the geotextile was slightly different compared to the first experiment. In this longer-term operation, insects and other random materials were occasionally observed, additionally to the previously mentioned greywater contents. The geotextile filter had to be replaced every 7–10 days, which means it only filtered 112–160 cm, compared to the 250–300 cm observed in the preliminary tests. Some type of slime was observed on the surface of the geotextile but not measured in any way; this could, however, indicate some bacterial activity.

Figure 14.5 shows the progression in infiltration time for a single dose (5.66 cm) on the surface layer of the ISFs, as an indirect measurement of its hydraulic permeability. A clear increase in infiltration time was observed for three of the cases (0.30 mm with and without geotextile and the 0.60 mm without geotextile). Differences started to be noticeable after just two weeks. ISF without geotextile and

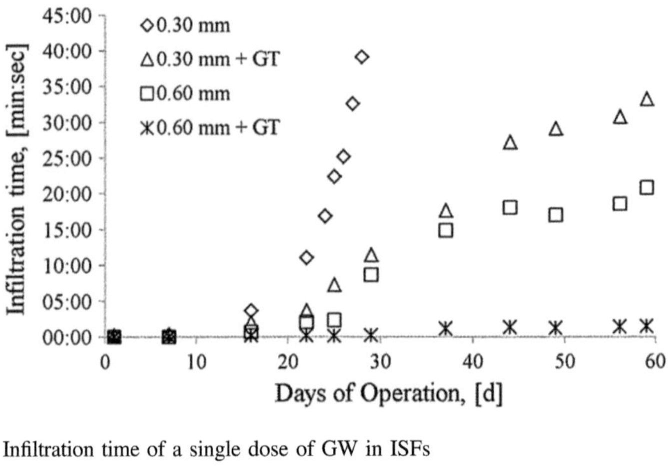

Fig. 14.5 Infiltration time of a single dose of GW in ISFs

fine sand was clogged after 29 days, after an almost exponential increase in the infiltration time in its last week. On day 28th, the total daily dose of 16 cm did not completely infiltrate from the previous day; the column was stopped at the end of day 29th, letting it to rest for one more day and then proceeded to disassembling. Loss on ignition test was performed immediately. The ISF with same media but geotextile did experience sustained increase in infiltration time, although on a slower rate.

For the case of 0.60 mm sand with geotextile pretreatment, the decrease in permeability was the least severe, increasing from 17 s in the initial day to 92 s in day 60. Although there was an increment in the infiltration time, the increase rate was significantly slower compared to the case without geotextile.

The remaining three ISFs were dismantled after 60 days of operation, and the LOI tests were performed. Figure 14.6 shows the difference in weight loss in each of the layers. The most notable difference is observed in the top layer (0 m to 0.05 m depth). ISF of media 0.60 mm with geotextile had a weight loss of only 0.49%, whereas the one without filter was 0.77%. ISFs with media 0.30 mm with and without geotextile had 0.89 and 0.92% weight loss. The LOI values from below surface were similar for all ISFs.

Furthermore, following experiments showed that after operating for more than 8 months in similar conditions, no clogging was observed in media size $d_{10} = 0.60$ mm (data not shown). This confirms the ability of geotextile to extend lifetime of ISFs over a continuous usage.

At depth 100 cm, the quality of effluent from ISFs was high in terms of both SS (less than 1 mg L^{-1} in all cases) and organic matter (COD removal over 90%). The quality of the effluents did not significantly vary between the columns of same media size (Table 14.4).

The profile obtained in the last two weeks for the three remaining columns through samplings in the ports provided information about the degradation of organic components (Figs. 14.7 and 14.8). COD reduction was drastic in the upper

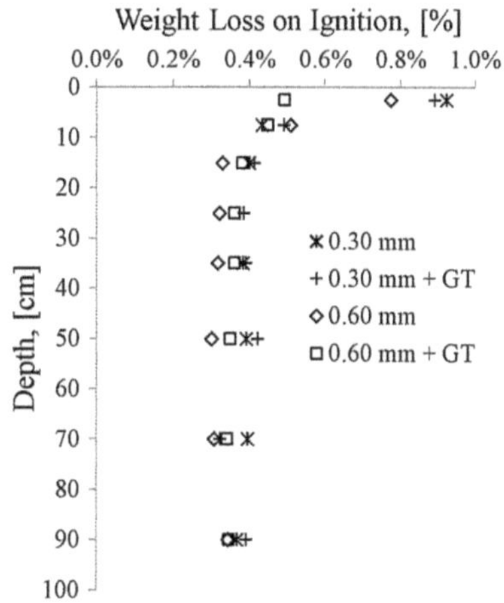

Fig. 14.6 Accumulation of organic material in ISFs

Table 14.4 Effluent quality

Parameter	GW	0.30 mm	0.30 mm +GT	0.60 mm	0.60 mm +GT
COD (mg L^{-1})	357.4 ± 72.0	30.5 ± 18.9	27.5 ± 16.2	32.6 ± 14.2	33.7 ± 18.9
SS (mg L^{-1})	99.9 ± 14.7	<1	<1	<1	<1
TN (mg L^{-1})	7.5 ± 2.9	1.9 ± 1.4	2.9 ± 2.4	3.5 ± 1.5	3.0 ± 1.9
TP (mg L^{-1})	7.5 ± 1.5	1.2 ± 0.2	1.2 ± 0.3	1.1 ± 0.2	0.9 ± 0.1
LAS (mg L^{-1})	25	3.0 ± 1.2	3.0 ± 1.5	3.5 ± 1.0	3.9 ± 1.2

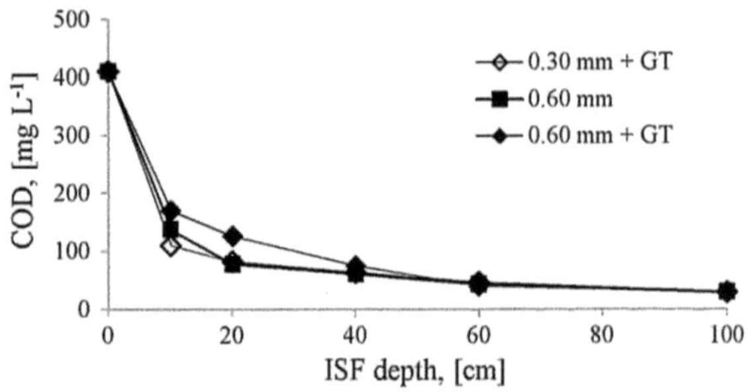

Fig. 14.7 COD reaction in ISFs

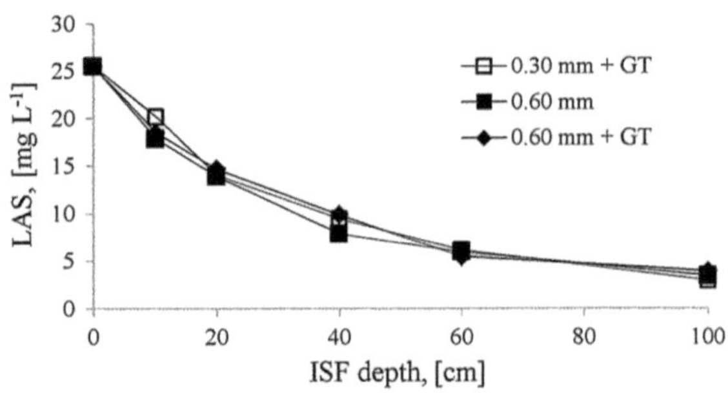

Fig. 14.8 LAS reaction in ISFs

10 cm, whereas LAS removal had a more linear relationship to depth, reaching overall efficiencies of 88% and 84–86% for fine and medium sand, respectively.

14.4 Discussion

As shown in the results from the first phase, the geotextiles have a remarkable ability to remove SS and COD in particulate form. Moreover, the particle size can be controlled. In the case of a traditional sedimentation tank, the removal of SS depends on the retention time, something difficult to control in the case of sudden and intermittent discharges. Additionally, the lack of maintenance in such systems can cause a sharp increase in SS exiting the chamber, a common occurrence. Geotextile filtration appears to be a good option to control the size of particles, although the maintenance required for current method of use seems a downside. In a previous study by Marino (2006), a geotextile was evaluated in removal of organic particles under horizontal flow, the material was able to separate the large elements and when the flow stopped, the retained material subsequently slurred, giving an idea of geotextile working as a simple mesh and separating particles of a given size; however, the nature of particles in greywater is different in the sense that particles attach to the fibers of the geotextile, giving little chance to reuse the filter under the current usage conditions. The inclusion of geotextile in a self-cleaning backwash system setup as shown in Silva Vieira et al. (2013) could enhance its lifetime and reduce the frequency of maintenance. In that scheme, the geotextile was installed in an inverted siphon setup, performing an up-flow filtration rather than top to bottom as in this study. This setup allowed the retained SS to settle in the bottom of the siphon after the influent stopped to be loaded. Furthermore, a magnetic valve located in the bottom part of the siphon opened under certain head pressure, allowing the automatic backwash when the accumulated liquid equilibrated the strength of the magnet. Although the influent in that study was rainwater, the same principle applied for greywater system could be a good option to be examined for a longer lifetime of geotextile.

Thermal-bonded side of geotextile materials NT1800, 2000, and 3000 seemed to be beneficial on the initial sieving capture of SS, and certainly the structure of the geotextile was more robust since the fibers were bonded together. Besides this characteristic, another important factor in removing particles seems to be the density of the materials. Despite having a theoretical smaller AOS and similar thickness, the model 4161N performance in removal of SS and COD was slightly below models NT2000 and NT3000, and above NT1800, an order which correlates to the densities shown in Table 14.1.

The improvement in performance over the course of loading greywater to geotextiles is undoubtedly due to the accumulation of trapped particles in the fibers of the filter and a logical reduction of porosity, as well as the formation of a cake layer. This is clearly shown when the SS load of larger particles was higher (Fig. 14.4). The removal rates jumped to near 80%, although with a reduced lifetime.

The effect of geotextile on the permeability of the ISFs was evident, although the experimental conditions were somewhat near worst-case scenario for the two columns without pretreatment. However, it was clear that the short-term clogging was avoided by pretreatment. Geotextile reduced the SS load, thus reducing the particulate accumulation in surface, which was not only important regarding the total volume but also the type and size of particles trapped; the geotextile removed not only particles, but also fibers originated in clothes, which had previously been suggested by Spychala and Blazejewski (2003) to contribute to the acceleration of clogging in sand filters. These type of SS probably have a higher chance to breakthrough a sedimentation tank, most likely if those are low-density synthetic fabrics, something that adds the complication of no biodegradation. Geotextile has a remarkable ability to trap such materials and similar (hair) within its complex array of fibers, compared to a simple metallic mesh. On the long run, this process could extend lifetime of a sand filter.

The quality of effluent in ISFs did not show any significant difference and, as expected, the top layer of the filters was the most important for the removal of COD and SS, mainly due to the capture of the particulate portion by sieving. The near linear pattern in LAS concentration to depth shows that removal may be initially influenced by adsorption process followed by biodegradation.

Biofilm growth is a process inherent to this type of bioreactors, but if the system works under appropriate conditions and resting times are allowed periodically, the system can have a stable performance for long periods of time. Clogging by deposition of large and non-degradable materials on the other hand is an avoidable process.

14.5 Conclusions

Geotextiles showed to be able to remove more than 50% of SS in greywater. Moreover, the removal of specific particles goes below the theoretical AOS. Reduction of large particles by pretreatment not only reduces the amount of accumulated particles but showed the potential to remove COD up to 30%. The materials with one side thermal-bonded as well as higher density showed best performance. The materials were able to filter an overall maximum of 200–250 cm of greywater before failure in the continuous use, and only 112–160 cm in the intermittent setup, meaning that if used in the same area ratio to the ISFs surface, a large amount of geotextile is needed to be replaced often. Using geotextile filter as pretreatment did not affect significantly the overall quality of ISFs effluent and positively reduced the accumulated volatile solids in the top layers of the media, effectively avoiding early clogging in fine media and decrease of permeability in medium sand during the experimental period.

The results of the experiment show that geotextile filter may have potential for extending the initial lifetime of ISFs. The downside is the periodical maintenance and the material expense needed every 7–10 days in the current usage form. It is recommended to explore alternative usage techniques to address this matter. Promising alternatives could be the usage of layered packed geotextile and/or the up-flow layout shown in some rainwater-oriented systems, but a cost-effectiveness assessment is needed for evaluating the real value of integrating such pretreatment in ISF systems.

References

Almeida MC, Butler D, Friedler E (1999) At-source domestic wastewater quality. Urban Water 1 (1):49–55

American Public Health Association, American Water Works Association, Water Environment Federation (2012) Standard methods for the examination of water and wastewater, 22nd edn. Washington, USA

Charchalac Ochoa SI, Ushijima K, Hijikata N, Funamizu N (2015) Treatment of domestic greywater by geotextile filter and intermittent sand filtration bioreactor. J Water Reuse Desalin 5(1):39–49. https://doi.org/10.2166/wrd.2014.042

Christova Boal D, Eden RE, McFarlane S (1996) An investigation into greywater reuse for urban residential properties. Desalination 106:391–397. https://doi.org/10.1016/0011-9164(96)00134-8

Darby JL, Tchobanoglous G, Asri No M, Maciolek D (1996) Shallow intermittent sand filtration: performance evaluation. Small Flows J 2:3–14

Eriksson E, Auffarth K, Henze M, Ledin A (2002) Characteristics of grey wastewater. Urban Water 4:85–104

Friedler E (2004) Quality of individual domestic greywater streams and its implication for on-site treatment and reuse possibilities. Environ Technol 25:997–1008

Friedler E, Hadari M (2006) Economic feasibility of on-site greywater reuse in multi-storey buildings. Desalination 190:221–234. https://doi.org/10.1016/j.desal.2005.10.007

Funamizu N, Mizukubo T, Lopez Zavala MA, Takakuwa T (2002) Fractioning gray-water in the differentiable onsite-wastewater treatment system. In: Proceedings of XXVIII Congreso Interamericano de Ingeniería Sanitaria y Ambiental, Cancún, México

Gajurel DR, Li Z, Otterpohl R (2003) Investigation on the effectiveness of source control sanitation concepts including pre-treatment with Rottebehaelter. Water Sci Technol 48:111–118

Hua GF, Zhu W, Zhao LF, Huang JY (2010) Clogging pattern in vertical-flow constructed wetlands: insight from a laboratory study. J Hazard Mater 180:668–674. https://doi.org/10.1016/j.jhazmat.2010.04.088

Huelgas A, Funamizu N (2010) Flat-plate submerged membrane bioreactor for the treatment of higher-load graywater. Desalination 250:162–166. https://doi.org/10.1016/j.desal.2009.05.007

Keraita B, Drechsel P, Konradsen F, Vreugdenhil RC (2008) Potential of simple filters to improve microbial quality of irrigation water used in urban vegetable farming in Ghana. J Environ Sci Health Pt A: Environ Sci Eng Toxic Hazard Subst Control 43:749–755. https://doi.org/10.1080/10934520801959948

Korkut EN (2003) Geotextiles as biofilm attachment baffles for wastewater treatment. Ph.D. Dissertation. Faculty of Civil Engineering, Drexel University

Kotha KK (2001) Modification of waste water treatment using geotextiles. MSc Thesis. Faculty of Civil Engineering, Texas Tech University

Langergraber G, Haberl R, Laber J, Pressl A (2003) Evaluation of substrate clogging processes in vertical flow constructed wetlands. Water Sci Technol 48:25–34

Leverenz HL, Tchobanoglous G, Darby JL (2009) Clogging in intermittently dosed sand filters used for wastewater treatment. Water Res 43:695–705. https://doi.org/10.1016/j.watres.2008.10.054

López Zavala MA, Funamizu N, Takakuwa T (2002) Onsite wastewater differentiable treatment system: modeling approach. Water Sci Technol 46:317–324

Managaki S, Kojima S, Harada A, Nanada N, Tanaka H, Takada H (2005) Development of analytical method of linear alkylbenzenesulfonates and their degradation intermediates by high performance liquid chromatography equipped with tandem mass spectrometry and its application to major Japanese rivers. J Japan Soc Water Environ 28:621–628

Marino RJ (2006) Geotextile filter treatment of combined sewer discharges. Ph.D. Dissertation. Faculty of Civil Engineering, Drexel University

Marino RJ, Walton J, Cheng SC, Martin JP (2006) Geotextile behavior relevant to filtering low density suspended organic particles. In: Proceedings from the 8th international geosynthetics conference, Yokohama, Japan. Millpress, Rotterdam, Netherlands

Morel A, Diener S (2006) Greywater management in low and middle-income countries, review of different treatment systems for households or neighborhoods. Swiss Federal Institute of Aquatic Science and Technology (EAWAG), Dübendorf

Mulligan CN, Davarpanah N, Fukue M, Inoue T (2009) Filtration of contaminated suspended solids for the treatment of surface water. Chemosphere 74:779–786

Mungray AK, Kumar P (2009) Fate of linear alkylbenzene sulfonates in the environment: a review. Int Biodeterior Biodegrad 63:981–987

Reed SC, Crites RW, Middlebrooks EJ (1995) Natural systems for waste management and treatment. McGraw-Hill, USA

Rodgers M, Mulqueen J, Healy MG (2004) Surface clogging in an intermittent stratified sand filter. Soil Sci Soc Am J 68:1827–1832

Rodgers M, Walsh G, Healy MG (2010) Different depth intermittent sand filters for laboratory treatment of synthetic wastewater with concentrations close to measured septic tank effluent. J Environ Sci Health Pt A: Toxic/Hazard Subst Environ Eng 46:80–85

Rolland L, Molle P, Liénard A, Bouteldja F, Grasmick A (2009) Influence of the physical and mechanical characteristics of sands on the hydraulic and biological behaviors of sand filters. Desalination 248:998–1007. https://doi.org/10.1016/j.desal.2008.10.016

Roy C, Auger R, Chénier R (1998) Use of non woven textile in intermittent filters. Water Sci Technol 38:159–166

Silva Vieira A, Weeber M, Ghisi E (2013) Self-cleaning filtration: a novel concept for rainwater harvesting systems. Resour Conserv Recy 78:67–73. https://doi.org/10.1016/j.resconrec.2013.06.008

Spychala M, Blazejewski R (2003) Sand filter clogging by septic tank effluent. Water Sci Technol 48:153–159

Standard D4491-99a (Reapproved 2004) Standard test methods for water permeability of geotextiles by permittivity. American Society for Testing and Materials USA

Ushijima K, Ito K, Ito R, Funamizu N (2013) Greywater treatment by slanted soil system. Ecol Eng 50:62–68. https://doi.org/10.1016/j.ecoleng.2012.04.028

Winter KJ, Goetz D (2003) The impact of sewage composition on the soil clogging phenomena of vertical flow constructed wetlands. Water Sci Technol 48:9–14

Zhao L, Zhu W, Tong W (2009) Clogging processes caused by biofilm growth and organic particle accumulation in lab-scale vertical flow constructed wetlands. J Environ Sci (China) 21:750–757. https://doi.org/10.1016/S1001-0742(08)62336-0

Part V
Agriculture

Chapter 15
Application of Human Urine in Agriculture

Moustapha Sene, Nowaki Hijikata, Ken Ushijima and Naoyuki Funamizu

Abstract Although it is known that human urine contains valuable plants nutrients, too much urine application may cause an accumulation of several ions in soil. Among them, excess of Na inhibits plant growth, while excess of nitrogen can build up in plant tissues. Therefore, the objectives of this study was to (1) evaluate when and how often human urine should be applied in agriculture as fertilizer solution, (2) examine the effects of uncontrolled application of human urine dose in agriculture, (3) and examine the effects of continuous application of uncontrolled human urine application dose in agriculture through continuous cultivations. The results revealed that application of urine at one time before cultivation is not effective for plant growth due to high ions intensity in low saline soil and high nutrients leached from soil. Besides, continuous application of extra urine volumes in the range of 2–3 times (U-2Vo and U-3Vo) higher than plant requirement (U-Vo) had no adverse effect on plant growth and yields under three-time continuous cultivations, since treated soils were far to be saline (EC < 2mS/cm; SAR < 13), and more than 39% of the total Na applied in all urine treatments was removed by plant from soil. However, continuous application of double (U-2Vo) and triple (U-3Vo) volumes of urine caused accumulation of N in plant and soil and decreased plant K/Na ratio (substitution of K^+ by Na^+). While in the adequate urine volume (U-Vo), there was no accumulation of *N* in soil, no increased of *N* in plant tissues,

M. Sene · N. Hijikata (✉)
Graduate School of Engineering, Hokkaido University,
kita13-nishi 8, Kita-ku, Sapporo, Hokkaido 060-8628, Japan
e-mail: hijikata.nowaki@gmail.com

K. Ushijima
Northern Regional Building Research Institute, Hokkaido Research Organization (HRO), Midorigaoka, Asahikawa 078-8801, Japan

N. Funamizu
Graduate School of Global Food Resources, Hokkaido University,
Kita-9, Nishi-9, Kita-ku, Sapporo 060-8589, Japan

N. Funamizu
Research Institute of Humanity and Nature, Motoyama Kamigamo,
Kita-Ku, Kyoto 603-8047, Japan

N. Funamizu (ed.), *Resource-Oriented Agro-sanitation Systems*,
https://doi.org/10.1007/978-4-431-56835-3_15

but plant K/Na ratio decreased. So, these results suggested that adequate application of human urine based on plant *N* requirement and management of salts from urine is required even in the adequate urine volume application for its sustainable reuse in agriculture.

Keywords Urine reuses · Uncontrolled application · Continuous cultivation Adverse effects · Salts management

15.1 Introduction

The global NPK fertilizer demand has increased through the years and is basically affected by population and economic growth, agriculture productions, price, and governance policies (FAO 2008a). This increase of fertilizer demand leads ineluctably to the instability of its price. For instance, in the mid-2008, this global nutrients consumption has decreased due mainly the economic crisis, and consequently, the phosphate rock commodity price has increased up to 700% from 2007 to 2008 (FAO 2009a). And high fertilizer price causes low yields and less agricultural productions, especially in marginal area of sub-Saharan Africa, where tools are limited and accessibility of fertilizer when needed is currently a great concern. More concerning is the global reduction of phosphorus production due to the insufficient phosphate rock reserve which is predicted to occur around 2030; and once the maximum reached the production will drop down year upon year and seeing a widening gap between supply and demand (Cordell 2008). In the other hand, it is already well known that human excreta contains valuable major plant nutrients such as nitrogen (N), phosphorus (P), and potassium (K). Therefore, to mitigate the global fertilizer demand and build a sustainable society based on resource recycling and low carbon society, the linkage between agriculture and sanitation is an holistic strategy for waste management, food security, and food production. Consequently, in recent year, source separation of black water has spurred a strong interest for both developed and developing world and is highlighted on the On-site Wastewater Differential Treatment System (OWDTS) and Ecological Sanitation (EcoSan), which are low-cost sanitation system, easy to implement, operate, and maintain. At the household level, faeces, urine, and gray water are properly separated and treated. The merits of this approach are mainly recovering and recycling nutrients, controlling micropollutants and pathogens and reducing wastewater flower (Otterpohl et al. 1997, 1998; Lopez et al. 2002). In this system, urine fraction contains the largest amount of nutrients in domestic wastewater with approximately 80% of the nitrogen (N), 55% of the phosphorus (P), and 60% of the potassium (K) (Höglund 2001). Thus, urine has a potential to be reused in agriculture as liquid fertilizer in order to replace industrial fertilizer (Kirchmann and Petterson 1995; Jönsson et al. 2000). Therefore in recent years, human urine has been compared successfully with commercial fertilizer, using diverse types of vegetables and cereals. However, urine contains some undesirous salts by plants and agricultural soil, (Krichmann and Pettersson 1995; Mnkeni et al. 2005, 2008), pathogens (Höglund 2001; Pradhan et al. 2007), pharmaceuticals (Winker et al. 2008, 2010), and hormones. Concerning the undesirous salts issues, too much urine application in agriculture land may cause accumulation of sodium (Na) in soil. The excess of Na in soil inhibits plant growth, while excess of N can build up in plant tissue and affect negatively the amount of sugar and vitamin in vegetables, causing therefore health and taste issues to consumer (Brady and Weil 1996; Turan and Sevimli 2005). Subsequently, to address salt issues and provide a better understanding of urine application in agriculture, this present research work was carried out with these following specific objectives to (1) evaluate when and how often human urine should be applied in agriculture as liquid fertilizer, (2) examine the effects of uncontrolled application of human urine dose in agriculture, (3) and examine the effects of continuous application of uncontrolled human urine application dose in agriculture.

15.2 Adequate Human Urine Application Pattern for Agriculture

Since human urine contains valuable plant nutrients, extensive research has been recently conducted to compare urine with commercial fertilizer using diverse type of vegetables and cereals: barley (Kirshmann and Pettersson 1995), wheat (Jönsson et al. 2004), maize (Guzha et al. 2005; Kassa et al. 2010; Dagerskog and Bonzi 2010), cabbage (Pradhan et al. 2007, 2010b) cucumber (Heinonen-Tanski et al. 2007), carrot, beetroot (Mnkeni et al. 2008), pumpkin, tomato, red beet, cauliflower, broadleaf mustard, potato, and radish (Pradhan et al. 2009a, b, 2010a, b). Sometimes, 1/4 of the total urine volume required for cultivation was applied 3 days before planting (cabbage, cauliflower mustard, etc.) and in other hand no application in prior to planting (cucumber, pumpkin, red beet, potato, etc.). In addition, Jönsson et al. (2004) reported that in crop cultivation, normal strategy of fertilizer application is once or twice per growing season and proposed if urine is applied only once, this should normally be carried out prior to or at the time of sowing/planting. Thereby, it has not been previously well addressed when and how often urine should be applied in soil as fertilizer. Therefore, it seemed reasonable to study human urine finding a better way for its application in soil. Pot test with Spinach (*Spinacia oleracea*) was conducted and then observed the plant growth and fertility effect under different urine application ways.

15.2.1 Pot Experiments

The experiment was conducted in greenhouse located in Hokkaido University in Japan (43° 04′ 11.6″N, 141° 20′ 22.4″E) from March to June 2011. Two compartments closed pot system was used in this experiment to analyze leached water. The inside compartment in the system was for cultivation, with a plastic pot (volume 0.7 L, top diameter 12.4 cm, bottom diameter 8.5 cm, and height 10.2 cm), and its bottom was covered by 150 μm nylon mesh to prevent root expansion and allowing water passing through. The external compartment was for collecting leached water, with a Wagner pot (cylindrical shape, diameter 11.0 cm and height 15.0 cm, defined as 1/10000 a size). Washed sandy soil, which had a low organic matter, was used as media to analyze nutrients (nitrogen and phosphorus) distribution in plant, soil, and leached water. Physical and chemical properties of the soil were shown in Table 15.1, and the method was described as below.

Within each experimental pot, six spinach seeds were germinated in petri dish with filter paper and incubated at 25°C for 2 weeks and then sown. After 2 weeks of sowing, the seedlings emerged and within each pot seedlings were reduced to two plants. Spinach was selected in this experiment due to its worldwide distribution, high marketability, and high nutrients requirement (Burt 2006). Synthetic urine (Wilsenach et al. 2007) was used in this work to have more general and precise effect of urine, since human urine fluctuates from person to another and depends mainly on diet and physical activity (Pradhan et al. 2010b). The synthetic urine contained 12.05 g/L of nitrogen, 0.96 g/L of phosphorus, and 2.04 g/L of potassium. Based on the total nitrogen required to grow Spinach (NPK 112-157-168 kg/ha) recommended by Kirkwyland and Thomas (2010), the composition of synthetic urine and the pot system size, the total volume of synthetic urine was designed to 29 ml and contains the following dose NPK 112.0-9.8-18.9 mg/pot. In practical, however, diluted (1/3) synthetic urine was performed. About 440 ml of irrigation water was applied in each pot in the first week. About 560, 560, 560, 700, 620, and 520 ml of irrigation were applied in the second, third, fourth, fifth, sixth, and seventh week, respectively. Totally 3.96 l of irrigation water was used per pot during all the cultivation period.

Table 15.1 Soil properties and nutrients contents before cultivation

Total nitrogen (mg-N/g-dry soil)	Total phosphorous (mg-P/g-dry soil)	Total carbon (mg-C/g-dry soil)	pH	EC (μS/cm)
0.21 ± 0.00	0.02 ± 0.00	1.01 ± 0.02	5.43 ± 0.18	49.5 ± 8.9
Porosity*	Water-holding capacity* (mL/100 g-dry soil)		Water drainage* (mL/s)	Moisture content*
0.31 ± 0.01	24.31 ± 2.58		0.90 ± 0.02	0.31 ± 0.07

Made by authors based on Sene et al. (2012)

Arithmetic mean values ± standard deviation ($N = 3$). *These values were determined by methods as described in Horisawa et al. (1999)

A total of seven treatments were set to know a better way of urine application in agricultural soil (a) 100% of urine equivalent to 29 ml was applied before cultivation; 60% of urine was applied before cultivation, and then remained urine (40%) was regularly applied (b) every week (c) every 2 days after 15th day sowing corresponding to the total emerged of plants in all pots; In similarly manner, 30% was applied before cultivation and then (d) every week (e) every 2 days; no urine applied before cultivation and then (f) every week (g) every 2 days. Totally 52 experimental pots were used for 7 treatments replicated 7 each, and 1 treatment without urine as control negative replicated 3 times.

The cultivation period of Spinach was designed for 70 days (Nishihara et al. 2001). However, the duration was shortened to 49 days due to weak plant growth. Therefore, the total input of nitrogen and phosphorus at the end of cultivation was (a) 112, 9.8 (b) 98, 7.8 (c) 94.8, 7.5 (d) 90, 7.1 (e) 89.2, 7.1 (f) 78.8, 6.3, and (g) 73.9, 5.9 mg/pot, respectively. Plants heights were measured weekly (from 1 week after urine started to be applied), and leaves area were measured after harvesting. In the plant height measurement, highest leaf in pot was selected. In the leaf area measurement, well grown three pots per treatment were selected, and then, two biggest external leaves were measured by paper copy method. Furthermore, fresh weight of shoots and roots were weighted separately. Lyophilized shoots and roots were also weighted as dry weight and then stored at −30 °C. Chemical properties (pH) and electrical conductivity (EC) in soil after harvests were shown in Fig. 15.1.

Total nitrogen and total carbon in soil were determined by Sumigraph NC-220F (Sumika Chemical Analysis Service, Japan) apparatus. Regarding total phosphorus in soil, soil samples were digested with nitrate-perchloric acid digestion method and then measured by ascorbic acid method using spectrophotometer at 880 nm

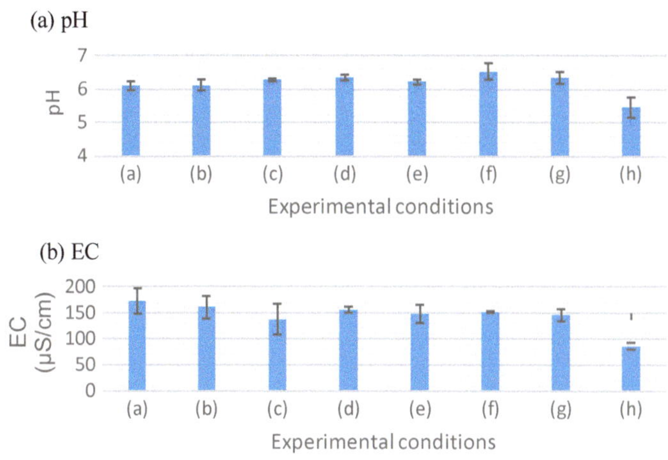

Fig. 15.1 Soil properties after cultivation, $N = 3$ in each experimental conditions. Experimental conditions: **a** 100% urine, **b** 60% before/week, **c** 60% before/2 days, **d** 30% before/week, **e** 30% before/2 days, **f** every week, **g** every 2 days, **h** control. Made by authors based on Sene et al. (2012)

(Standard Method 1989). pH and EC in soil were determined by 1:2.5 dilution method and 1:5 dilution method, respectively, with specific electrochemical probes. Measurements of porosity, water-holding capacity, and water drainage were performed following the methods described by Horisawa et al. (1999). In the case of plant, shoot and root were separately analyzed. Prior to nutrients analysis, dried shoot and root were grinded by mortar and muddler. Total nitrogen and total phosphorus in powdered plant were measured by same method with soil analysis described above. In leached water, total nitrogen and total phosphorus were analyzed by Hach Method 10071 and 8190, respectively. All data in the results were analyzed with three samples in each treatment, and its mean value and standard error were shown in results. The statistical analysis of each data was analyzed with Stat View software version 5.0.

15.2.2 Results

The plant heights at 35 days were (a) 3.9 ± 0.8, (b) 6.4 ± 0.6, (c) 6.6 ± 0.9, (d) 6.1 ± 1.4, (e) 6.6 ± 1.0, (f) 7.0 ± 0.6, (g) 6.6 ± 1.0, and (h) 3.3 ± 1.0 cm, respectively. Urine application at one time before seeding (a) was in the same level with the control (h) and significantly lower than (b), (c), (d), (e), (f), and (g) (Tukey–Kramer test, $p < 0.05$). Similarly to plant height, leave areas after cultivation in (a) was in the same level with that of control (h) and significantly lower than that of (b), (c), (d), (e), and (f) (Tukey–Kramer test, $p < 0.05$) (Fig. 15.2). Furthermore, shoot dry biomass of (a) was same level with the control (h), but both were significantly lower than treatments (c) and (g) (Tukey–Kramer test, $p < 0.05$). In the case of root dry weight, root dry biomass of (a) and the

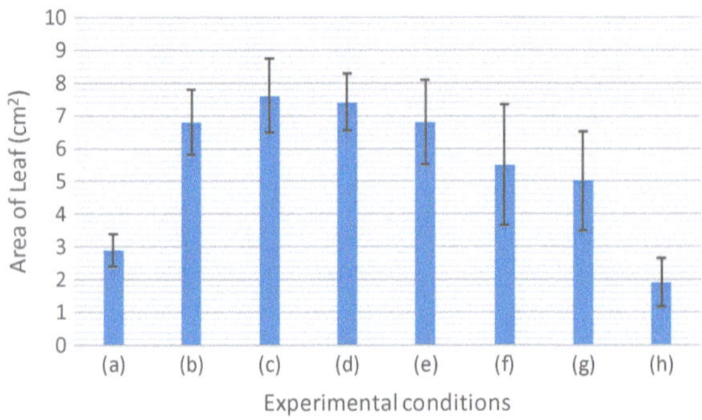

Fig. 15.2 Leaf area measurement of spinach plant after cultivationIn each pot, the two biggest external leaves were selected for three pots per treatment (arithmetic means ± SD). Experimental conditions: **a** 100% urine, **b** 60% before/week, **c** 60% before/2 days, **d** 30% before/week, **e** 30% before/2 days, **f** every week, **g** every 2 days, **h** control. Made by authors based on Sene et al. (2012)

control (h) were also in the same level, but both significantly lower than treatment (b). Therefore, from these results, it was clearly observed a low plant growth and production in 100% urine (a) applied at one time before cultivation. Besides, there is tendency that frequent urine application provided high growth of plant shoot.

To confirm nutrients effect in all treatment, nitrogen and phosphorus content in plant shoot was measured (Fig. 15.3). In the case of nitrogen, no significant difference was observed urine treatment (a)–(g), although control (h) (35.3 ± 1.3 mg/g plant DW) was significantly lower level than (a), (d), (f), and (g) (46.9 ± 2.9, 46.1 ± 2.5, 47.6 ± 1.2, and 46.5 ± 5.6 mg/g plant DW, respectively) (Tukey–Kramer test, $p < 0.05$). In the case of phosphorus, plant shoot concentration in (a) (0.5 ± 0.1 mg/g plant DW) was same level with control (h), but both were significantly lower than treatment (g) (1.6 ± 0.2 mg/g plant DW) (Tukey–Kramer test, $p < 0.05$). N/P ratio, which is one indicator of nutrient balance in plant, showed high value (about 25-100) in all treatment (Fig. 15.3). The value of (a) (95.2 ± 24.7) was significantly higher than that of (g) (28.4 ± 3.5) (Tukey–Kramer test, $p < 0.05$).

Nitrogen distribution in shoot of Spinach, root of Spinach, soil, and leached water is given in Fig. 15.4. The distribution in (a) was shoot 7.6%, root 1.1%, soil 35.9%, leached water 38.3%, and unknown part 17.1%. In the contrast, the distribution in (g) was shoot 49.3%, root 8.8%, soil 39.8%, leached water 0.1%, and unknown part 2.0%. Nitrogen distribution ratio of shoot part in (a) was significantly lower than that in other treatments, and the ratio of leached water in (a) was higher than that in other treatment. Similarly to nitrogen distribution, phosphorus distribution ratio in (a) was lower in shoot part (1.0%) and higher in leached water

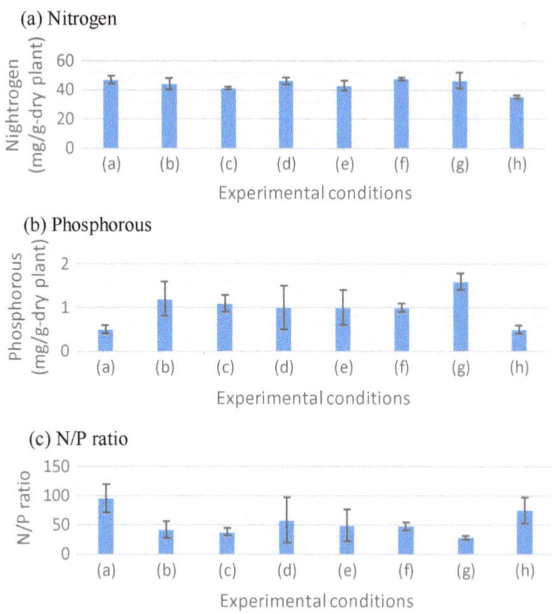

Fig. 15.3 Concentration of nitrogen and phosphorous in spinach shoot after cultivation. Experimental conditions: **a** 100% urine, **b** 60% before/week, **c** 60% before/2 days, **d** 30% before/week, **e** 30% before/2 days, **f** every week, **g** every 2 days, **h** control. Made by authors based on Sene et al. (2012)

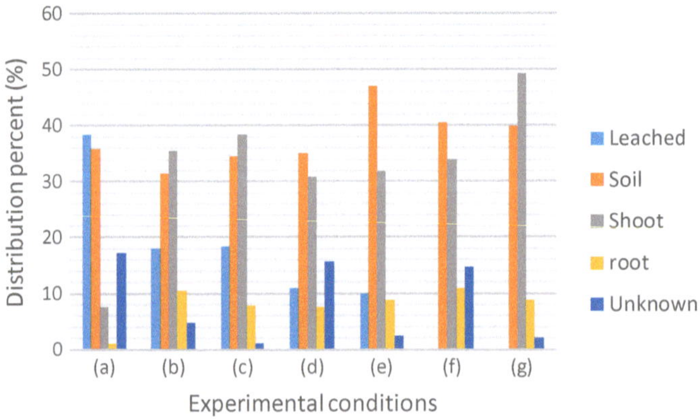

Fig. 15.4 Nitrogen distribution in each urine treatment after cultivation ($N = 3$). Made by authors based on Sene et al. (2012)

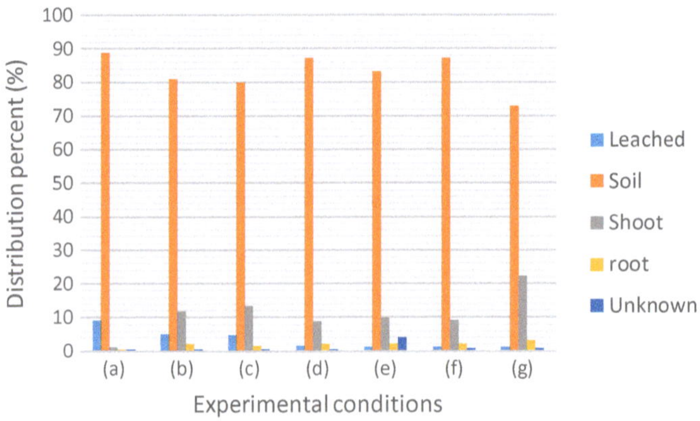

Fig. 15.5 Phosphorus distribution (%) in each urine treatment after cultivation ($N = 3$). Made by authors based on Sene et al. (2012)

(9.1%) than those in other treatments (Fig. 15.5). Phosphorus distribution ratio of shoot part in (g) was higher (22.2%), and the ratio of leached water was lower (1.0%). In all treatments, about 80% of phosphorus was remained in soil.

15.2.3 Discussions

In the present study, low growth of spinach in 100% urine applied at one time before seeding (a) and high growth in no urine applied before every 2 day (g) were observed. One reason for the difference might be phosphorus availability. It has been known that high N/P ratio ($N/P > 16$) in shoot indicates phosphorus limitation, generally speaking (Peter and Oleksyn 2004). In the present results, therefore, all spinach would suffer from phosphorus limitation. Especially, phosphorus limitation in (a) was severe ($N/P = 95.2 \pm 24.7$) and that in (g) was moderate ($N/P = 28.4 \pm 3.5$). This limitation leads us to consider that phosphorus availability might affect spinach growth in the present study.

It has been reported that phosphorus in urine was mainly existed as phosphate ion (Krichmann and Pettersson 1995) and the phosphate was easy to make an immobile form with Fe and Al under acidic soil condition (Schachtman et al. 1998) such as present study. The low phosphorus content in plant shoot of 100% urine might be associated to a low root activity compared to other treatments, since our original soil had low ions intensity and received a lot of ions in one time. Thus, that might cause some damages in the root zone and thereafter provided low plant growth and production in 100% urine application. Furthermore, frequent urine application such as treatment (g) (every 2 days urine application with no application before cultivation) caused high phosphorus uptake by Spinach, since available phosphorus was supplied frequently. In the present study, it was also observed that leached nitrogen was high ratio in 100% urine applied at one time before cultivation (a). This phenomenon was also considered as another possible reason for the low plant growth and production in (a), since nitrogen is major factor of plant growth. Pertnilla et al. (2007) found relatively low amount of leached N (6.3%) in clay soil by using stored urine as fertilizer for Wheat. One reason might be the difference in physical structure of soil because sandy soil used in the present study was characterized as low water-holding capacity (Hijikata et al. 2011). Second reason might be nitrogen form in synthetic urine and biological activity in soil. It has been known that urea in urine was changed to ammonium in storage urine (Müllegger et al. 2010). Retention capacity of ammonium form is higher than that of urea in soil due to the cation exchange capacity (CEC) of soil, although sandy soil has been regarded as low CEC. Generally speaking, urea applied in soil is easy to degrade to ammonium by enzyme from soil microorganism. However, activity of soil microorganism in the present study might be low in the early stage of cultivation, because our soil was washed and air-dried before cultivation. Therefore, if we consider the urine reuse in low water-holding capacity, low CEC, and low activity of soil microorganisms such as dryland agriculture (FAO 2008b), it was suggested that 100% urine applied at one time before seeding (a) is not adequate way and frequent urine application might be better from the view of nitrogen leaching. Spinach growth in all treatment in the present study was relatively lower than that of reference (Nishihara et al. 2001). One reason might be that Spinach is sensitive to acid condition (Burt 2006), and second reason might be unbalance of nitrogen and phosphorus in urine as mentioned above. This unbalance is in agreement with previous studies (Heinomen-Tanski et al. 2007; Pradhan et al. 2009a), which have shown that phosphorus in urine is low compared to chemical fertilizer. In urine reuse for agriculture, therefore, it is better to supplement with ash (Heinomen-Tanski et al. 2007) or use with compost (Hijikata et al. 2011), which has the ability to supply phosphorus and soil modification.

From this investigation, it was concluded that urine application at one time before cultivation was not effective for plant growth and production because of high ions intensity in low saline soil and high leached of nitrogen from soil. Furthermore, every 2 days application after germination showed relatively well plant growth but also high uptake of nitrogen and phosphorus. Thereafter, human urine application at frequent interval gave relatively favorable condition for spinach growth under sandy soil condition. However, P limitation in urine was palpably manifest in this work. Consequently, urine used as fertilizer with ash and compost might be a better vision to promote high plants yield in sandy soil.

15.3 Effects of Extra Human Urine Volume Application

Although it has been known that human urine could be reused as nutrients for plant, too much urine application in agricultural land may cause accumulation of several ions in soil. Among these ions, excess of sodium (Na) in soil inhibits plant growth by disrupting the water uptake in the root, dispersing soil particles, restricting root growth, and/or interfering with the uptake of competitive nutrients (Asano et al. 2007; Franzen 2007; Rosen et al. 2008; Lee 2012). While, excess of nitrogen possibly decreases amount of sugar and vitamins in vegetables and builds up nitrate concentration in the plant tissues (Brady and Weil 1996; Turan and Sevimli 2005). These phenomena are associated health and taste issues for consumers. Therefore, the objective of this study was to investigate the effect of extra urine volume application in plant and soil. Pot test with Komatsuna (*Brassica rapa var. perviridis*) was conducted in greenhouse and then observed the plant growth and nutrients effect in both plant and soil under different urine volume applications.

15.3.1 Pot Experiments

Pot test was conducted in the greenhouse at Hokkaido University, Japan from August 6–September 9, 2011. Inside the greenhouse, the following temperature range 19–23 °C min. (31–43 °C max.), humidity range 27–58% min. (61–88% max.), monthly mean global radiation range 7.2–21.8 MJ/m^2, and sunshine duration range 1.2–10.3 h were recorded. Two compartments closed pot system was used in this experiment to analyze leached water (see Sect. 15.2.1). Washed sandy soil mixed with small gravels (6:3 soil:gravel ratio) to keep good aeration of soil and facilitate root development was set as media (0.84-Kg soil pot). Initially, the soil pH was 5.4, and therefore before cultivation, the soil was limed by adding 2 g of $CaMg (CO_3)_2$ in each pot as recommended by Fujiwara and Narimatsu (2006). The physical and chemical characteristics of the soil are shown in Table 15.2. Komatsuna (*Brassica rapa var. perviridis*), a leafy vegetable commonly consumed

Table 15.2 Physical and chemical properties of bulk soil before cultivation

Total nitrogen (mg-N/g-dry soil)	Total phosphorous (mg-P/g-dry soil)	Total carbon (mg-C/g-dry soil)	pH
0.18 ± 0.01	0.28 ± 0.03	1.38 ± 0.17	7.50 ± 0.09
EC (μS/cm)	Potassium (mg-K/g-dry soil)	Sodium* (mg-Na/g-dry soil)	SAR value
49.4 ± 4.70	1.91 ± 0.36	0.60 ± 0.04	0.21 ± 030

Made by authors based on Sene et al. (2013a)

Mean values ± standard deviation with five replications; *this value was total sodium extracted with nitrate-perchloric acid digestion

in Japan, was selected in this work because it presents high growth rate and can be harvested in short term (35 days) after sowing in warm climates (Thao et al. 2008). Within each experimental pot, six Komatsuna seeds were germinated in petri dish with filter paper and incubated at 25 °C for 1 day and then sown. Seven days after seedling emerged, the seedlings were reduced to two plants per pot.

Synthetic urine (Wilsenach et al. 2007) was used in this work to obtain more general and precise effects of extra urine volume application in plant and soil, because the composition of urine fluctuates from one person to another and depends mainly on diet and physical activity (Pradhan et al. 2010b). The synthetic urine contained 12.05 g/L of nitrogen, 0.96 g/L of phosphorus, 2.04 g/L of potassium, and 2.84 g/L of sodium. Based on the requisite nitrogen for Komatsuna growth (140 kg/ha) recommended by Fujiwara and Narimatsu (2006), the composition of synthetic urine and the pot system size, diluted (1/3) synthetic urine volume was designed as U-Vo (34.86 ml), and containing the following dose of N (140 mg/pot), P (11.12 mg/pot), K (23.73 mg/pot) and Na (33.05 mg/pot). In other term U-Vo was defined as the adequate urine volume required for 35 days cultivation of Komatsuna. To know the effects of extra human urine volume on plant and soil, three different urine treatments volumes [U-Vo, U-2Vo (NPK 280-22.24-47.46, Na 66.1 mg/pot) and U-3Vo (NPK 420-33.36-71.19, Na 99.15 mg/pot)] were set in one hand. In the other hand, one modified Hoagland's solution volume [V-H (NPK 140-9.41-31.53, Na 0.0 mg/pot)] and one non-fertilizer treatment [C (NPK 0-0-0, Na 0.0 mg/pot] were set, respectively, as positive (same N with plant requirement and without minors compounds except KH_2PO_4) and negative controls. Totally, 25 experimental pots were prepared for this experiment, with four fertilizer treatments and one control (five cases replicated five times each). All fertilizers were regularly applied just after the seedlings reduced to two plants per pot at days 7, 14, 21, and 28 after sowing as recommended by Sene et al. (2012).

Tap water was used as irrigation water. Before the experiment, the theoretical daily amount of water needed for irrigation (24.6 ml/pot) was determined by multiplying the crop coefficient of Komatsuna in the development stage (Kc = 0.60) and the potential evapotranspiration (ETo = 4.1 mm/day) and the pot size system (1/10000a). The ETo value was computed by using ETo calculator version 3.1 (FAO 2009b), and the input data (solar radiation, temperatures max and min, humidity max and min) were derived from Sapporo weather station 47412 (Japan) for the cover period of August 6 to September 2010. In practical, the theoretical daily irrigation was modified to 50 ml/pot. However, based on the daily observation, the modified daily irrigation water was slightly modified during the cultivation period. The application of irrigation water was done by a graduated pet bottle (50 mL between two graduations where the total volume was 0.5 L) equipped with a plastic shower head which had allowed spraying water uniformly inside the pot. The total volumes of water applied in pot during the whole cultivation period in each treatment were 1.85 (U-Vo), 1.95 (U-2Vo), 1.90 (U-3Vo), 1.90 (U-Vo), and 1.66 L (C). In contrast, 51.4 (U-Vo), 50.56 (U-2Vo), 49.16 (U-3Vo), 50.46 (V-H), and 70.47 mL (C) were leached from experimental pots. Plants heights were measured every week and from one week after sowing. In the plant height measurement, the highest leaf in each pot was selected. At 35 days after sowing, Komatsuna shoots and roots were separately harvested. The fresh samples were lyophilized and weighted as dry weight and then stored at −30 °C.

15.3.2 Results

Komatsuna plants grew well in all treatments except control as shown in Fig. 15.6. The weekly plant height measurement for all treatments was not statistically different among them (ANOVA, $p > 0.05$), but all were significantly higher than the control C at 21 and 28 days after sowing (Tukey–Kramer, $p < 0.05$). Similarly to plant height, shoot dry weight (g/pot) obtained after cultivation in all treatments did not have any significant difference among them. Similar observations were also found in root dry biomass. Therefore, from these results, it was observed that extra urine volume application had no inhibition effects on plant height growth, shoot, and root dry biomasses under our current "experimental conditions."

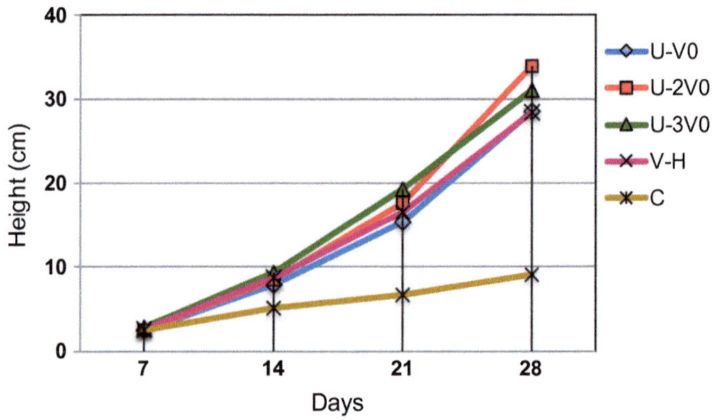

Fig. 15.6 Change in time course of Komatsuna height during the first cultivation Made by authors based on Sene et al. (2013a)

The chemical contents in plant shoot and root are summarized in Tables 15.3 and 15.4. Concentration of nitrogen in plant shoot increased significantly (Tukey–Kramer, $p < 0.05$) with increasing urine volume application in pot. Roots nitrogen contents were significantly higher in U-2Vo and U-3Vo (both in same level) than U-Vo treatment. In the case of phosphorus, no significant difference was observed in plant shoot and root for urine treatments U-Vo, U-2Vo, and U-3Vo. Na concentrations in Komatsuna shoot and root were significantly lower in treatment U-Vo than other urine conditions (U-2Vo and U-3Vo, both same level), but all of them were obviously higher than modified Hoagland's solution (V-H). In contrast, the concentrations of K in Komatsuna shoot and root for U-2Vo and U-3Vo (both in same level) were statistically lower than U-Vo, while modified Hoagland's solution, V-H, was significantly higher than all urine conditions.

Nitrogen concentration in cultivated soil was statistically higher in treatment U-3Vo than U-Vo, U-2Vo, and V-H, which were in the same level (Table 15.5). Phosphorus concentrations in remained had no significant effect with urine volume application. Concentrations of Na in soil after cultivation were in the same level for all treatments. However, soil SAR value in urine treatment U-3Vo was significantly higher than other conditions. Nitrogen distribution in Komatsuna shoot, root, soil, and leached water is given in Table 15.6. The distribution of nitrogen showed that nitrogen in shoot increased with increasing urine volumes application in the experimental pot. Concerning soil data, the amounts of nitrogen remained in soil after cultivation in the cases of U-Vo and U-2Vo were lower than that in the original soil.

Na distribution ratio in shoot increased with increasing urine volume application in soil (Table 15.7). However, more than 96% of the total amount of Na applied in pot remained in soil after cultivation. It should be noted that the amount of Na uptaken by plants is equivalent to 51–56% of the total input of Na by urine and irrigation water.

15.3.3 Discussion

It was observed that extra human urine application volume did not show any specific difference on Komatsuna plant height, shoot dry weight, and root dry weight. One possible reason might be a salt sensitivity of the plant species. It has been reported that extra urine application caused an inhibition effect of spinach and cabbage (both moderately salts sensitive) dry biomass at 800 mg-N/Kg soil levels (Mnkeni et al. 2005) and carrot (salt-sensitive) biomass and yield at 400 mg-N/Kg soil level, whereas no depressed effect was found in beetroot (salt tolerant) at the similar level (Mnkeni et al. 2008). In the case of our study, Komatsuna, one family of cabbage, is defined as salt moderately sensitive and the highest urine application (U-3Vo) was 500 mg-N/Kg soil, lower than spinach and cabbage cases (both moderately salts sensitive) but slightly in the same level with carrot, salt-sensitive plant. Another possible reason could be a low Na applied through urine volume

Table 15.3 Effect of urine volume application on N, P, Na, Ca, Mg, and K concentrations (mg/g-dry plant) in Komatsuna shoots cultivated in sandy soil; $N = 5$ in each treatment

Treatment	Shoot						
	N	P	Na	Ca	Mg	K	K/Na
U-Vo	42.45 ± 2.33c	0.66 ± 0.19ab	14.28 ± 1.22b	19.87 ± 1.45a	5.25 ± 0.30a	41.56 ± 4.76a	2.91 ± 0.11a
U-2Vo	63.75 ± 6.22b	0.93 ± 0.34a	20.60 ± 2.47a	18.53 ± 1.30a	4.94 ± 0.40a	28.40 ± 3.72b	1.40 ± 0.07b
U-3Vo	80.40 ± 1.83a	0.96 ± 0.22a	17.75 ± 1.16a	14.31 ± 1.05b	3.77 ± 0.30b	24.75 ± 1.81b	1.40 ± 0.15b
V-H	31.74 ± 2.43d	0.49 ± 0.14b	5.08 ± 1.21c	23.03 ± 3.74	6.03 ± 0.83	50.03 ± 8.94	9.97 ± 0.72
N-C	10.45 ± 1.73e	0.34 ± 0.12b	1.92 ± 0.34d	16.28 ± 2.54	4.85 ± 0.53	22.31 ± 4.70	11.59 ± 0.88

Reprinted from Sene et al. (2013b), with permission from TJPRC

Means ± SD; different letters indicate significant difference (Tukey–Kramer test, $p < 0.05$). In the case of Ca ($p = 0.0037$), Mg ($p = 0.003$), K ($p = 0.0011$), and K/Na ($p = 0.0003$) parameters, all five treatments included were significantly different in nonparametric test (Kruskal–Wallis)

Table 15.4 Effect of urine volume application on N, P, Na, Ca, Mg, and K concentrations (mg/g-dry plant) in Komatsuna roots cultivated in sandy soil; $N = 5$ in each treatment

Treatment		Root					
	N	P	Na	Ca	Mg	K	K/Na
U-Vo	23.22 ± 1.26b	0.47 ± 0.19	12.82 ± 1.34b	9.91 ± 1.45b	2.64 ± 0.32	8.19±1.34a	0.64±0.12a
U-2Vo	32.78 ± 2.22a	0.61 ± 0.43	16.38 ± 1.39a	17.02 ± 2.21a	2.40 ± 0.23	4.68±0.74b	0.28±0.02b
U-3Vo	32.65 ± 4.09a	1.03 ± 0.25	14.88 ± 2.64a	13.56 ± 3.06a	2.92 ± 0.82	3.61±0.59c	0.11±0.05c
V-H	19.58 ± 2.10b	0.46 ± 0.27	8.82 ± 1.17c	23.00 ± 2.53	3.75 ± 0.20	13.92±1.20	1.59±0.08
N-C	11.93 ± 1.23c	0.47 ± 0.36	2.34 ± 0.36d	11.22 ± 0.80	3.93 ± 0.42	15.88±1.58	5.44±0.51

Reprinted from Sene et al. (2013b), with permission from TJPRC

Means ± SD; different letters indicate significant difference (Tukey–Kramer test, $p < 0.05$). In the case of Ca ($p = 0.0007$), Mg ($p = 0.005$), K ($p = 0.0002$), and K/Na ($p = 0.0001$) parameters, all five treatments included were significantly different in nonparametric test (Kruskal–Wallis)

Table 15.5 Soil physical and chemical properties after Komatsuna cultivation in sandy soil; $N = 5$ in each treatment

Treatment	Total N (mg-N/g-dry soil)	Total P (mg-P/g-dry soil)	Total sodium (mg-Na/g-dry soil)	SAR value
U-Vo	0.16 ± 0.02b	0.24 ± 0.04	0.70 ± 0.05	0.46 ± 0.04b
U-2Vo	0.17 ± 0.01b	0.25 ± 0.01	0.72 ± 0.05	0.52 ± 0.06b
U-3Vo	0.22 ± 0.01a	0.29 ± 0.03	0.76 ± 0.03	0.72 ± 0.08a
V-H	0.14 ± 0.02bc	0.24 ± 0.03	0.68 ± 0.04	0.34 ± 0.01
N-C	0.12 ± 0.01c	0.20 ± 0.03	0.67 ± 0.07	0.38 ± 0.03

Made by authors based on Sene et al. (2013a)
Mean values ± standard deviation; different letters indicate significant difference (Tukey–Kramer test, $p < 0.05$)

Table 15.6 Nitrogen distribution in each treatment after cultivation ($N = 5$)

N distribution (mg/pot)	U-Vo	U-2Vo	U-3Vo	V-H
Input				
Original soil	148.90	148.90	148.90	148.90
Urine	140	280	420	140
Irrigation water	8.66	9.13	8.89	8.89
Seeds	0.74	0.74	0.74	0.74
Total input	298.30	438.77	578.53	298.53
	(100%)	(100%)	(100%)	(100%)
Total output				
Shoot	100.85 ± 7.74	157.51 ± 52.67	240.78 ± 37.06	85.65 ± 9.92
	(33.81%)	(35.90%)	(41.62%)	(28.69%)
Root	12.57 ± 3.68	13.62 ± 5.09	14.56 ± 5.58	11.73 ± 2.28
	(4.21%)	(3.10%)	(2.52%)	(3.93%)
Soil remained	133.62 ± 13.83	143.82 ± 11.48	182.1 ± 8.60	121.53 ± 14.64
	(44.79%)	(32.78%)	(31.48%)	(40.71%)
Leached	0.25 ± 0.01	0.06 ± 0.04	0.08 ± 0.01	0.09 ± 0.01
	(0.08%)	(0.01%)	(0.01%)	(0.03%)
"Unknown"	51.01	123.75	141.01	79.53
	(17.10%)	(28.20%)	(24.37%)	(26.64%)

Reprinted from Sene et al. (2013b), with permission from TJPRC
Mean values ± standard deviation (SD); original soil means soil before cultivation

Table 15.7 Sodium (Na) distribution in each treatment after cultivation (N = 5)

Na distribution (mg/pot)	U-Vo	U-2Vo	U-3Vo	V-H
Input				
Original Soil	508.47	508.47	508.47	508.47
Urine	33.05	66.10	99.15	0.00
Irrigation	53.78	56.69	55.23	55.23
Seeds	0.002	0.002	0.002	0.002
Total input	595.30	631.26	662.86	563.71
	(100%)	(100%)	(100%)	(100%)
Total output				
Shoot	33.90 ± 3.02	56.19 ± 6.24	60.04 ± 6.80	13.56 ± 2.65
	(5.69%)	(8.90%)	(9.06%)	(2.41%)
Root	6.77 ± 1.39	7.16 ± 3.70	7.50 ± 0.88	5.21 ± 0.37
	(1.14%)	(1.13%)	(0.98%)	(0.92%)
Soil remained	590.20 ± 39.10	610.51 ± 43.72	636.74 ± 25.38	573.61 ± 34.96
	(99.14%)	(96.71%)	(96.06%)	(101.76%)
Leached	0.22 ± 0.04	0.27 ± 0.10	0.18 ± 0.07	0.18 ± 0.05
	(0.04%)	(0.04%)	(0.03%)	(0.03%)
"Unknown"	−35.01	−42.88	−40.62	−28.86
	(−6.01%)	(−6.79%)	(−6.13%)	(−5.12%)

Reprinted from Sene et al. (2013b), with permission from TJPRC
Mean values ± standard deviation (SD); original soil means soil before cultivation

application in our experimental conditions. Especially, soil EC in the present study was approximately 10 times lower than that given in the literature. It was pointed out that soil whose EC is below 4 mS/cm is considered to be non-saline soil (FAO 1998). Besides, it has been reported that SAR value below 13 is desirable level of sodium in soil and may not affect plant growth (Davis et al. 2007). Considering this EC and SAR references values, soils in urine treatments U-Vo, U-2Vo, and U-3Vo were far to be saline (Table 15.5). Thus, the low level of sodium in soil may probably justify the none inhibition effects of urine volume application observed in Komatsuna growth, shoot, and root dry matters.

The decreased uptake of K and an increased uptake of Na in plant shoot and root for all urine treatments were observed, and remarkably, the ratio of K/Na was significantly declined in three-time fold the basic urine volume application (U-3Vo) as shown in Tables 15.3 and 15.4. These findings support a result of low K^+ and high Na^+ uptake in red beet with urine cultivation (Pradhan et al. 2010a) and high level of residual K^+ in urine fertilized soil after tomato cultivation (Pradhan et al. 2009b). One possible explanation for the phenomena could be the substitution of

K^+ by Na^+, which is well known as the primary effects of plant responses to salinity and leads to nutritional imbalances (Alfocea et al. 1996; Gorham et al. 1997; Dasgan et al. 2002; Tuna et al. 2007).

It was observed that excess of urine volume application affected nitrogen concentration in plant shoot and root, and this finding was clearly shown in shoot part (Table 15.3). This agrees with the observations of Mnkeni et al. (2005, 2008), who have reported an increase of nitrogen concentration in plants tissues of spinach, cabbage, carrot, and beetroot by increasing human urine application rate in experimental pot. Unfortunately, the level of nitrate in Komatsuna could not be studied in this work, since when excess of N is applied to the soil, nitrate concentration in vegetables (cabbage, carrot, okra) is often high and its metabolites may cause public health issues (Turan and Sevimli 2005; Mubashir et al. 2010). Subsequently, care should be taken regarding this high nitrogen contents in shoot by using more urine volume than plant requirement. Similar to nitrogen concentration in shoot, the nitrogen uptake rate in plant shoot increased with the increasing urine volume application (Table 15.6). Furthermore, it was also observed from the N-balance, a none accumulation of nitrogen remained soil after cultivation in adequate (U-Vo) and double (U-2Vo) application of urine volumes in comparison to the original soil. Therefore, adequate human urine application did not accumulate nitrogen in soil. One probable reason might be the remarkably observed "Unknown" part of nitrogen which shown high values where excess volumes of urine were applied (Table 15.6). One possible explanation for this "Unknown" of nitrogen might be volatilization, because nitrogen in urine is applied in urea form, which is easy to degrade by enzyme from soil microorganism to ammonium and then converted to ammonia gas. This process occurs more slowly in acidic soil than in alkaline one (Staines et al. 2011). Moreover, it has been known that this process is faster in soil where urea concentration is high and soil pH included in the range of 6.5–9.5 (Cabrera et al. 1991), and at the pH 7.5, the concentration of ammonia at equilibrium is higher enough that much of N can be lost to the atmosphere in a few days (On-Farm 2008). So considering, the high amount of urea applied in soil (U-2Vo and U-3Vo) and our soil pH (Table 15.7), it was suggested that extra urine volume application in soil (U-2Vo and U-3Vo) is not adequate way and application of urine volume based on the plant requirement (U-Vo) might be better in the viewpoint of volatilization. The nitrogen loss from human urine fertilized soil has been showed as 6–7% in a report of Kirchmann and Pettersson (1995) and was smaller than those obtained in the present results (17–28%). One reason might be that our pot test was conducted in summer where the maximum temperature in greenhouse reached 43 °C, and it is well documented that ammonia volatilization is high in warm temperature, and increasing temperature markedly increased ammonia volatilization (Ernst and Massey 1960; Jones et al. 2013).

Excess of urine volume application at the level of 2–3 times higher than plant volume requirement had no effect on plant phosphorus contents in shoots and roots. Likewise in plant shoot and root, phosphorus concentration in remained soil was not affected by extra urine volume application. This might probably be due to the low phosphorus contents in human urine (Pradhan et al. 2009a, b; Sene et al. 2012).

From these results, it was concluded that application of human urine volume in the range of 2–3 times higher than plant requirement had no inhibition effects on Komatsuna growth, shoot biomass, and root biomass in one-time cultivation, since treated soils were far to be saline when considering the references values given by FAO 1998 (EC < 4 mS/cm; SAR < 13). However, application of extra urine volume (U-2Vo & U-3Vo) increases nitrogen concentration in plant tissues. Besides, human urine application causes nutritional imbalance (substitution uptake of K^+ by Na^+), which is severe in double (U-2Vo) and triple (U-3Vo) of urine volume applications. In the case of soil, more than 50% the total sodium applied through urine and irrigation water in all urine treatments remained in soil after one-time cultivation. Subsequently, further applications of urine may have a potential to increase accumulated sodium ion in soil and thereafter affect adversely plant growth and production. Furthermore, application of double (U-2Vo) or triple (U-3Vo) volume of urine increase N lost from soil and triple volume of urine causes a accumulation of nitrogen in soil. Therefore, we suggest that application of adequate amount of urine based on plant nitrogen requirement might be a better option for its sustainable use in agriculture.

15.4 Effects of Continuous Application of Extra Human Urine Volume on Plan and Soil

For the long-term view related to a sustainable agricultural production, uncontrolled urine application in agriculture land may cause simultaneously soil salinity, sodium (Na) accumulation, and nitrogen accumulation in soil. It has been reported that extra urine application inhibited a growth of cabbage, spinach, and carrot accompanied with increase of soil EC (Mnkeni et al. 2005, 2008). Furthermore, sodium is contained in the urine at the level of 1–9 g/L (Krichmann and Pettersson 1995; Mnkeni et al. 2008; Pradhan et al. 2010b). The excess of sodium accumulation in soil has a potential to be detrimental to plant growth, by disrupting the water uptake in the root zone, interfering with completive nutrients and/or dispersing soil particles as a result of breaking down soil aggregates (Asano et al. 2007; Franzen 2007; Allotey et al. 2008; Lee 2012). Regarding nitrogen, additionally, high volume of urine application increases nitrogen concentration in plant tissues (Mnkeni et al. 2008; Sene et al. 2013a, b). The high concentration of nitrogen in soil might affect negatively amount of sugar and vitamins in vegetables and build up in the plant tissues causing therefore health and taste issues to consumers (Brady and Weil 1996; Turan and Sevimli 2005). It has been known that nitrogen lost from soil during cultivation was promoted in urine fertilization (Kirchmann and Pettersson 1995) and in high rate of urine application (Sene et al. 2013a). Thereby, it has not been addressed the effects of continuous application of excess urine volume on plant and soil through continuous cultivations. Therefore, the objective of this study was to investigate the effects of continuous application of extra urine volume in plant and soil. Pot test with Komatsuna (*Brassica rapa var. perviridis*), was conducted in greenhouse for continuous cultivations and then observed plant growth and nutrients effects in both plant and soil under continuous application of different urine volumes.

15.4.1 Pot Experiments

The experiments were carried out during three successive cultivation periods (August–September 2011, May–June 2012, and July–August 2012) in the greenhouse at Hokkaido University, Japan. The recorded climate data in each cultivation period are summarized in Table 15.8. After the first cultivation period (see Sect. 15.3), no adverse effect was observed on Komatsuna growth through application of urine volume at the rate of 2–3 times higher than plant requirement. Therefore, the experiments were pursued using the previous same experimental pots, the same sandy soil, and same plant (Komatsuna), and then second and third cultivations were performed successively. Komatsuna (*Brassica rapa var. perviridis*) was continuously cultivated three times (35 days each), since it is resistant to diseases for continuous cultivation (Fjellstrom and Williams 1997; Morikawa and Saigusa 2001). The total volume of water applied in experimental pot during the first, second, and third cultivation period and theirs chemical contents are shown in Table 15.9.

Table 15.8 Climate data inside the greenhouse during the three successive cultivation periods

Cultivation periods	Temperature (°C)	Humidity (%)	Sunshine duration (h)	Solar radiation (MJ/m^2)
August–September 2011	18.8–42.9	27.3–88.5	4.6–5.8	7.2–21.8
May–June 2012	12.9–37.7	16.0–72.0	4.7–6.8	13.1–27.3
July–August 2012	17.3–42.4	17.7–77.7	4.3–6.0	11.1–25.1

Reprinted from Sene et al. (2013b), with permission from TJPRC

The sunshine duration (h) and solar radiation (MJ/m^2) were obtained from Sapporo Weather Station No 47412 (Lat 43° 03.6′N Lon 141° 19.7′E), while the temperature and humidity were recorded inside the greenhouse using a wireless thermometer and humidity sensor

Table 15.9 Total irrigation volume and total applied amount of N, P, and Na derived from irrigation water during each cultivation periods

Treatments	Total volume (L/pot)	Total N (mg/pot)	Total P (mg/pot)	Total Na (mg/pot)
First cultivation period				
U-Vo	1.85	8.66	0.30	53.78
U-2Vo	1.95	9.13	0.31	56.79
U-3Vo	1.90	8.89	0.30	55.23
V-H	1.90	8.89	0.30	55.23
N-C	1.66	7.77	0.27	48.26
Second cultivation period				
U-Vo	1.97	9.22	0.32	57.27
U-2Vo	1.97	9.22	0.32	57.27
U-3Vo	1.97	9.22	0.32	57.27
V-H	1.97	9.22	0.32	57.27
N-C	1.70	7.96	0.27	49.42
Third cultivation period				
U-Vo	1.35	6.32	0.22	39.24
U-2Vo	1.35	6.32	0.22	39.24
U-3Vo	1.30	6.08	0.21	37.79
V-H	1.35	6.32	0.22	39.24
N-C	0.95	4.45	0.15	27.62

First cultivation (33 days), second cultivation (34 days), and third cultivation (34 days); U-Vo, urine volume Vo applied
U-2Vo, urine volume 2Vo applied; U-3Vo, urine volume 3Vo applied; modified Hoagland's solution applied
N-C, non-fertilizer control

15.4.2 Results

Shoot and root dry biomasses obtained from first, second, and third cultivation are summarized in Table 15.10. Komatsuna shoot dry weight obtained after the first cultivation step did not differ significantly among all fertilizer treatments compared to the N-C. In the second cultivation step, shoot dry weight for all treatments were statistically difference among them except U-Vo and V-H (both same level), and the highest values were observed in the highest rate of urine application (U-2Vo and U-3Vo). Concerning, the third cultivation, shoot dry biomass production in urine treatments U-Vo, U-2Vo, and U-3Vo and V-H did not show any significant different among them. Furthermore, root dry matter productions in all set conditions were not statistically different among them, except the N-C in the first, second, and third cultivation. Therefore, from these results, it was observed that continuous application of extra urine volume had no depressed growth on plant height, shoot, and root dry weights under our present experimental conditions.

Table 15.10 Komatsuna dry biomass obtained from different urine volume application in sand soil for continuous cultivation; $N = 5$ in each treatment

Treatments	Parameters	
	Shoot dry weight (g/pot)	Root dry weight (g/pot)
First cultivation period		
U-Vo	2.38 ± 0.23a	0.54 ± 0.14a
U-2Vo	2.49 ± 0.88a	0.42 ± 0.19a
U-3Vo	3.10 ± 0.43a	0.45 ± 0.17a
V-H	2.71 ± 0.33a	0.60 ± 0.07a
N-C	0.34 ± 0.21b	0.16 ± 0.10b
Second cultivation period		
U-Vo	2.97 ± 0.32c	0.58 ± 0.11x
U-2Vo	3.61 ± 0.72b	0.64 ± 0.10x
U-3Vo	4.82 ± 0.41a	0.58 ± 0.06x
V-H	3.21 ± 0.38c	0.66 ± 0.12x
N-C	0.22 ± 0.07d	0.04 ± 0.03y
Third cultivation period		
U-Vo	1.44 ± 0.28x	0.34 ± 0.07x
U-2Vo	1.92 ± 0.60x	0.31 ± 0.09x
U-3Vo	1.30 ± 0.62xy	0.25 ± 0.21x
V-H	1.71 ± 0.35x	0.36 ± 0.08x
N-C	0.09 ± 0.05y	0.04 ± 0.03y

Reprinted from Sene et al. (2013b), with permission from TJPRC

Means ± SD; different letters within a column (a, b, c, d) indicate honestly significant difference (Tukey–Kramer test, $p < 0.05$); different letters within a column (x, y) indicate at least significant difference (Dunnett T3 test, $p < 0.05$)

Figure 15.7 is the observed nitrogen concentration in shoot, showing that shoot nitrogen concentration increased significantly (Tukey–Kramer, $p < 0.05$) with increasing extra urine volume application. On the other hand, phosphorus concentrations in plant shoot did not show any specific change in urine treatments U-Vo, U-2Vo, and U-3Vo in the first, second, and third cultivation as shown in Fig. 15.8.

K/Na ratio of Komatsuna shoot shown in Fig. 15.9 decreased from the first to second cultivation in all treatments, and then, the decrease was modelated from the second to third cultivation. Additionally, in each of cultivation step, plant shoot K/Na ratios were significantly higher in all modified Hoagland's solution (V-H) than other conditions.

Soil physicals and chemicals properties after each of the cultivation period are summarized in Table 15.11. Within each cultivation, nitrogen concentrations in remained soil increased significantly in extra urine volumes application (U-2Vo, U-3Vo) cases, while the concentration of N remained soils in the adequate urine

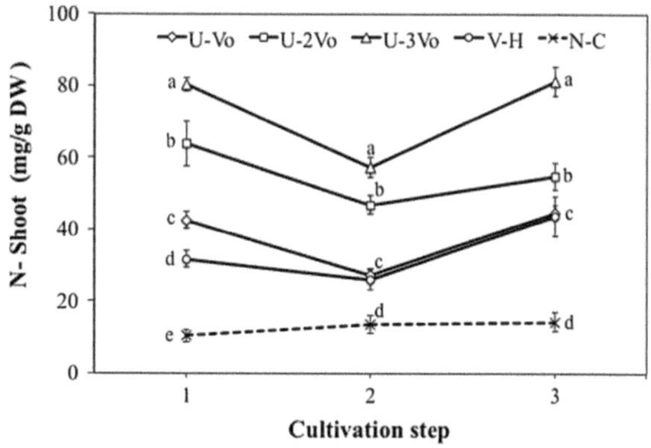

Fig. 15.7 Effect of continuous application of extra urine volume application on nitrogen concentration in Komatsuna shoot (mg/g-dry plant). Different letters within cultivation step (a, b, c, d, e) indicate honestly significant difference (Tukey-Kramer test, $p < 0.05$)

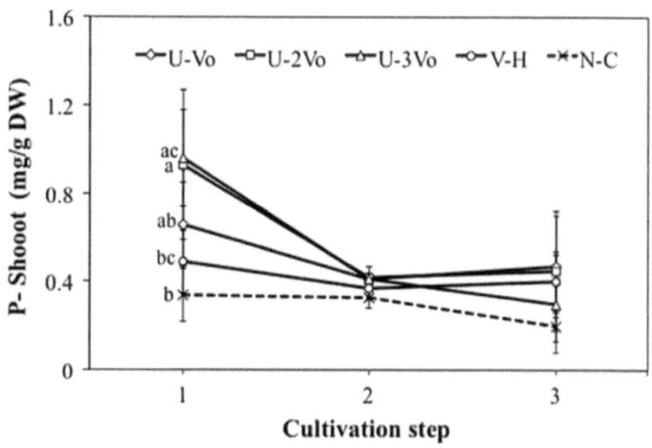

Fig. 15.8 Effect of continuous application of extra urine volume application on phosphorous concentration in Komatsuna shoot (mg/g-dry plant). Different letters within cultivation step (a, b, c, d, e) indicate honestly significant difference (Tukey-Kramer test, $p < 0.05$)

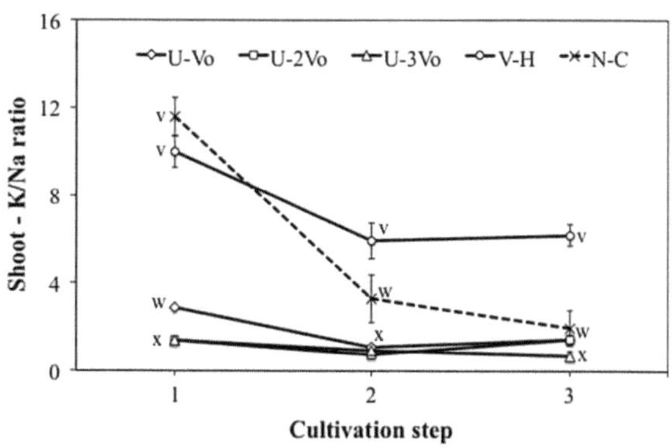

Fig. 15.9 Effect of continuous application of extra urine volume application on Komatsuna shoot K/Na ratio. Different letters within cultivation step (v, w, x) indicate at least significant difference (Dunnett T3 test, $p < 0.05$)

volume application (U-Vo) and V-H was statistically in the same level from the first cultivation step to the third one. Phosphorus concentration in soil after the first and second cultivation in all fertilized treatment (U-Vo, U-2Vo, U-3Vo, and V-H) did not show any statistical difference among them, but in the third cultivation, U-3Vo was significantly higher than other treatments. Concentrations of sodium in cultivated soil for U-Vo, U-2Vo, and U-3Vo were not significantly different (Tukey–Kramer, $p < 0.05$) in all cultivation periods except in the third one, where urine treatments U-2Vo and U-3Vo were in the same level, but both were statistically greater than U-Vo. Nevertheless, sodium concentration in soil and soil SAR were gradually increased during three-time cultivation in all urine treatments. Soil electrical conductivity (EC) increased from the first cultivation to the third one in all treatment, and this increase was faster in extra urine volume application cases. However, soil EC in U-Vo and V-H was not significantly different (Tukey–Kramer, $p < 0.05$) within each cultivation step. Conversely to the increase of soil EC from the first to third cultivation, soil pHs decreased in urine treatments and were statistically different among them after the second and third cultivation, but this decrease was faster in extra urine volumes applications.

15.4.3 Discussion

The results show that continuous application of extra urine volume did not cause any depressed growth on Komatsuna height, shoot dry biomass, and root dry biomass after three-time cultivation. It has been reported a depressed growth of spinach and cabbage (both salt moderately plant) dry biomasses at higher application rate of human urine (800 mg-N/Kg soil level) which has been caused by salinity (EC = 18.8 mS/cm) of the treated soil in comparison to the set control (EC = 2 mS/cm) (Mnkeni et al. 2005). In our study, Komatsuna, one family of cabbage, is defined as salt moderately plant and the highest urine application volume (U-3Vo) in the original soil (EC = 0.05 mS/cm) from the first to the third cultivation was 1500 mg-N/Kg soil level, which was higher than that applied by Mnkeni et al. (2005), but EC level of our treated soil was 0.84 mS/cm, and this was till far to be saline condition given by FAO (1998) where soil is rated by EC value as follows: 0–2 mS/cm: none saline soil; 2–4 mS/cm: slightly saline; 4–8 mS/cm: moderately saline; 8–16 mS/cm: strongly saline; >16 mS/cm: very strongly saline. Therefore, it was suggested that application of extra human urine volume in saline soil might accelerate plant inhibition more strongly, and its application in non-saline soil may delay the increase of soil EC and eventually attenuate the adverse effects of saline soil on plant growth. On the other hands, even though soil EC (Table 15.11) was low in comparison with the references ranges provided by FAO (1998), it was obviously observed that soil EC increased during three-time cultivations in all set conditions of this present study.

However, among all treatments, this increase of soil EC was always high in extra urine applications (U-2Vo and U-3Vo) and low in the adequate urine volume (U-Vo) from one cultivation step to another. Therefore, some salt management plans such as leaching requirement or cultivation of clean up plants should be applied even in the case of low EC soil (U-Vo), when urine is continuously reused in several years scale.

Table 15.11 Soil physical and chemical properties after different urine volume application in sandy soil for continuous cultivation of Komatsuna. $N = 5$ in each treatment

Treatments	Parameters					
	pH	EC (μS/cm)	Total N (mg/g-dry)	Total P (mg/g-dry)	Total Na (mg/g-dry)	SAR values
Zero cultivation period						
Initial soil	7.50 ± 0.09	49.4 ± 4.70	0.18 ± 0.01	0.28 ± 0.03	0.60 ± 0.04	0.210 ± 0.03
First cultivation period						
U-Vo	7.60 ± 0.10b	61.0 ± 4.9y	0.16 ± 0.02b	0.24 ± 0.04ab	0.70 ± 0.05	0.46 ± 0.04xy
U-2Vo	7.69 ± 0.17ab	88.8 ± 37.7xy	0.17 ± 0.01b	0.25 ± 0.01ab	0.72 ± 0.05	0.52 ± 0.06x
U-3Vo	7.75 ± 0.08ab	180.3 ± 23.1w	0.22 ± 0.01a	0.29 ± 0.03a	0.76 ± 0.03	0.72 ± 0.08w
V-H	7.85 ± 0.06a	96.8 ± 9.3x	0.14 ± 0.02bc	0.24 ± 0.03ab	0.68 ± 0.04	0.34 ± 0.01z
N-C	7.62 ± 0.10b	88.2 ± 7.0x	0.12 ± 0.01c	0.20 ± 0.03b	0.67 ± 0.07	0.38 ± 0.03yz
Second cultivation period						
U-Vo	7.58 ± 0.15b	132.0 ± 11.2c	0.20 ± 0.01c	0.41 ± 0.03a	0.73 ± 0.06ab	0.61 ± 0.08ab
U-2Vo	6.96 ± 0.06c	200.8 ± 43.6b	0.23 ± 0.02b	0.42 ± 0.05a	0.77 ± 0.09a	0.55 ± 0.04ab
U-3Vo	6.54 ± 0.05d	372.6 ± 40.3a	0.27 ± 0.01a	0.41 ± 0.03a	0.80 ± 0.07a	0.64 ± 0.06a
V-H	7.90 ± 0.14a	158.2 ± 20.4bc	0.20 ± 0.01c	0.37 ± 0.05ab	0.64 ± 0.05b	0.49 ± 0.09b
N-C	7.69 ± 0.06b	132.1 ± 19.6c	0.11 ± 0.02d	0.33 ± 0.05b	0.61 ± 0.08b	0.54 ± 0.03a
Third cultivation period						
U-Vo	7.43 ± 0.13b	223.4 ± 41.2c	0.18 ± 0.02y	0.44 ± 0.03b	0.74 ± 0.07b	0.73 ± 0.04x
U-2Vo	6.78 ± 0.18c	438.8 ± 60.2b	0.26 ± 0.03x	0.44 ± 0.03b	0.80 ± 0.04a	0.72 ± 0.02x
U-3Vo	6.50 ± 0.12d	868.0 ± 83.0a	0.42 ± 0.04w	0.51 ± 0.03a	0.88 ± 0.08a	0.94 ± 0.06w
V-H	7.87 ± 0.15a	294.6 ± 44.4c	0.17 ± 0.01y	0.39 ± 0.05b	0.63 ± 0.06c	0.51 ± 0.02z
N-C	7.97 ± 0.18a	229.8 ± 29.4c	0.11 ± 0.02z	0.29 ± 0.02c	0.66 ± 0.05c	0.63 ± 0.03y

Reprinted from Sene et al. (2013b), with permission from TJPRC

Means ± SD; different letters within a column (a, b, c, d) indicate honestly significant difference (Tukey–Kramer test $p < 0.05$); different letters within a column (w, x, y, z) indicate at least significant difference (Dunnett T3 test, $p < 0.05$)

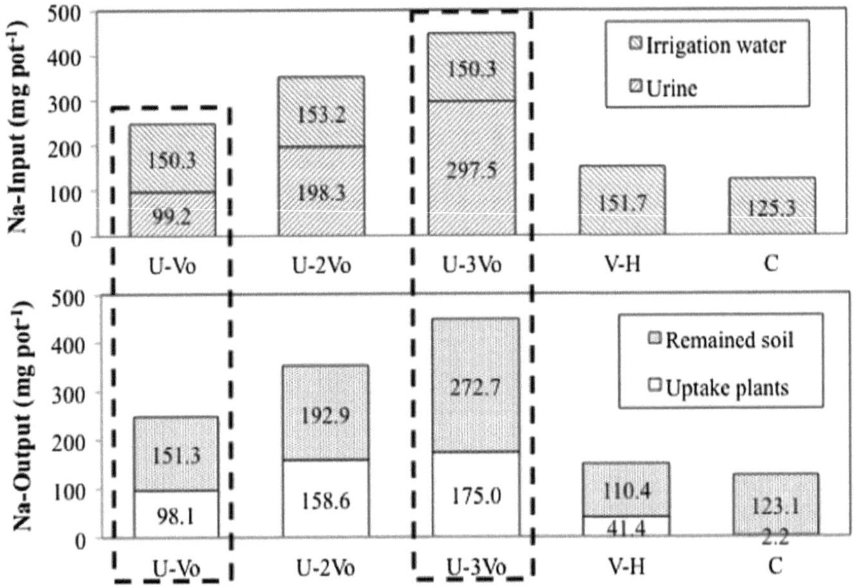

Fig. 15.10 Total input (irrigation water and urine) and output (remained in soil and uptake plants) of Na under the first, second, and third cultivation from continuous application of urine

Figure 15.10 shows the Na balance after three-time cultivations with continuous application of extra urine volume. Na was supplied through urine and irrigation water in the present study. The total amount of Na removed by plan increased with increasing the application of urine.

It was observed that soil pH decreased in all urine treatments after the second and third cultivations, and this decrease was significantly faster in the highest urine applications rates (U-2Vo and U-3Vo) (Table 15.11). One explanation might be a high level of nitrate contents in the extra urine soil conditions after continuous cultivation of Komatsuna. It is widely known that nitrogen in urine is in urea form, which can be easily degraded by urease enzymes to ammonium and ammonia, and this process released one hydroxide ion (OH^-). Since our soil is well aerated, ammonium is nitrified (NO_3^-) by soil nitrifying bacteria and this reaction is accompanied by a release of two protons ions ($2H^+$). The uptake of nitrate by plant roots in the soil solution emits one hydroxide ion (OH^-). Globally, two protons and two hydroxides ions are released in soil, which mean the net effect of pH on the soil is small (Höglund 2001). However, in the case of our experimental conditions, high urea concentration was applied in extra urine treatments soil and nitrate uptake by plant was a limited, since the uptake reached the plateau. Thereafter, no release of hydroxide ions from roots and more protons and/or nitrate remained in soil subsequently causing soil acidification in extra urine treatments.

Continuous application of extra urine in the range of 2–3 times higher than plant requirement had no effect on phosphorus concentration in Komatsuna shoot after three-time cultivations. Similar observations were also found in soil after all cultivations step except in the third one, where the highest urine application rate (U-3Vo) was significantly increased compared to other treatments (Table 15.11). Therefore, these results confirm also the low level of phosphorus contents in human urine (Pradhan et al. 2009a; Sene et al. 2012), but long-term use of urine may have potential to accumulate or improve soil phosphorus.

From this investigation, it was illustrated that continuous application of extra human urine volume in the range of 2–3 higher than plant requirement had no inhibition effect on Komatsuna growth, shoot dry biomass, and root dry biomass after three-time cultivations, since under three-time cultivations, soil EC of all urine-treated soils was lower than the soil salinity rating based on EC given by FAO (EC < 2mS/cm), and the removal Na applied from urine and irrigation water reached more than 39% in all urine fertilizer treatments.

However, urine fertilization causes nutritional imbalance, but this phenomenon was not accelerated under three-time cultivations with double (U-2Vo) and triple (U-3Vo) volumes urine application. Besides, N contents in plants tissues were not accelerated under three-time cultivations, but its level was constantly high in the highest urine applications volumes. Furthermore, soil EC increased from the first to the third cultivation in all set treatments, and this increase was significantly high in the extra urine volumes application. Contrast to EC, pH decreased in extra urine conditions and might probably due to a high level of nitrate contents soil conditions after continuous cultivation. Moreover, continuous application of excess amount of urine (U-2Vo and U-3Vo) accumulated nitrogen and sodium in soil; while adequate urine volume (U-Vo) application caused no accumulation of N in soil, but more than 60% of Na applied through urine and irrigation water remained in soil after three-time cultivations. Besides, the amount of N lost from soil was likely be to related to the difference in climate conditions and not to the continuous cultivation with extra urine, because N lost from soil was lower in the second cultivation period than in the first one. Nevertheless, the N lost was constantly high in extra urine applications volumes after each cultivation time. Therefore, from these results, we suggest that application of human urine volume should be based on plant N requirement, and management of salts from urine is required even in the adequate urine volume application for its sustainable reuse in agriculture.

References

Alfocea FP, Balibrea ME, Cruz AS, Estan MT (1996) Agronomical and physiological characterization of salinity tolerance in a commercial tomato hybrid. Plant Soil 180:251–257. https://doi.org/10.1007/BF00015308

Allotey DFK, Asiamah RD, Dedzoe CD, Nyamekye AL (2008) Physico-chemical properties of three—salt-affected soils in the lower Volta Basin and management strategies for their sustainable utilization. West Afr J Appl Ecol 12:1–14. https://doi.org/10.4314/wajae.v12i1.45776

Asano T, Burton FL, Leverenz HL, Tsuchihashi R, Tchobanoglous G (2007) Water reuse: issues technologies and applications. McGraw-Hill, New York, USA, pp 959–962

Brady NC, Weil RR (1996) The nature and properties of the soils, 14th edn. Prentice Hall, New York, USA

Burt J R (2006) English spinach. Review Farmenote 37/1997, Department of Agriculture-Western. Australia (WA_AGRIC)

Cabrera ML, Kissel DE, Bock BR (1991) Urea hydrolysis in soil: effects of urea concentration and soil pH. Soil Biol Biochem 23:1121–1124. https://doi.org/10.1016/0038-0717(91)90023-D

Cordell D (2008) The story of phosphorus: missing global governance of a critical resource. Paper prepared for SENSE earth systems governance, Amsterdam, 24–31st August

Dagerskog L, Bonzi M (2010) Opening mind and closing loops productive sanitation initiatives in Burkina Faso and Niger. EcoSan Club Sustain Sanitation Pratice 3:4–11

Dasgan YH, Aktas H, Abak K, Cakmak I (2002) Determination of screening techniques to salinity tolerance in tomatoes and investigation of genotype response. Plant Sci 163:695–703. https://doi.org/10.1016/S0168-9452(02)00091-2

Davis JG, Waskom RM, Bauder TA, Cardon GE (2007) Managing sodic soils. Colorado State University Extension. Crop Series/ Soil, Fact sheet No. 0.504

Ernst JW, Massey HF (1960) The effects of several factors on volatilization of ammonia formed from urea in the soil. Soil Sci Soc Am Proc 24:87–90

FAO (1998) Salt-affected soils and their management. FAO Soils Bulletin 39. Food and agriculture Organization of the United Nations, FAO, Rome, Italy

FAO (2008a) Current world fertilizer trends and outlook to 2011/12. Food and Agriculture Organization of the United Nation, FAO, Rome, Italy, pp 1–57

FAO (2008b) Water and cereals in drylands. The Food and Agriculture Organization of the United Nations and Earthscan. ISBN 978-92-5-1060520

FAO (2009a) Current world fertilizer trends and outlook to 2013. Food and Agriculture Organization of the United Nation, FAO, Rome, Italy, pp 1–14

FAO (2009b) ETo calculator version 3.1. FAO Land and Water Digital Media Series N°36. Food and Agriculture Organization of the United Nation, FAO, Rome, Italy

Fjellstrom RG, Williams PH (1997) Fusarium yellows and turnip Mosaic virus resistance in Brassica rapa and B. juncea. HortScience 32(5):927–930

Franzen D (2007) Managing saline soils in North Dakota. NDSU Extension Service, North Dakota State University, Fargo, ND, 58105, SF-1087 (Revised), pp 1–11

Fujiwara S, Narimatsu J (2006) Fertilizing and cultivation of vegetables-from property of nutrient absorbance to practical fertilizing method. Rural Culture Association, Japan. ISBN 4540052195

Gorham J, Bridges J, Dubcovsky J, Dvorak J, Hollington PA, Luo MC, Khan JA (1997) Genetic analysis and physiology of a trait for enhanced K +/Na + discrimination in wheat. New Phytol 137:109–116. https://doi.org/10.1046/j.1469-8137.1997.00825.x

Guzha E, Nhapi I, Rockstrom J (2005) An assessment of the effect of human urine faeces and urine on maize production and water productivity. Phys Chem Earth 30:840–845. https://doi.org/10.1016/j.pce.2005.08.028

Heinonen-Tanski H, Sjoblom A, Fabritius H, Karinen P (2007) Pure human urine is a good fertilizer for cucumbers. Biores Technol 98:214–217. https://doi.org/10.1016/j.biortech.2005.11.024

Hijikata N, Yamauchi N, Yabui K, Ushijima K, Funamizu N (2011) Characterization of several agricultural wastes as a matrix of composting toilet—from fecal degradation to reuse as a soil conditioner. Proceedings of 8th IWA international symposium on waste management problems in agro-industries, pp 317–324

Höglund C. (2001) Evaluation of microbial health risks associated with the reuse of source–separated human urine. Doctoral thesis, Royal Institute of Technology (KTH) and Swedish Institute for Infectious Disease Control (SMI), Stockholm

Horisawa S, Sunagawa M, Tamai Y, Matsuoka Y, Miura T, Terazawa M (1999) Biodegradation of nonlignocellulosic substances II: physical and chemical properties of sawdust before and after use as artificial soil, Japan. Wood Res Soc 45:492–497. https://doi.org/10.1007/BF00538959

Jones C, Brown BD, Engel R, Horneck D, Olson-Rutz K (2013) Factors affecting nitrogen Fertilizer Volatilization. Montana State University, Extension, EB0208 new February, 1–6

Jönsson H, Vinneras B, Höglund C, Stenström TA, Dalhammar G, Kirchmann H (2000) Recycling source separated human urine (Kallsorterad Humanurin). VA-Forsk Report 2000-1, VAVA AB, Stockholm, Sweden (In Sedish, English summary)

Jönsson H, Richert Stintzing R, Vinneras B, Salomon E (2004) Guidelines on the use of urine and faeces in crop production. EcoSanRes publication. Stockholm Environmental Institution (SEI), pp 2–18

Kassa K, Meinzinger F, Zewdie W (2010) Experience from the use of human urine in Arba Minch, Ethiopia. Sustain Sanitation Pratice 3:12–17 EcoSan Club

Kirchmann H, Pettersson S (1995) Human urine—Chemical composition and fertilizer use efficiency. Fert Res 40:149–154. https://doi.org/10.1007/BF00750100

Kirkwyland M, Thomas EG (2010) Production guide for organic Spinach. NYS IPM Publication No. 139

Lee TS (2012) Irrigation system and practices in challenging environments, 169–190. InTech, Janeza Trdine 9,51000 Rijeka, Croatie

Lopez Zavala MA, Funamizu N, Takakuwa T (2002) Onsite wastewater differential treatment system: modeling approach. Wat Sci Tech 46(6–7):317–324

Mnkeni PNS, Austin A, Kutu F R (2005) Preliminary studies on the evaluation of human urine as a source of nutrients for vegetables in the Eastern Cape Province, South Africa. In: Ecological sanitation: a sustainable integrated solution. Proceedings of the 3rd international ecological sanitation conference, Durban, South Africa, 23–26 May 2005, pp 418–426. http://conference2005.ecosan.org/papers/mnkeni_et_al.pdf

Mnkeni PNS, Kutu FR, Muchaonyerwa P (2008) Evaluation of human urine as a source of nutrients for selected vegetables and maize under tunnel house conditions in the Eastern Cape, South Africa. Waste Manag Res 26:132–139. https://doi.org/10.1177/0734242X07079179

Morikawa CK, Seigura M (2001) Effects of three types of fertilizer on acidification of Andisol and Growth of Komatsuna (Brassica campestris var. Maruha). Tohoku J Agric Res 52(1–2):021–028

Mubashir M, Malik SA, Khan AA, Ansari TM, Wright S, Brown MV, Islam KR (2010) Growth yields and nitrate accumulation of irrigated carrot and okra in response to nitrogen fertilization. Pak J Bot 42:2513–2521

Müllegger E, Langergraber G, Lechner M (2010) sustainable sanitation practice: use of urine. EcoSan Club (3)

Nishihara E, Inoue M, Kondo K, Takahashi K, Nakata N (2001) Spinach yield and nutritional quality affected by controlled soil water matric head. Agric Wat Manag 51:217–229

On-Farm Network (2008) Understanding soil nitrogen dynamics. UPDATE, Agronomic research and economic performance. Advanced Agriculture Performance, http://www.isafarmnet.com/onfarmupdate/oct08updateHIREZ.pdf

Otterpohl R, Grottker M, Lange J (1997) Sustainable water and waste management in urban area. Wat Sci Tech 35:121–133. https://doi.org/10.1016/S0273-1223(97)00190-X

Otterpohl R, Albold A, Olddenburg M (1998) Differentiating of water resources and waste in urban areas. In: Foo EL, Senta TD (eds) Proceedings of the internet conference on integrated bio-systems in zero emissions applications, 114. http://www.ias.unu.edu/proceedings/icibs

Pertnilla T, Mattsson B, Jonsson H (2007) Environmental impact of wheat production using human urine and mineral fertilisers—a scenario study. J Clean Prod 15:52–62. https://doi.org/10.1016/j.jclepro.2005.04.019

Peter B, Oleksyn J (2004) Global patterns of plant leaf N and P in relation to temperature and latitude. PNAS 101(30):11001–11006. https://doi.org/10.1073/pnas.0403588101

Pradhan SK, Nerg AM, Sjoblom A, Holopainen JK, Heinonen-Tanski H (2007) Use of human urine fertilizer in cultivation of cabbage (Brassica oleracea)—impact on chemical, microbial and flavor quality. J Agric Food Chem 55:8657–8863

Pradhan SK, Pitkanen S, Heinonen-Tansk H (2009a) Fertilizer value of urine in pumpkin (Cucurbita maxima L.) cultivation. J Agric Food Sci 18:57–68

Pradhan SK, Holopainen JK, Heinonen-Tanski H (2009b) Stored human urine Supplemented with wood ash as fertilizer in tomato (Solanum lycopersicum) cultivation and its impacts on fruit yield and quality. J Agric Food Chem 57:7612–7617. https://doi.org/10.1021/jf9018917

Pradhan SK, Holopainen JK, Weisell J, Heinonen-Tanski H (2010a) Human urine and wood ash as plant nutrients for red beet (Beta vulgaris) cultivation: impacts on yield quality. J Agric Food Chem 58:2034–2039. https://doi.org/10.1021/jf9029157

Pradhan SK, Piya RC, Heinomen-Tanski H (2010b) Eco-sanitation and its benefits: an experimental demonstration program to raise awareness in central Nepal. Environ Dev Sustain 13:507–518. https://doi.org/10.1007/s10668-010-9273-5

Rosen CJ, Bierman PM, Eliason RD (2008) Soil test interpretation and fertilizer management for laws, turf, garden, and landscape plants. Regents of University of Minnesota, Department of Soil, Water and Climate, pp 1–48

Schachtman DP, Reid RJ, Ayling SM (1998) Phosphorus uptake by plants: from soil to cell. Plant Physiol 116:447–453. https://doi.org/10.1104/pp.116.2.447

Sene M, Hijikata N, Ushijima K, Funamizu N (2012) Adequate Human urine application pattern for agriculture. Int Res J Agric Sci Soil Sci 2(1):038–045

Sene M, Hijikata N, Ushijima K, Funamizu N (2013a) Effects of extra human urine volume application in plant and soil. Int Res J Agric Sci Soil Scie 3(6):182–191

Sene M, Hijikata N, Ushijima K, Funamizu N (2013b) Effects of continuous application of extra human urine volume on plant and soil. Int J Agric Sci Res 3(3):75–90

Staines M, Morris R, Tess C, Bolland M, Russell B, Guthridge L, Lucey J, Bennett D (2011) Nitrogen for intensively grazed dairy pastures. Greener Pasture 2, Bulletin 4815, ISSN: 1833-7236, Western Australia Agriculture authorities

Standard Methods (1989) For the examination of water and wastewater. 17th edition

Thao HTB, Yamakawa T, Shibata K, Sarr PS, Myint AK (2008) Growth response of Komatsuna (Brassica rapa var. peruviridis) to root and pholiar application of phosphate. Plant Soil 308:1–10. https://doi.org/10.1007/s11104-008-9598-0

Tuna LA, Kaya C, Ashraf M, Altunlu H, Yokas I, Yagmur B (2007) The effects of calcium sulphate on growth membrane stability and nutrient uptake of tomato plants grown under salts stress. Environ Exp Bot 59:173–178. https://doi.org/10.1016/j.envexpbot.2005.12.007

Turan M, Sevimli F (2005) Influence of Different nitrogen sources and levels ion content of cabbage (Brassica oleracea var.capitate). New Zeal J Crop Hort Sci 33:241–249

Wilsenach JA, Schuurbiers CAH, Loosdrecht MCMV (2007) Phosphate and Potassium recovery from source separated urine through struvite precipitation. Wat Res 41:458–466. https://doi.org/10.1016/j.watres.2006.10.014

Winker M, Tettenborn F, Faika D, Gulyas H, Otterpohl R (2008) Comparison of analytical and theoretical pharmaceutical concentrations in human urine in Germany. Wat Res 42:3633–3640. https://doi.org/10.1016/j.watres.2008.06.002

Winker M, Clemens J, Reich M, Gulyas H, Otterpohl R (2010) Ryegrass uptake of carbamazepine and ibuprofen applied by urine fertilization. Sci Total Environ 408:1902–1908. https://doi.org/10.1016/j.scitotenv.2010.01.028

Chapter 16
On-site Use of Reclaimed Greywater

Nowaki Hijikata

Abstract Greywater is a non-toilet component of household wastewater and potentially reusable water source for irrigation due to higher proportion volume in wastewater and lower organic load. The reuse would have economic benefits an offer an attractive option particularly in dried area where water resource is fragile, because the reuse has advantages for reducing potable water use, expanding agricultural activities, increasing crop yields by additional inputs of irrigation amount and fertility elements. On the other hands, planning and managing of the reuse would need special and case-by-case cares with a consideration of cost-effectiveness in each local condition, since greywater may potentially contain undesirable factors for plant and soil as well as pathogens for human, and the quality is depended on culture, lifestyle family constitution, water infrastructures. A case study in rural area in Burkina Faso showed an example of the planning approach for low-income targets in dried area.

Keywords Economic benefit · Soil management · Phyto-toxicity
Environmental risk · Pathogenic risk

16.1 Economic Benefit of On-site Greywater Reuse

Greywater is a non-toilet component of household wastewater that originates predominantly from laundry and bathrooms of residential buildings. Greywater constitutes about 60% of domestic wastewater (DHWA 2002) and a potentially reusable water resource for irrigation of household lawns and gardens (Al-Jayyousi 2003). The reuse particularly offers an attractive option in arid and semi-arid zone, where water is scarce and water resource is fragile due to unstable rainfall. The reuse could potentially reduce potable water use by up to 50% in dried area

N. Hijikata (✉)
Graduate School of Engineering, Hokkaido University, Kita13-Nishi 8,
Kita-ku, Sapporo City 060-8628, Japan
e-mail: hijikata.nowaki@gmail.com

N. Funamizu (ed.), *Resource-Oriented Agro-sanitation Systems*,
https://doi.org/10.1007/978-4-431-56835-3_16

(DHWA 2002). In Cyprus, a study on greywater reuse indicates 36% reduction in water bills when household greywater is reused (Morel and Diener 2006). The reuse would also expand opportunities of agricultural activity in poverty area where public water service is limited. It has been preliminary speculated that the annual benefit from agricultural products derived from on-site greywater may reach 106-240 EURO per one family for rural small-scale farmer in Burkina Faso (Hijikata et al. 2012). Accounted for by increased product yields, as well as reduced water and fertilizer cost, it may reach an average annual benefit of USD 376 in urban, Jordon (Faruqui and Al-Jayyousi 2002). Source separation of greywater from urine and feces has certain advantage that it would reduce these organic loads (Lopez Zavala et al. 2002) and source control of greywater would reduce pathogenic risks (Maimon et al. 2010). Thus, the separation simplifies each treatment which leads to reduced overall investment for sanitation facilities (Winblad and Simpson-Hebert 2004). In Israel, the return on investment of on-site greywater management system (comprising a recycled vertical flow constructed wetland and garden irrigation system) approximately three years and regarded as economically attractive (Gross et al. 2007a).

16.2 Quantity of Greywater

The generated amount of greywater widely varies a range of 20–225 l/p/d depending on lifestyles, living standard, habitant, number of household members, ages of the member, existing water supply service and infrastructure. The volume in low-income countries with water scarcity and rudimentary form of water supply can be as low as 20–30 l/p/d (Morel and Diener 2006). The literature indicates a typical greywater amount of 90–120 l/p/d with piped water in houses without water shortage. The greywater amount greatly influences the available garden area, particularly in the area with water scarce and simple form of water supply. A design of garden size can be calculated with following equations:

$$A = (G + I)/\mathrm{IR}_{\mathrm{max}} \tag{16.1}$$

$$\mathrm{IR}_t = (\mathrm{ET}_0 \times K_c \times 30 - P_e)/30 \tag{16.2}$$

where A is available area (m^2), G is the available amount of greywater (L/day), I is an available irrigation resource except the greywater and precipitation (L/day), $\mathrm{IR}_{\mathrm{max}}$ is a maximum irrigation requirement in the planed cultivation period (L/m^2/day).

A calculation for a daily irrigation requirement (IR_t) is followed by a guide of FAO (1986) and simply expressed in an Eq. (16.2), where IR_t is irrigation requirement at a cultivation day t (L/m^2/day = mm/day), ET_0 is a reference evapotranspiration (mm/day), K_c is a crop factor referred from the guide, and P_e is monthly effective precipitation (mm/month). The monthly effective precipitations

(P_e) and reference evapotranspiration by Penman-Monteith method (ET_0) are available using CROPWAT 8.0 software by FAO. The precipitation and climate data for many cities in the world can be obtained from CLIMEWAT 2.0 database by FAO.

In the case of rural area, Burkina Faso, it was speculated that the generated greywater amount covers increased garden area of 21 m^2/house, and the area corresponds to 30–40% of existed vegetable garden (Ushijima et al. 2012). Although the available area is approximately decreased half from rainy season to dry season cultivation due to increase of evapotranspiration, the greywater reuse would give a gardening opportunity to family even in dry season without irrigation facilities and piped water supply service.

16.3 Quality of Greywater

Composition of greywater also depends on several factors such as lifestyle and water culture. Greywater may contain soaps, food particles, grease, oil, lint, hair, pathogens personal care products, and domestic chemicals (Morel and Diener 2006). Greywater may also contain high levels of detergents, which is consisted of surfactants, builders, bleaches, thickening agent, and fluorescent brightener (Warne and Schifko 1999). Although several nutrients present in greywater such as nitrogen and phosphorus can be beneficial for plant growth, balanced concentration of nutrient is required to avoid nutrient efficiency or adverse effect on plant. Additionally, the nutrient compositions potentially cause clogging for irrigation facilities such as drip irrigation kit or irrigation pipeline. Besides, pathogens in greywater highly influence public health. Important parameters for agricultural reuse, particularly agricultural facilities, soil, plant, environmental risks, and public health, were summarized in Table 16.1. The following section aims at illustrating selected parameters which believed to be relevant for the design and of appropriate management strategies for the reuse.

16.3.1 Suspended Solids

Food, oil, and soil particle from kitchen sinks or hair and fibers from laundry can lead to high solids content in greywater. These particles cause turbidity in the greywater and may result in physical clogging of pipes, pumps, and filters which are used in treatment processes and irrigation facilities. Especially, non-biodegradable and persistent fibers from clothing, hair, powdered detergents, and soap, as well as colloids are the main reasons for physical clogging.

Table 16.1 Important parameters of greywater for agricultural reuse

Parameters	Possible impacts
Total suspended solids	Inducing physical clogging to irrigation pipes and equipment [1]
pH	Corrosion of equipment, damage to biota, precipitate of divalent cations in soil, indirect inhibition of nutrient uptake by plants [2–5]
Electrical conductivity	Reduction of plant productivity, possible changes in soil properties [1, 3, 6, 7]
Sodium adsorption ratio	Exacerbation of soil erosion, changing of soil hydraulic conductivity, reduction of water permeability in soil unbalanced nutrient uptake by plant [1, 3, 4, 6,7]
Nitrogen	Increase of lodging and insect damage possibility for plant, increase of nitrate concentration in leafy vegetable and nitrate contamination in groundwater through leachate, in excess concentration [1, 8]
Phosphorus	Eutrophication if excess concentrations reach surface water, inducing biological clogging of equipment [1]
Cl^-	Excess accumulation in soil may adversely affect plants [3, 9]
Boron	Excess accumulation in soil may be toxic to plant [3, 9, 10]
Oil and grease	Accumulation change soil hydraulic conductivity [11]
Surfactant	Accumulation may change soil hydraulic conductivity [11]. Toxicity to plant in high concentration and influence on soil enzymes and nitrification bacteria in low concentration [4, 12–15]
Xenobiotics	Toxicity to biota [2, 3]
E. coli and *Fecal coli*	Indication of pathogens contamination and increase of health risk probability for worker and consumer [3, 16, 17]

[1] Morel and Diener (2006), [2] Eriksson et al. (2002), [3] Maimon et al. (2010), [4] Hijikata et al. (2014a), [5] Pinto et al. (2010), [6] Al-Hamaiedeh and Bino (2010), [7] Mecham and Qian (2005), [8] Haruta and Nakaya (2011), [9] Gross et al. (2005), [10] Gross et al. (2007b), [11] Travis et al. (2010), [12] Garland et al. (2004), [13] Misra et al. (2010), [14] Jensen et al. (2001), [15] Elsgaard et al. (2001), [16] Ottoson and Stenstrom (2003), [17] WHO (2006)

16.3.2 Nitrogen and Phosphorus

Literatures have shown that on-site raw greywater contains total nitrogen (TN) of 25–45 mg/L in Israel (Gross et al. 2007a), 37 mg/L in Malaysia (Martin 2005) and 8.8–42.5 in global summary (WHO 2006). In the case of Burkina Faso, treated greywater still contained TN of 10–77 mg/L and it takes an account of approximately 20 g/m^2 (200 kg/ha) nitrogen application when the greywater is applied 6 L/day/m2 during 3-month cultivation (Hijikata et al. 2014a). The applied TN value corresponds to nitrogen requirement of normal vegetable. This indicates that the greywater reuse possibly reduces an amount of nitrogen fertilizer usage, but also that excess nitrogen may be applied in the field if users don't know the information. The excess nitrogen application potentially has a negative impact on plants yield and human health. For instance, paddy rice has been known as nitrogen sensitive plant. The excess nitrogen supply from irrigation may cause lodging by rank growth, lack of grain filling, and weaken by insect tolerance (Haruta and Nakaya 2011). In soybean case, it may also cause decrease of yield by rank growth. In vegetable case, excess nitrogen fertilization possibly causes high concentration of nitrate and nitrite in the leaf and fruit, which may cause disruption of taste and threat health of consumer (Turan and Sevimlib 2005; Mubashir et al. 2010; Pradhan et al. 2009). Haruta and Nakaya (2011) have pointed that study on the nitrogen fertility efficiency by the reuse of greywater has been still limited. Especially, nitrogen form, e.g., organic nitrogen, ammonium, or nitrate, affects nitrogen nutrient efficiency in plant. Therefore, impact of nitrogen concentration and its form in greywater on these nitrogen sensitive plants should be further studied.

In the case of phosphorus, the element in greywater could be beneficial for plant growth, certainly. Most of terrestrial plant tends to be a lack of phosphorus in dry-field cultivation. Approximately 80% of applied phosphorus is remained in soil as unavailable form for plant, since the phosphorus are easily incorporated with Ca (in alkaline) or Fe and Al (in acidic) in soil condition (Schachtman et al. 1998). Sene et al. (2012) showed frequent phosphate ion supply leaded efficient phosphorus absorption by plant root. This fact let us expect that contentious phosphate supply from greywater would be also effective for plant. On the other hands, it has been noted that high phosphorus concentration potentially cultures algae in storage tank and this would be a cause of biological clogging in irrigation facilities (Morel and Diener 2006).

16.3.3 pH

According to FAO (1985) and USEPA (2004), greywater pH should show in the range of 6.5–8.4 for easier bioprocess treatment and to avoid negative impacts on soil and plants when reused. pH of greywater is usually optimal and varies in the range of pH 6.5–8.4 (Morel and Diener 2006). In laundry greywater cases, however, 9.3–10 pH value was observed (Christova-Boal et al. 1996), as a result of sodium hydroxide based and bleach in detergent and soap. Additionally, continuous greywater irrigation potentially increases soil pH (Table 16.2) and causes a nutrient imbalance for plant (Table 16.3) (Misra and Sivongxay 2009; Pinto et al. 2010; Hijikata et al. 2014a). The soil pH value between 6 and 7.5 is usually considered optimal for growth of many plants and microbial health of soil. The pH greatly affects the availability of a range of soil nutrients. Particularly, soil taxonomy has defined moderate alkaline soil as pH > 8.2 which potentially leads salinity problem and sever alkaline soil as pH > 9.0 which leads nutrient imbalance of plant. In the high pH soil, essential element for plant nutrient such as P, Ca, Mg, Fe, Zn, Mn are easily precipitated. It prevents the mineral uptake by root in the soil. Therefore, it would be recommended to be taken care of the soil pH management when the soil pH rises more than 8.2 (Pinto et al. 2010). To decrease the soil pH and to promote soil pH buffering action, amendment of gypsum ($CaSO_4$), and compost would be one of solution.

Table 16.2 Soil chemical properties after two times pot cultivation with laundry detergent irrigation

	TW	Laundry detergent treatments			
		LD0.1	LD0.3	LD1	LD3
pH	7.21 ± 0.03a	7.29 ± 0.08a	7.34 ± 0.11a	7.83 ± 0.10b	9.24 ± 0.03c
EC (mS/cm)	1.08 ± 0.32	0.90 ± 0.24	0.81 ± 0.27	1.10 ± 0.23	1.20 ± 0.28
Water extracted Na (mg/g)	0.25 ± 0.03a	0.32 ± 0.17a	0.27 ± 0.04a	0.48 ± 0.26a	1.13 ± 0.31b
SAR	0.14 ± 0.01a	0.18 ± 0.10a	0.16 ± 0.02a	0.28 ± 0.16a	0.66 ± 0.17b

LD0.1, LD0.3, LD1, and LD3 showed different concentrations of laundry detergent (0.1, 0.3, 1 and 3 mg/L) irrigated in the soil
A group of alphabetical letters (a, b, c) showed significant difference ($p < 0.05$, Tukey-HSD test)

Table 16.3 Major cations concentration in komatsuna shoot after two times cultivation with laundry detergent irrigation

	TW	Laundry detergent treatments			
		LD0.1	LD0.3	LD1	LD3
K (mg/g)	20.2 ± 0.6a	17.3 ± 1.0b	16.3 ± 1.6b	14.5 ± 0.7bc	12.6 ± 2.0c
Ca (mg/g)	15.6 ± 0.7e	14.2 ± 1.1e	14.0 ± 2.4ef	12.2 ± 0.2e	9.3 ± 0.6f
Mg (mg/g)	5.7 ± 0.1e	5.2 ± 0.2ef	5.4 ± 0.6ef	4.8 ± 0.1f	4.7 ± 0.1f
Na (mg/g)	2.6 ± 0.4f	3.4 ± 0.3ef	3.3 ± 0.4f	6.9 ± 0.9e	22.0 ± 8.5f
K/Na	7.8 ± 1.2a	5.2 ± 0.7b	5.0 ± 0.7b	2.1 ± 0.3c	0.7 ± 0.3c
Ca/Na	6.0 ± 1.0a	4.2 ± 0.6a	4.3 ± 1.0a	1.8 ± 0.2b	0.5 ± 0.2b

LD0.1, LD0.3, LD1, and LD3 showed different concentrations of laundry detergent (0.1, 0.3, 1, and 3 mg/L) irrigated in the soil
A groups of alphabetical letters (a, b, c) showed significant difference ($p < 0.05$, Tukey-HSD test)
A group of alphabetical letters (e, f) showed significant difference ($p < 0.05$, Dunnet t-test)

16.3.4 Salinity and Sodium Adsorption Ratio

Greywater contains salts, indicated as electrical conductivity (EC). EC value represents salinity of all dissolved ions, including negatively charged ions and positively charged ions. The EC in greywater is typically in the range of 300–1500 mS/cm, but 2700 mS/cm of greywater was observed as a higher case (Morel and Diener 2006). Salinity of greywater normally not problematic, but it became a hazard when the greywater is reused as irrigation. High EC of irrigation water may reduce yield potential. Furthermore, long-term reuse of saline greywater potentially leads increase of salinization of topsoil (Mecham and Qian 2005; Al-Hamaiedeh and Bino 2010). This problem easily occurs especially on clay and loamy soils with low percolation rates in arid climate due to high evapotranspiration.

Hazard of sodium accumulation in soil is regarded as different problem from salinity accumulation, and it affects soil physical property and imbalance of plant nutrition. The sodium hazard is indicated as sodium adsorption ratio (SAR), which quantifies the proportion of sodium (Na^+) to calcium (Ca^{2+}) and magnesium (Mg^{2+}). The SAR values of greywater are within a typical range of 2–10, depending mainly on the laundry powder used by the household (Gross et al. 2005). The water quality influenced SAR value of irrigated soil (Table 16.2) (Al-Hamaiedeh and Bino 2010; Hijikata et al. 2014b). The sodium needs special concern when applied to loamy soils in poor calcite or calcium/magnesium. High SAR potentially causes the degradation of well-structured soils (compacted soil); it may inhibit aeration and water permeability. In the high SAR soil, additionally, nutrient absorption by plant root is disrupted and substitute uptake of K and Na, which is regarded as primary saline stress and represented as high Na/K ratio in leaf, is potentially occurred (Table 16.3) (Tuna et al. 2007; Sene et al. 2013a; Hijikata et al. 2014b).

It would be difficult to remove these mineral ions, which affects to greywater salinity and sodium factors, with normal and economical on-site greywater treatment facilities. Thus, the management should be concerned with the view of soil and agronomy. It has been known that water harvesting technique, construction of drainage canal under fields, amendment of gypsum and cultivation of cleaning clops such as sorghum and forage are effective for leaching of salinity ions (Qadir et al. 2005). A choice of salt tolerant vegetables and crops is also one of solution to avoid yield decrease by the salinity (Mnkeni et al. 2008). The overall design of soil management and crop rotation is recommended when the saline greywater is reused as irrigation in long-term.

16.3.5 Oil and Grease

Greywater may contain significant amounts of oil and grease (O&G) originating mainly kitchen sinks and dishwashers. The concentrations in greywater ranging between 37 and 78 mg/L and averaging 8–35 mg/L (Christova-Boal et al. 1996), while values as high as 230 mg/L were observed in Jordan for mixed greywater (Al-Jayyousi 2003) and 1000–2000 mg/L in restaurant wastewater was also observed (Crites and Tchobanoglous 1998). The O&G potentially congeal and cause mats on the surface of settling tanks, on the interior of pipes and other surface. This may cause a shutdown of treatment and disposal units such as infiltration trenches or irrigation fields (Morel and Diener 2006). Furthermore, the O&G may affect physical properties of irrigated soil such as water droplet penetration time, water capillary, and zeta potential, in long-term or high load irrigation (Travis et al. 2008, 2010, 2012). These changes may not cause adverse effect on plant, directory. However, water permeability in soil greatly influences leaching capacity of salinity factors and sodium. Thus, the accumulation of O&G in field soil may decrease efficiency of salinity leaching by rainfall or sufficient irrigation.

16.3.6 Surfactant

Greywater contains surfactant, which is the main component of household cleaning products. Surfactant consists of a hydrophilic head and hydrophobic tail. The double function plays significant role to decrease the surface tension of water, make an emulsion with oily stains in water, and keep them dispersed (Morel and Diener 2006). Thus, the compound allows to clean hands, clothes, and dishes more effectively. Most of common surfactants in household cleaning chemicals are anionic surfactant such as LAS (Liner-alkylbenzene sulfonate) and SLES (sodium laureth sulfate), nonionic as AE (alcohol ethoxylate), amphoteric as CAPB (cocamidopropyl betaine). The anionic surfactants in greywater are able to be determined by mainly methylene blue method and specific quantification of the each surfactant is determined by GC/MS (Ding et al. 1998), LC/MS (Garland et al. 2004; Huelgas and Funamizu 2010; Ushijima et al. 2013). Anionic surfactant (as MBAS) in domestic raw greywater was varied at the range of 4.1–10 mg/L (Travis et al. 2008, 2010), 4.7–15.6 mg/L (Gross et al. 2007a), 37 mg/L (Friedler 2004) and 0.7–70 mg/L (Wiel-Shafran et al. 2006). Morel and Diener (2006) has reviewed the general range of 1–60 mg/L (averaging 17–40 mg/L) and the surfactants usage of 3.5–10 g MBAS/person/day, while 45–170 mg MBAS/L has been also reported as the highest case (Al-Jayyousi 2003).

It has been known that high concentration of surfactant has direct adverse effect on plant growth. The significant growth reduction has been reported at 250 ppm of igepon irrigation with lettuce (Budenheim et al. 1997), 300 ppm of AE and CAPB with wheat (Garland et al. 2004), 150 mg/L industrial grade LAS with tomato (Misra et al. 2010), and 200 mg/L of LAS with komatsuna (Hijikata et al. 2014b). 10% effective concentration (EC10) of the LAS irrigation for komatsuna growth has been also estimated at 135 mg LAS/L (Hijikata et al. 2014b). These effective concentration values by surfactants are higher than general concentration in raw greywater, described in above. Furthermore, it has been reported that a continuous cultivation with LAS irrigation did not accelerate the growth inhibition (Fig. 16.1)

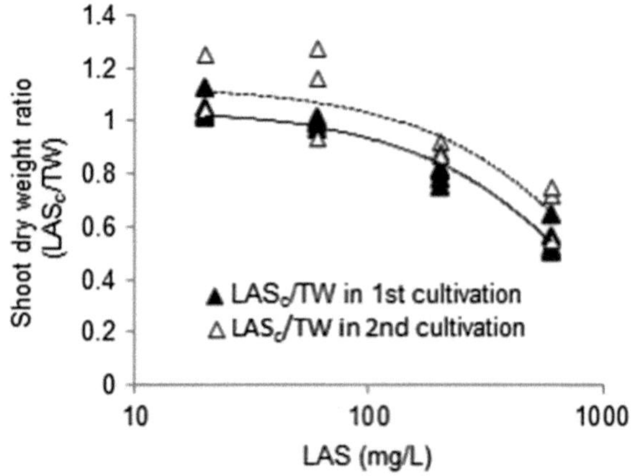

Fig. 16.1 Dry weight ratio of Komatsuna shoot cultivated with LAS irrigation

due to high LAS degradation in the soil condition (Hijikata et al. 2014b). This may be supported by references that the LAS degradation is promoted in the existence of viable plant root (Mortensen et al. 2001), and a number of bacteria able to grown with biodegradable surfactant are increased after hydroponic culture with the surfactant (Garland et al. 2000). Therefore, it is suggested that adverse effect for plant growth by the anionic surfactant may rarely observe in general on-site greywater reuse.

In the case of LAS which is widely consumed in the world (Cirelli et al. 2008), phytotoxicological data has been well summarized and EC10 soil value of various terrestrial plants is ranged between 50 and 200 mg/kg soil (Jensen et al. 2001, 2007; Ying 2006). On the other hands, the anionic surfactant influences soil microorganisms (Jensen et al. 2001; Sanchez-Peinado et al. 2009), nitrification bacteria (Brandt et al. 2002) and soil enzyme activities (Elsgaard et al. 2001; Brandt et al. 2002; Sanchez-Peinado et al. 2009) in lower concentration at the range of 5–20 mg/kg soil. These parameters play a significant role of biochemical reaction in soil, particularly in organic or low-input chemical fertilizer agriculture. Involving these eco-toxicological views, cut-off value of 5 mg LAS/kg soil (Jensen and Folker-Hansen 1995) and predicted no observed effective concentration (PNEC) of 4.6 mg LAS/kg soil (Jensen et al. 2001) has been proposed, when sludge compost is amended. These facts indicate that greywater reuse also potentially influence the bacteria and biochemical reactions in soil. Because continuous 20 mg LAS/L irrigation leads 6.2 mg LAS/kg soil at the end of cultivation (Fig. 16.2) (Hijikata et al. 2014b) and the value is within an expectable concentration of general domestic greywater.

One of difficulty of the phytotoxic and ecotoxic assessment of surfactants in agricultural field is a variance of the biodegradability depending on a difference of soil environmental condition. Biodegradability of LAS in soil is influenced by applied LAS concentration, population of microflora, concentration of divalent cations, soil organic matter, and existence of plant root (Cirelli et al. 2008) and has widely half-life time with a range of 7–33 days (Ying 2006). It is also remarked that young komatsuna root growth was significantly inhibited in 20 mg LAS/L in petri dish germination test, which test is regarded as non-biodegradable condition

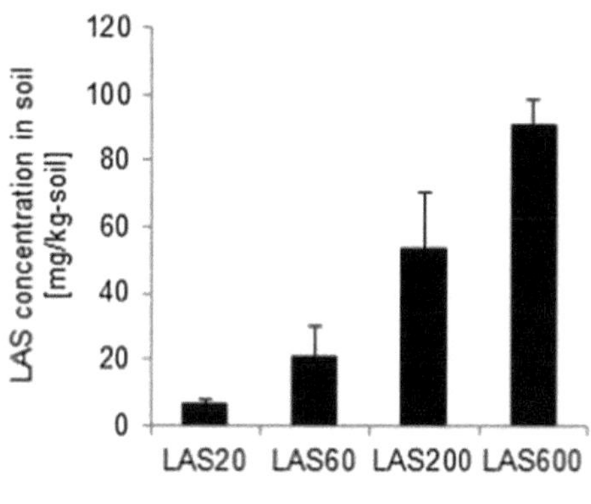

Fig. 16.2 LAS concentration in soil after two times cultivation with LAS irrigation

(Hijikata et al. 2011). These values would be excluded from the situation of on-site greywater reuse and dried-field cultivation, however, the surfactant biodegradability in the cultivation condition needs to be careful.

16.3.7 Microorganisms

Microorganisms, particularly pathogenic risk, are significant parameter in on-site greywater reuse for irrigation, since the food and worker's safety are one of the most interest points for farmers and consumers. Greywater is often considered by the public as safer than blackwater. However, it is well established that it can still pose considerable health risks if not used appropriately (Dixon et al. 1999; Eriksson et al. 2002; Ottoson and Stenstrom 2003; Maimon et al. 2010). Pathogens, such as viruses, bacteria, protozoa, and parasites, may derive from three main sources: fecal contamination, food handling, and opportunistic pathogens, such as those found on the skin or respiratory organs (Maimon et al. 2010). It has been regarded that pathogens from fecal origin mainly pose the main health risks, and thus most of risk assessment and guideline are considered to the fecal pathway (Ottoson and Stenstrom 2003; WHO 2006; Maimon et al. 2010). On the other hands, a few studies have investigated the risk from washing of vegetables and raw meat, such as *Salmonella* spp. (Diaper et al. 2001). Opportunistic pathogens usually associated with skin and ears insults, such as *Pseudomonas aeruginosa* and *Staphylococcus aurous*, are occasionally found in greywater (Winward et al. 2008; Gross et al. 2007b; Gilboa and Friedler 2008).

Fecal contamination of greywater is traditionally expressed by fecal indicators such as thermotorelant fecal coliforms. Average concentration of the fecal coliform in greywater is 10^4–10^6 cfu/100 ml (Morel and Diener 2006; WHO 2006). In the higher concentration case, however, 10^7–10^8 cfu/100 ml was also observed (Al-Jayyousi 2003; Dallas et al. 2004). In greywater systems, re-growth of enteric bacteria including the fecal indicators has been observed (Ottoson and Stenstrom 2003). It has been pointed that these bacterial indicator numbers may lead to an overestimation of fecal contamination in the case of greywater (Ottoson and Stenstrom 2003; Morel and Diener 2006; WHO 2006).

Target value of the fecal indicators in reused irrigation has been well described in a guideline of wastewater irrigation by WHO (2006) and a report by Maimon et al. (2010), the values are generally assessed by a quantitative microbial risk assessment (QMRA) consisting hazard identification, exposure assessment, dose-response characterization, and risk characterization (Shuval et al. 1997; Haas et al. 1999). The WHO guideline has stated that the target concentration of *E. coli* for unrestricted irrigation was 10^3–10^4 cfu/100 ml and 10^3–10^4 reduction was adequate greywater treatment. The values is leaded from rotavirus infection risk for lettuce consumer with an exposure scenario; 100 g eaten raw per person every 2 days, 10–15 ml of wastewater remaining on the crop. Maimon et al (2010) have assumed a case of on-site greywater irrigation reuse and stated that safety *E. coli* concentration was 10^2–10^4 cfu/100 ml. The value are leaded from rotavirus infection risk with the worst exposure scenario for workers; 1 ml routine indirect ingestion from touching plants and lawns with the frequency of 90 events per annual. These flagship values indicate that the use of untreated greywater may be unsafe (Maimon et al. 2010). However, as remarked by Dixon et al. (1999) and Maimon et al. (2010), the smaller reuse cycle is the lower the pathogen risk. Thus, reusing greywater from a single house system is much safer than reusing greywater on a neighborhood-scale system and most regulatory programs allow restricted use of untreated greywater within the context of a single household property (Maimon et al. 2010).

Even in the single household scale, however, pathogens contamination, survival or re-growth in greywater conveyance system may pose unnecessary risks to the direct users (Ottoson and Stenstrom 2003). Therefore, it may need to promote suitable treatments. It could be noted that several greywater treatment systems were found to reduce fecal indicators such as *E. coli* and coli phage concentration to low (Gross et al. 2007b; Gilboa and Friedler 2008; Ushijima et al. 2013). In order to minimize human contact with potentially hazardous bacteria, complementary approach such an establishment of "barriers" related to source control scheme and tiered approach, in which there are different requirement for different type of reuse schemes, has also stated in several regulatory programs (Maimon et al. 2010).

16.4 A Case Study of On-site Greywater Reuse in Rural Area, Burkina Faso

Many research papers about practical attempts of on-site greywater reuse have been published and showed successfully its availability. The following section was described the case study in rural area of Burkina Faso, where locates in sub-Sahel region. Particularly, planning, preliminary income estimation, and monitoring were mentioned.

16.4.1 Background

In Burkina Faso, where is located in sub-Sahel region, the proportion of population using an improved sanitation facility was 33% in urban area and only 6% in rural area at the time of 2008 (WHO and UNICEF 2010). The under-five mortality rate is 169 per 1000 live birth and diarrhea occupies 19% of cause of these deaths. Water and sanitation are urgent issues in the country. On the other hands, the country belongs to the semi-arid zone and the main soil type is Lixisols. It is possible to undertake agriculture but the productivity is low and fragile due to unstable rainy water, low-fertility soils and low accessibility to chemical fertilizer for small-scale farmers who consist more than 80% of the population. Therefore, improvement of sanitation and agricultural condition is urgent issue in the country. Comprehensive approach of resource-oriented sanitation linked with agriculture such as compost, urine, and greywater reuse has been proposed.

According to a feasibility study in a rural area in the country, Ubritenga commune (Ushijima et al. 2012), main water resources are deep well, shallow well, rainfall and dam lake. Water from deep well is mainly used for human daily life such as drinking, cooking, and shower. The averaged water usage is 20–30 L/day/capita. Shallow well is mainly used for animal, laundry water, and gardening irrigation. However, the well rarely existed in the villages and the residents cannot use the well in dry season due to low level of groundwater. Rainfall is main water resource of crop fields and house lawn garden but limited from middle of June to end of September. There is a small dam lake near village but far from resident houses, normally. The water is used for animals, cash vegetable in dry season. This means, agricultural activities are limited by water resources in the area. Three types of the activities are existed, (i) crop field near house, (ii) vegetable garden in-house lawn, (iii) cash vegetable fields near dam lakes. Income source of the rural people was vegetable production and the cultivation was conducted without adequate amount of chemical fertilizer and composts.

16.4.2 Planning of On-site Greywater Reuse

For economical sustainability and local acceptability, value-chain base approach was taken for a planning of an on-site greywater reuse in the area, which approach are consisted following steps: (i) survey a local vegetable distribution (ii) analyze vegetable marketability, (iii) design a crop rotation in a garden (iv) design a size of the garden, and (v) estimate maximum annual income from vegetables.

16.4.2.1 Vegetable Distribution

In a current situation, farmers have two routes to sell their vegetables. One is sale in local market and second is direct sale to a broker (Fig. 16.3). In the former case, the farmer transports the vegetable productions to local market by bicycle or donkey cart. Then, the vegetables are sold to broker, who comes from capital city, with negotiation. While some parts of the productions are sold to retailers in the local market through the broker, the most of parts are mainly transported to wholesale market in capital city by the broker with bus. The vegetables are repacked to small unit to sale for retailer in capital city. Retailer comes to the wholesale market and buys in the vegetables. In the later case, individual broker directly comes in farmland and buys in farmers vegetables at the place.

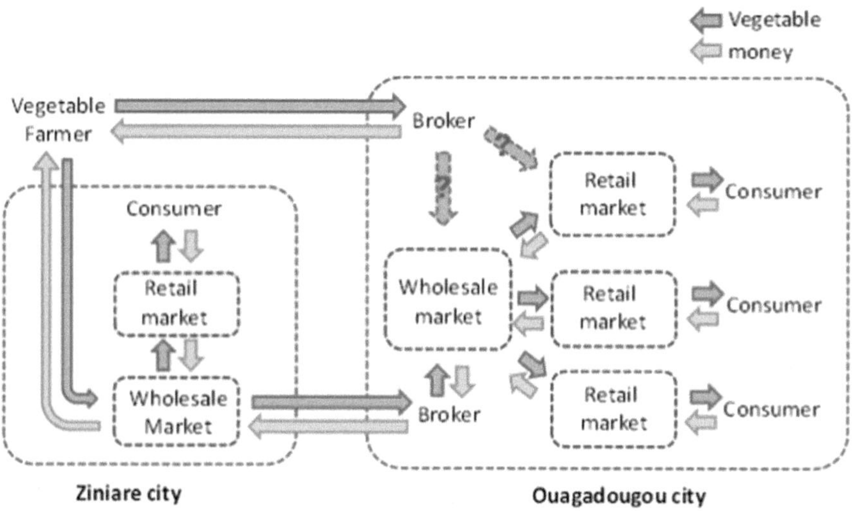

Fig. 16.3 Distribution network of vegetable near Ouagadougou, Burkina Faso

16.4.2.2 Marketability of Vegetables

The sale price from farmer to broker, which is income for farmer in other words, was questioned to broker in local market in Ziniare and farmer in Ouagadougou in different season. The higher marketability products were tomato, onion, and zucchini (Table 16.4). Carrot and cabbage were also traded at relatively higher price around capital city. Most of vegetables are sold as a unit base using 50 kg soil bags which volume is about 40 L. In the tomato case, wood box is normally used as a sold unit and small basket is also used in local market. In the case of carrot and cabbage around capital city, broker comes to farmers land and buys in the products as a cultivation area or piece. But we were seldom to see these two vegetables in Ziniare market.

According to a seed company in capital city, tomato and onion are exported to neighbor countries, where are Ghana, Niger, Togo, Cort d'voal. Therefore, the capacity of needs for tomato and onion would be larger. Onion, chili pepper, and gumbo are preferable for small state famer, because refrigeration technology is not diffused in this area but these products are able to storage by natural drying.

Expectable income from vegetable garden was preliminarily estimated from expectable yield of vegetables and the wholesale price (farmer to broker) with a consideration of filling rate. The estimation showed that tomato, onion, zucchini, and carrot were well value and would be a good income source (Table 16.4). From the view of marketability and storage, tomato, onion, carrot, chili pepper, and okra are selected as a candidate vegetable production as an income source of agro-sanitation system.

Table 16.4 Vegetable price from farmer to broker in market

	Wholesale price (farmer to broker)			Expectable maximum yield			Expectable maximum income (FCFA/m^2)
	FCFA	Unit	Volume (L)	kg/m^2	Plant/ m^2	Filling volume (L/m^2)	
Tomato	3000	Small basket					
	3500	Middle basket					
	5000	Large basket					
	15,000–20,000	Tomato box	200	8.0		11.2	980
Onion	7500–10,000	50 kg bag	40	4.8		6.7	1470
Okra	4000	50 kg bag	40	2.0		3.5	350
Egg plant	2500–4000	50 kg bag	40	4.8		6.7	544
Chili pepper	5000–6000	50 kg bag	40	3.2		5.6	770
Sweet pepper	3000	50 kg bag	40	4.8		6.7	502
Zucchini	7000–12,500	50 kg bag	40	4.8		6.7	1638
Carrot	10,000	10 m^2			30		1000
Cabbage	50–100	1 number			5		375

16.4.2.3 Design of Crop Rotation

For a sustainable agricultural productivity and stable income, design of crop rotation is required. One crop rotation model was designed, considered with vegetable marketability, irrigation requirements, general crop rotation rule, and local crop cultivation style (Fig. 16.4). In the model, three rotations garden was proposed; one rotation was vegetables for sale, second was sorghum because of Na-phytoremediation and third was gumbo due to high demand of local people. In the vegetables rotation, better rotation was proposed as tomato or chili pepper from April to August, onion from August to December, and cucumber, zucchini, or carrot from January to March as a low irrigation vegetable. All sanitary products, which are treated greywater, urine, and compost, are applied in the vegetable rotation, since the rotation is expected as main income source. In the sorghum cultivation as a second rotation, it is assumed that sorghum is cultivated by local technique with rain-fed and no-sanitary products. During the cultivation, it is expected that accumulated sodium in soil is removed to sorghum shoot and leached

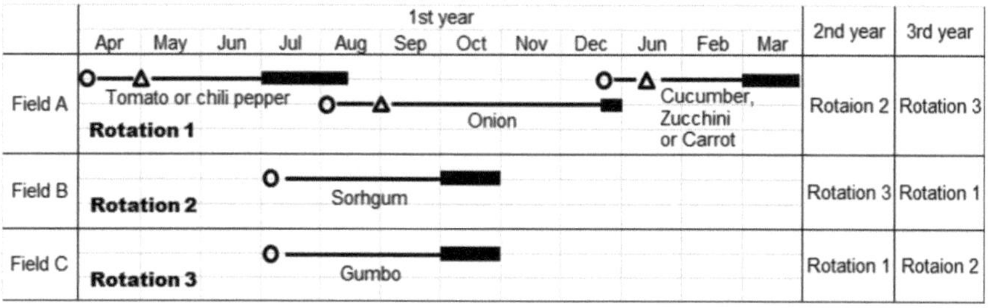

Fig. 16.4 One model of crop rotation in Ziniare

with rain (Qadir et al. 2005). In gumbo cultivation as a third rotation, it was assumed that gumbo was cultivated by local agricultural technique with rain-fed, urine, and compost. The gumbo has been cultivated as self- consumption in pilot-families, but also it was sold in local market. Therefore, the gumbo is expected as both self-consumption and income source.

16.4.2.4 Estimation of Available Gardening Size

A potential amount of greywater and irrigation requirement in the climate was assessed to estimate an available gardening size for greywater and urine reuse, following a logic described above (see Sect. 16.2).

Potential amount of greywater was estimated by a question-based survey with three pilot families (KD1, KD2, and BK1) who committed to install sanitary facilities and use as a pilot model (Table 16.5). The result showed that total water usage in the pilot families was 1120, 160, and 200 L/day/family, and the mean value of the usage per capita was 28 L/day/capita. Potential amount of greywater irrigation in the pilot family was estimated to 391, 85, and 185 L/day/family, respectively. The value was ranging 11–23 L/day/capita and averaging 17 L/day/capita.

Irrigation requirement for vegetables cultivation was calculated by FAO method (FAO 1986) with climate data of Ouagadougou station. The result, for instance, showed that tomato cultivation in June–September (rainy season) and October–

Table 16.5 Water usage and potential amount of irrigation in three pilot-families, Ziniare

	Family persons	Water use			Irrigation potential	
		Total (L/day/family)	Bathing (L/day/family)	Laundry (L/day/family)	(L/day/family)	(L/day/capita)
KD1	8	160	72	13	85	11
KD2	11	200	165	20	185	17
BK1	24	1120	336–672	36	391–707	23

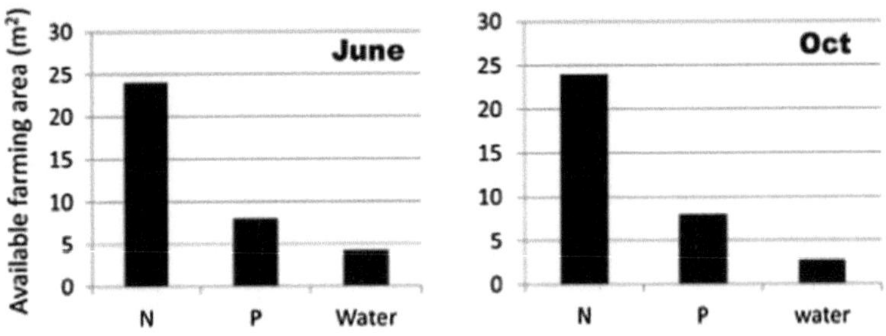

Fig. 16.5 Balance of nitrogen and phosphorus from urine and potential irrigation water originated from one person for four month, assumed as tomato 3-month cultivation from June and October

December (dry season) under the Ouagadougou climate required 4.0 and 6.1 L/day/m^2, respectively. If 17 L/day/capita of greywater was obtained in everyday, 4.3 m^2 (rainy season) and 2.8 m^2 (dry season) of gardening area would be available. N and P originated from general urine (Mullegger et al. 2010), which are contained an amount obtained from one person during the four month's cultivation period, covers 24 and 7.5 m^2, respectively (Fig. 16.5). These results indicated that greywater amount limits the available cultivation area in the pilot-garden under the climate.

Assumed as the designed crop rotation model (Sect. 16.4.2.3), available gardening areas in each pilot family was estimated in Table 16.6. The estimated area in BK1 was larger and approximately 100 m^2 with tomato in April–August and 50 m^2 with carrot in January–March. The area in KD1 was smaller and 20 m^2 with tomato in April–August and 10 m^2 with carrot in January–March.

16.4.2.5 Preliminary Income Estimation

A maximum annual income in the pilot family, which is considered proposed crop rotation model involved tomato, onion, zucchini, and okra cultivation, was also estimated with the expectable income from vegetables and available cultivation area in the pilot (Table 16.7). The result showed that BK1 potentially had 337,000

Table 16.6 Estimated available cultivation area in each pilot family with several vegetables in designed crop rotation

	Available cultivation area (m^2)				
	Tomato	Chili pepper	Onion	Carrot	Cucumber
	(April–August)	(April–August)	(August–December)	(January–March)	(January–March)
KD1	20.3	21.1	15.1	11.0	12.2
KD2	47.0	48.7	34.8	25.3	28.2
BK1	99.0	102.6	73.4	53.4	59.4

Table 16.7 Maximum income estimation per year in pilot family

	Expectable maximum income [FCFA][a]				Expectable annual income[b]			
	Tomato (April–August)	Onion (August–December)	Zucchini (January–March)	Okra (July–September)	Income/family		Income/capita	
					FCFA	EURO	FCFA	EURO
KD1	19,898	22,147	19,945	7107	69,097	106	8637	13
KD2	46,015	51,215	46,123	16,434	159,787	246	14,526	22
BK1	97,005	107,966	97,233	34,645	336,849	518	14,035	22

[a]Calculated with available farming area (Table 16.6) and estimated vegetable price (Table 16.4)
[b]Sum of expectable income of tomato, onion, zucchini, and okra

FCFA (518 Euro) incomes from 23 persons'. BK1 consists of 23 persons and 3 family groups. Thus, the maximum income in one family group was estimated to 170 Euro. Besides, KD1 and KD2 family might have 69,000 FCFA (106 Euro) and 160,000 FCFA (246 Euro). Therefore, shower room with greywater treatment system and urine collection system potentially gives 8–11 members' family 100–240 Euro as a maximum gain and it corresponded to 11–22 Euro/year/capita.

16.4.3 Installing of On-site Treatment Facilities

To adjust the expectable income level, simple and low-cost greywater treatment system, slanted soil system (Ushijima et al. 2013) were selected for the pilot. A urine diversion shower room which connects to the treatment system and urine collection tank was installed in the pilot site (Fig. 16.6a–c). The local people traditionally defecates urine in shower room. Vegetable garden (approximately 25–50 m^2) was also installed a side of the shower room in their house lawn (Fig. 16.6c, d). Then, quality of the urine and greywater, soil chemical properties in the gardens, and acceptability of the whole systems were monitored.

16.4.4 Monitoring of Pilot-Garden

16.4.4.1 Quality of Urine and Greywater in Pilot-Garden

Chemical properties of urine were shown in Table 16.8. N, P, and K in source-separated storage urine of 6 pilot families were 3.6 ± 0.7, 0.7 ± 0.3 and 0.6 ± 0.3 g/L, respectively. In the case of minor cations, only boron and iron was detected. Others found these chemical properties varied from 1.7–8.4 g-N/L, 0.3–0.8 g-P/L, 0.9–2 g-K/L, and 2.3–9.0 g-Na/L (Kirchmann and Pettersson 1995; Mnkeni et al. 2008; Pradhan et al. 2010; Mullegger et al. 2010). The present results were similar and most of values were inside of these ranges. Differences among

Fig. 16.6 Pictures of installed shower room connected with greywater treatment facility and house lawn garden (photo by Hijikata). **a** Outside of shower room connected with slanted soil treatment system, **b** inside of the shower room, **c** slanted soil system, collected urine and house lawn garden, **d** house lawn garden with urine fertilization and greywater irrigation

samples reflect family lifestyle, food culture, or differences in rainfall contribution. Similar to the urine, chemical properties of greywater also varied particularly in EC, SAR, COD, total N and Na (Table 16.9). The range of greywater parameters was also followed by previous report (Morel and Diener 2006; Al-Hamaiedeh and Bino 2010) excepting TN and microbial indicators. In the TN case, the higher value might be caused by urine contamination in greywater due to traditional behavior of pilot area, or caused by a lack of maintenance. However, relatively high bacteria were observed. This might be caused by lack of on-site maintenance. Therefore, simplified maintenance should be further investigated.

From the survey with pilot family, the total input of irrigation was estimated at 2.6–9.0 (mean 5.7 ± 2.7) L/m^2/day. From the date and chemical composition in urine and greywater, total mineral input from urine and greywater was speculated (Fig. 16.7). Major cations input from greywater were higher than that from urine. A large amount of Na was derived from greywater. Even the combined application

Table 16.8 Composition of urine in pilot site

	Min	Max	Mean	STDEV
pH	8.6	9.0	8.8	0.1
EC (mS/cm)	20.7	37.1	28.1	7.1
Total N (g/L)	2.4	4.4	3.6	0.7
Total P (g/L)	0.4	1.2	0.7	0.3
Na (g/L)	0.8	2.3	1.3	0.7
K (g/L)	0.4	1.2	0.6	0.3
Ca (mg/L)	2.0	26.3	10.9	9.1
Mg (mg/L)	2.0	32.8	10.1	11.5
B (mg/L)	0.5	2.0	1.1	0.6
Fe (mg/L)	n.d.	1.3	0.3	1.7
Zn (mg/L)	n.d.	n.d.	–	–
Mn (mg/L)	n.d.	n.d.	–	–
Cu (mg/L)	n.d.	n.d.	–	–
E. coli (cfu/ml)	n.d.	n.d.	–	–
Total coliform (cfu/ml)	n.d.	n.d.	–	–

Table 16.9 Composition of greywater in pilot site

	Min	Max	Mean	SD
pH	7.1	8.4	7.8	0.4
EC (μS/cm)	473	1523	849	281
COD (mg/L)	42	195	79	46
Total N (mg/L)	10	77	33	19
Total P (mg/L)	0.1	7.6	2.1	2.2
K (mg/L)	21.6	45.0	32.7	10.2
Ca (mg/L)	28.4	42.6	35.7	6.9
Mg (mg/L)	3.7	22.2	12.4	5.8
Na (mg/L)	23.4	106.2	61.2	23.6
SAR	0.7	3.0	1.8	0.7
E. coli (cfu/ml)	4.E+01	2.E+04	3.E+03	6.E+03
Total coliform (cfu/ml)	8.E+02	3.E+05	5.E+04	9.E+04

of urine and greywater, total phosphorus was generally insufficient for okra and eggplant. Sene et al. (2012) showed low phosphorus content of plant in urine reuse. Therefore, it was suggested that additional materials such as wood ash (Pradhan et al. 2010) or human feces compost from composting toilet (Guzha et al. 2005; Hijikata et al. 2015) would be beneficial for vegetable production.

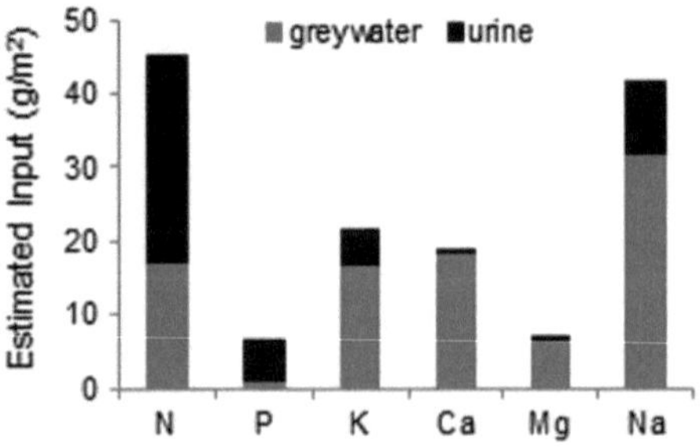

Fig. 16.7 Estimation of total input from urine and greywater in pilot site (reprinted from Hijikata et al. (2014a) with permission from the copyright holders, Japanese Association for Arid Land Studies)

16.4.4.2 Salt Monitoring in Soil of Pilot Garden

Soil EC in KD2 and BK1 site increased significantly from before cultivation to after cultivation (Fig. 16.8a). EC at other sites did not change significantly. High urine application elevates soil EC (Mnkeni et al. 2008; Sene et al. 2013b), but smaller urine applications did not alter soil EC more than chemical fertilizer (Pradhan et al. 2009, 2010). Furthermore, greywater irrigation also increases soil EC (Wiel-Shafran et al. 2006; Al-Hamaiedeh and Bino 2010). Therefore, the combination application of urine and greywater has the potential to increase soil EC value even during the rainy season.

Soil SAR in pilot-garden responded differently than soil EC with two sites (KD1 and KD2) observing a decrease in SAR over the course of the experiment (Fig. 16.8b). All other locations did not observe a change in SAR. Soil SAR has been shown to increase with large amount of urine (Sene et al. 2013b) and long-term greywater irrigation (Al-Hamaiedeh and Bino 2010) with less irrigation and rainfall than this experiment.

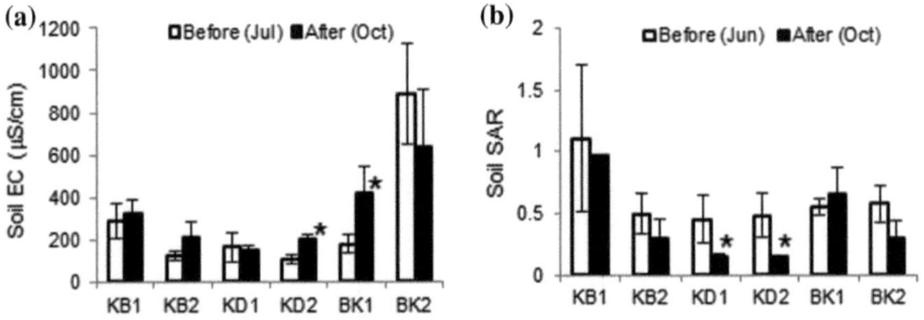

Fig. 16.8 Soil EC (**a**) and SAR (**b**) with urine fertilization and greywater irrigation in pilot site (reprinted from Hijikata et al. (2014a) with permission from the copyright holders, Japanese Association for Arid Land Studies)

16.4.4.3 Acceptability by Pilot-Families

From communication-based question survey, the pilot family seemed to feel an efficiency and availability of the reuse. Especially, they recognized a worth that they can cultivate and obtain vegetables even in dry season without rainfall (see Fig. 16.9). On the other hands, some of families failed the dry season cultivation due to lack of greywater irrigation (unbalance of cultivation area and amount of greywater) and feeding damage by their own livestock. This means, agronomic technical supports would be necessary for the beginning of the reuse. Furthermore, while cash vegetables were successfully cultivated, some families preferred the proposed crop rotation to continuous Okra cultivation since their favorite food in the area. This may be caused by lack of connection with cash vegetables market and/or their prioritization between direct income from cash vegetable and daily certisefaction from self-consumption vegetables. The behavior might indicate that a value for life of quality also become a good incentive for the reuse, not only direct income value.

Fig. 16.9 Pictures of house lawn garden with greywater irrigation (photo by Hijikata). **a** Greywater reuse enables gardening in dry season, **b** Okra cultivation with greywater irrigation in dry season, **c** greywater reuse hastens cultivation period, **d** chili pepper cultivation with greywater for selling

16.5 Conclusion

This chapter is described on-site use of reclaimed greywater on agricultural field such as household lawn garden, particularly the economic benefits, important parameters for safety and sustainable agriculture and a practical case study in rural area, sub-Sahel region. The reuse has economic advantages in dried area and lower-income area where limits public water service. An amount of greywater significantly influenced available gardening area and it would affect the economic benefit. While several target values of enteric microbial parameters for the safety reuse has been shown, a health risk of single household scale reuse is considered to be lower than neighborhood-scale collection reuse. Although several low-cost and simple technologies for greywater reclamation have been successfully developed and enable to reduce some important parameters such as suspended solids, total nitrogen, oil, grease, surfactants, and enteric microbial parameters, it may be difficult for these technologies to reclaim factors affecting alkalinity, salinity, and sodicity in soil. Therefore, soil and tillage management would be helpful options for sustainable and stable agricultural benefits. To maximize the reuse benefits, design of crop rotation with a consideration of agricultural value chains and food culture in each local is worth to be investigated. This may positively contribute a local acceptability for the reuse.

While potential impacts of greywater on plant, soil, and health risks are mentioned in the chapter, it is recommended that these reclaiming treatments and cares would be taken from a view of cost-effectiveness and labor-effectiveness in each local condition with a long-term speculation.

References

Al-Hamaiedeh H, Bino M (2010) Effect of treated grey water reuse in irrigation on soil and plants. Desalination 256:115–119. https://doi.org/10.1016/j.desal.2010.02.004

Al-Jayyousi OR (2003) Greywater reuse: towards sustainable water management. Desalination 156:181–192. https://doi.org/10.1016/S0011-9164(03)00340-0

Brandt KK, Pedersen A, Sorensen J (2002) Solid-phase contact assay that uses a lux-marked *Nitrosomonas europaea* reporter strain to estimate toxicity of bioavailable linear alkylbenzene sulfonate in soil. Appl Environ Microbiol 68:3502–3508. https://doi.org/10.1128/AEM.68.7.3502-3508.2002

Budenheim D, Wignarajah K, Berry W, Wydeven T (1997) Phytotoxic effect of grey water due to surfactants. J Am Soc Hortic Sci 122:792–796

Christova-Boal D, Eden RE, Macfarlane S (1996) An investigation into greywater reuse for urban residential properties. Desalination 106:391–397. https://doi.org/10.1016/0011-9164(96)00134-8

Cirelli AF, Ojeda C, Castro MJJ, Salgot M (2008) Surfactants in sludge-amended agricultural soils: a review. Environ Chem Lett 6:136–148. https://doi.org/10.1007/s10311-008-0146-1

Crites R, Tchobanoglous G (1998) Small and decentralized wastewater management systems. McGraw-Hill, New York

Dallas S, Scheffe B, Ho G (2004) Reedbeds for greywater treatment—case study in Santa Elena-Monteverde, Costa Rica. Cent Am Ecol Eng 23(1):55–61. https://doi.org/10.1016/j.ecoleng.2004.07.002

DHWA (2002) Draft guidelines for reuse of greywater in Western Australia. Department of Health, Perth Australia 37

Diaper C, Dixon A, Bulier D, Fewkes A, Parsons SA Strathern M, Stephenson T, Stritt J (2001) Small scale water recycling systems—risk assessment and modelling. Water Sci Technol 43 (10):83–90

Ding WH, Lo JH, Tzing SH (1998) Determination of linear alkylbenzenesulfonates and their degradation products in water samples by gas chromatography with ion-trap mass spectrometry. J Chromatogr A 818:270–279. https://doi.org/10.1016/S0021-9673(98)00550-0

Dixon AM, Butler D, Fewkes A (1999) Guidelines for greywater re-use: health issues. Water Environ J 13(5):322–326. https://doi.org/10.1111/j.1747-6593.1999.tb01056.x

Elsgaard L, Petersen SO, Debosz K (2001) Effect and risk assessment of linear alkylbenzene sulfonates in agricultural soil. 1. Short-therm effects on soil microbiology. Environ Toxicol Chem 20:1656–1663. https://doi.org/10.1002/etc.5620200806

Eriksson E, Auffarth K, Hnze M, Ledin A (2002) Characteristics of grey wastewater. Urban Water 4(1):85–104. https://doi.org/10.1016/s1462-0758(01)00064-4

FAO (1985) Water quality for agriculture. Irrigation and drainage paper 29 Rev. 1. Food and Agriculture Organization of the United Nation, Rome 174

FAO (1986) Irrigation water management. Training manual no. 3

Faruqui N, Al-Jayyousi O (2002) Greywater reuse in urban agriculture for poverty alleviation—a case study in Jordan. Water Int 27(3):387–394

Friedler E (2004) Quality of individual domestic greywater streams and its implication for on-site treatment and reuse possibilities. Environ Technol 25(9):997–1008. https://doi.org/10.1080/09593330.2004.9619393

Garland JL, Levine LH, Yorio NC, Adams JL, Cook KL (2000) Greywater processing in recirculating hydroponic system: phytotoxicity, surfactant degradation, and bacterial dynamics. Water Res 34:3075–3086. https://doi.org/10.1016/S0043-1354(00)00085-3

Garland JL, Levine LH, Yorio NC, Hummerick ME (2004) Response of greywater recycling systems based on hydroponic plant growth to three classes of surfactants. Water Res 38:1952–1962. https://doi.org/10.1016/j.watres.2004.01.005

Gilboa Y, Friedler E (2008) UV disinfection of RBC-treated light greywater effluent: kinetics, survival and regrowth of selected microorganisms. Water Res 42(4–5):1043–1050. https://doi.org/10.1016/j.watres.2007.09.027

Gross A, Azulai N, Oron N, Ronen Z, Arnold M, Nejidat A (2005) Environmental impact and health risks associated with greywater irrigation: a case study. Water Sci Technol 52:161–169

Gross A, Shmueli O, Ronen Z, Raveh E (2007a) Recycled vertical flow constructed wetland (RVFCW)—a novel method of recycling greywater for irrigation in small communities and households. Chemosphere 65(5):916–923. https://doi.org/10.1016/j.chemosphere.2006.06.006

Gross A, Kaplan D, Baker K (2007b) Removal of chemical and microbiological contaminants from domestic greywater using a recycled vertical flow bioreactor (RVFB). Ecol Eng 31:107–114. https://doi.org/10.1016/j.ecoleng.2007.06.006

Guzha E, Nhapi I, Rockstrom J (2005) An assessment of the effect of human feces and urine on maize production and water productivity. Phys Chem Earth 30:840–845. https://doi.org/10.1016/j.pce.2005.08.028

Haas CN, Rose JB, Gerba CP (1999) Quantitative microbial risk assessment. Wiley, New York

Haruta S, Nakaya T (2011) Present condition and evolvability of reclaimed wastewater application on paddy field water. Land Environ Eng 79(11):813–816

Hijikata N, Nakaoka E, Ushijima K, Funamizu N (2011) Phytotoxicity assay of several grey waters for reuse as agricultural irrigation. In: Proceeding of IWA Xi'an conference, Xi'an

Hijikata N, Ushijima K, Ito R, Sawadogo B, Funamizu N (2012) Design of vegetable garden for income estimation toward to agro-sanitation business model, a case study in Burkina Faso. In: Proceedings of dry toilet conference, Tampere, Finland

Hijikata N, Fujii T, Sangare D, Sou M, Ushijima K, Funamiuz N (2014a) Salts monitoring and management for human urine fertilization and treated greywater irrigation in sub-Sahel region. J Arid Land Study 24(1):85–88

Hijikata N, Nakaoka E, Sawadogo B, Ochoa EC, Ushijima K, Funamizu N (2014b) Evaluation of laundry detergent irrigation on continuous vegetable cultivation for stable domestic greywater reuse. Agricultural water and management (submitted)

Hijikata N, Yamauchi N, Ishiguro M, Ushijima K, Funamizu N (2015) Suitability and availability of biochar as a matrix for improving performance of composting toilets. Waste Manage Res 33 (4):313–321. https://doi.org/10.1177/0734242X15572179

Huelgas A, Funamizu N (2010) Flat-plate submerged membrane bioreactor for the treatment of higher-load greywater. Desalination 250:162–166. https://doi.org/10.1016/j.desal.2009.05.007

Jensen J, Folker-Hansen P (1995) Soil quality criteria for selected organic compounds. Working report no. 47, Danish Environmental Protection Agency, Denmark

Jensen J, Løkke H, Holmstrup M, Krogh PH, Elsgaad L (2001) Effects and risk assessment of liner alkylbenzene sulfonates in agricultural soil. 5. Probabilistic risk assessment of liner alkylbenzene sulfonates in sludge-amended soils. Environ Toxicol Chem 20:1690–1697. https://doi.org/10.1002/etc.5620200810

Jensen J, Smith SR, Krogh PH, Versteeg DJ, Temara A (2007) European risk assessment of LAS in agricultural soil revisited: species sensitivity distribution and risk estimates. Chemosphere 69:880–892. https://doi.org/10.1016/j.chemosphere.2007.07.029

Kirchmann H, Pettersson S (1995) Human urine—chemical composition and fertilizer use efficiency. Fert Res 40:149–154. https://doi.org/10.1007/BF00750100

Lopez Zavala MA, Funamizu N, Takakuwa T (2002) Onsite wastewater differentiable treatment system: modelling approach. Water Sci Technol 46:317–324

Maimon A, Tal A, Friedler E, Gross A (2010) Safe on-site reuse of greywater for irrigation—a critical review of current guidelines. Environ Sci Technol 44:3213–3220. https://doi.org/10.1021/es902646g

Martin C (2005) Ecological sanitation greywater demonstration project at Hui Sing garden, Kuching, Sarawak, Malaysia, Urban Environmental Management system (UEMS). Project Natural Resources and Environment Board (NREB), Sarawak

Misra RK, Sivongxay A (2009) Reuse of laundry greywater as affected by its interaction with saturated soil. J Hydrol 366:55–61. https://doi.org/10.1016/j.jhydrol.2008.12.010

Misra RK, Patel JH, Baxi VR (2010) Reuse potential of laundry greywater for irrigation based on growth, water and nutrient use of tomato. J Hydrol 386:95–102. https://doi.org/10.1016/j.jhydrol.2010.03.010

Mnkeni PNS, Kutu FR, Muchaonyerwa P (2008) Evaluation of human urine as a source of nutrients for selected vegetables and maize under tunnel house conditions in the Eastern Cape, South Africa. Waste Manage Res 26:132–139. https://doi.org/10.1177/0734242X07079179

Morel A, Diener S (2006) Greywater management in low and middle-income countries, review of different treatment systems for housholds or neighbourhoods. Swiss Federal Institute of Aquatic Science and Technology (Eawag), Dubendorf, Swizerland

Mortensen GK, Egsgaard H, Ambus P, Jensen ES, Gron C (2001) Influence of plant growth on degradation of linear alkylbenzene sulfonate in sludge-amended soil. J Environ Qual 30:1266–1270. https://doi.org/10.2134/jeq2001.3041266x

Mubashir M, Malik SA, Khan AA, Ansari TM, Wright S, Brown MV, Islam KR (2010) Growth, yield and nitrate accumulation of irrigated carrot and okra in response to nitrogen fertilization Pak. J Bot 42(4):2513–2521

Mullegger E, LAngergraber G, Lechner M (2010) Sustainable sanitation practice, use of urine, EcoSan club 3

Ottoson J, Stenstrom TA (2003) Fecal contamination of greywater and associated microbial risks. Water Res 37(3):645–655. https://doi.org/10.1016/S0043-1354(02)00352-4
Pinto U, Mahesgwari BL, Grewal HS (2010) Effect of greywater irrigation on plant growth, water use and soil properties. Resour Conserv Recycl 54:429–435. https://doi.org/10.1016/j.resconrec.2009.09.007
Pradhan SK, Pitkänen S, Heinonen-Tanski H (2009) Fertilizer value of urine in pumpkin (*Cucurbita maxima* L.) cultivation. Agri Food Sci 18:57–68
Pradhan SK, Holopainen JK, Weisell J, Heinonen-Tanski H (2010) Human urine and wood ash as plant nutrients for red beet (*Beta vulgaris*) cultivation: impacts on yield quality. J Agric Food Chem 58:2034–2039. https://doi.org/10.1021/jf9029157
Qadir M, Noble AD, Oster JD, Schubert S, Ghafoor A (2005) Driving forces for sodium removal during phytoremediation of calcareous sodic and saline–sodic soils: a review. Soil Use Manage 21:173–180. https://doi.org/10.1111/j.1475-2743.2005.tb00122.x
Qian YL, Mecham B (2005) Long-term effects of recycled wastewater irrigation on soil chemical properties on golf course fairways. Agron J 97:717–721. https://doi.org/10.2134/agronj2004.0140
Sanchez-Peinado MM, Rodelas B, Martinez-Toledo MV, Gonzalez-Lopez J, Pozo C (2009) Response of soil enzyme to liner Alkylbenzene Sulfonate (LAS) addition in soil microcosms. Soil Biol Biochem 41:69–76. https://doi.org/10.1016/j.soilbio.2008.09.019
Schachtman DP, Reid RJ, Ayling SM (1998) Phosphorus uptake by plants: from soil to cell. Plant Physiol 116:447–453. https://doi.org/10.1104/pp.116.2.447
Sene M, Hijikata N, Ushijima K, Funamizu N (2012) Adequate human urine application pattern for agriculture. Int Res J Agric Sci Soil Sci 2(1):038–045
Sene M, Hijikata N, Ushijima K, Funamizu N (2013a) Effects of continuous application of extra human urine volume on plant and soil Int. J Agri Sci Res (IJASR) 3(3):75–90
Sene M, Hijikata N, Ushijima K, Funamizu N (2013b) Effects of extra human urine volume application in plant and soil. Int Res J Agric Sci Soil Sci 3(6):182–191
Shuval H, Lampert Y, Fattal B (1997) Development of a risk assessment approach for evaluating wastewater reuse standards for agriculture. Water Sci Technol 35:15–20. https://doi.org/10.1016/S0273-1223(97)00228-x
Travis MJ, Weisbrod N, Gross A (2008) Accumulation of oil and grease in soils irrigated with greywater and their potential role in soil water repellency. Sci Total Environ 394:68–74. https://doi.org/10.1016/j.scitotenv.2008.01.004
Travis MJ, Weil-Shafran A, Weisbrod N, Adar E, Gross A (2010) Greywater reuse for irrigation: effect on soil properties. Sci Total Environ 408:2501–2508. https://doi.org/10.1016/j.scitotenv.2010.03.005
Travis MJ, Weisbrod N, Gross A (2012) Decentralized wetland-based treatment of oil rich farm wastewater for reuse in an arid environment. Ecol Eng 39:81–89. https://doi.org/10.1016/j.ecoleng.2011.11.008
Tuna LA, Kaya C, Ashraf M, Altunlu H, Yokas I, Yagmur B (2007) The effects of calcium sulphate on growth, membrane stability and nutrient uptake of tomato plants grown under salts stress. Environ Exp Bot 59:173–178. https://doi.org/10.1016/j.envexpbot.2005.12.007
Turan M, Sevimlib F (2005) Influence of different nitrogen sources and levels on ion content of cabbage (*Brassica oleracea* var. capitate) New Zealand. J Crop Hortic Sci 33(3):241–249. https://doi.org/10.1080/01140671.2005.9514356
US EPA (2004) Guidelines for water reuse. U.S. Environmental Protection Agency, Washington DC
Ushijima K, Hijikata N, Ito R, Funamizu N (2012) Effect estimation of dry-toilet application for rural farmer family in Burkina Faso. J Arid Land Stud 22(1):99–102
Ushijima K, Ito K, Ito R, Funamizu N (2013) Greywater treatment by slanted soil system. Ecol Eng 50:62–68. https://doi.org/10.1016/j.ecoleng.2012.04.028
Warne MSJ, Schifko AD (1999) Toxicity of laundry detergent components to a freshwater Cladoceran and their contribution to detergent toxicity. Ecotoxicol Environ Safety 44:196–206. https://doi.org/10.1006/eesa.1999.1824

WHO (2006) Guidelines for the safe use of wastewater, excreta and greywater. In: WHO Ed
WHO and UNICEF (2010) Progress on sanitation and drinking-water 2010 update
Wiel-Shafran A, Ronen Z, Weisbrod N, Adar E, Gross A (2006) Potential change in soil properties following irrigation with surfactant rich greywater. Ecol Eng 26:348–354. https://doi.org/10.1016/j.ecoleng.2005.12.008
Winblad U, Simpson-Hebert M (2004) Ecological sanitation, revised and enlarged edn. SEI, Stockholm
Winward GP, Avery LM, Franzer-Williams R, Pidou M, Jeffery P, Stephenson T, Jefferson B (2008) A study of the microbial quality of greywater and an evaluation of treatment technologies for reuse. Ecol Eng 32(2):187–197. https://doi.org/10.1016/j.ecoleng.2007.11.001
Ying GG (2006) Fate, behavior and effects of surfactants and their degradation products in the environment. Environ Int 32:417–431. https://doi.org/10.1016/j.envint.2005.07.004

Part VI
Cases

Chapter 17
Experiment of a Resource-Oriented Agro-sanitation System in Sahelian Rural Community: Case of Burkina Faso

Mariam Sou/Dakoure, Ynoussa Maïga, Amadou Hama Maïga, Maïmouna Traore Bologo and Seyram Sossou

Abstract This case study presents partial results of a research project in agro-sanitation. The study took place in rural area of Burkina Faso, a typical Sahelian country in West Africa. The objectives were (1) to design and implement an agro-sanitation system and (2) assess the willingness of users to reuse the sanitation by-products in small-scale subsistence farming. The sanitation system consists of collecting onsite greywater, urine, and feces which after separated treatment are reused as by-products. The greywater is treated by a slanted soil treatment system. Feces are converted in compost inside a composting toilet and urine treated by solar inactivation. General performances of the sanitation system are in progress at middle stage of the research project. However, the system is already a success in two points: (1) design simple and robust facilities adapted in a context of water shortage and not electricity; (2) better understand the social and economic criteria associated to the acceptability of users.

Keywords Agro-sanitation · Composting toilet · Slanted soil treatment system Urine · Small farming

17.1 Introduction

In sub-Saharan Africa, open defecation has decreased by 11% from 1990 to 2010, but due to population growth, absolute number increased to 33 million in the same period. The practice is more important in rural population (35%) than urban ones (8%) (Galan et al. 2013). In typical Sahelian countries like Burkina Faso,

---

M. Sou/Dakoure (✉) · Y. Maïga · A. H. Maïga · M. T. Bologo · S. Sossou
International Institute for Water and Environmental Engineering (2iE), Ouagadougou, Burkina Faso
e-mail: mariam.sou@2ie-edu.org

N. Funamizu (ed.), *Resource-Oriented Agro-sanitation Systems*,
https://doi.org/10.1007/978-4-431-56835-3_17

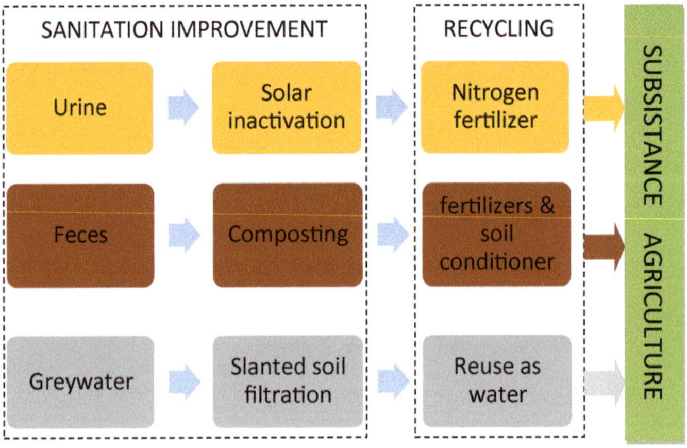

Fig. 17.1 Agro sanitation system

open defecation is still practiced by more than 70% of the population (ENA 2010). The related waterborne diseases mainly reach vulnerable part of population: children under 5 years old and women, mainly in rural areas. At the same time, this vulnerable population also suffers from malnutrition due to several reasons among which water shortage and lack of fertilizer to ensure their own subsistence agriculture. The purpose of Ameli-EAUR project is to improve sanitation condition while providing water and fertilizer to sustain subsistence agriculture to most vulnerable population in Sahelian countries. It is based on agro-sanitation concept, which consists of proposing an integrated model of sanitation with not only organic fertilizer (i.e., compost and urine) but also treated greywater (Fig. 17.1). Another innovation is the development of local farmers own capacity of treating and reusing themselves their own by-products (compost, urine, and greywater). Such system improves farmers' independence concerning by-products provision, contributing then to project sustainability.

The project was divided into four main components: (1) design and test robustness of composting toilet; (2) design and test robustness of greywater treatment plant; (3) collect, treat urine, and reuse total by-products (i.e., compost, treated urine, and treated greywater) in agriculture in respect of environmental and sanitary conditions (4) assess farmers acceptability to reuse sanitary by-products.

17.2 Experimental Sites

The study was conducted in the central part of Burkina Faso. A West African country dominated by the Sudano Sahelian climate. Years are regulated by two seasons: a short rainy one from June to September (4 months) with annual rainfall varying from 600 to 900 mm, and a long dry one from October to May (8 months).

One experimental site was identified in the semi-urban area of Ouagadougou (namely Kamboinsin) in the central region (Fig. 17.2a) and two experimental sites in rural area of the plateau central region around the city of Ziniaré (namely Kologoudiessé and Barkoundba) (Fig. 17.2b).

The two villages were selected based on the ethnic and religious group membership and the activities carried out by the inhabitants (Maiga et al. 2014). Indeed, in the studied region, "Mossi" and "Peul" are the main ethnic groups. More generally, "Mossi" is the main ethnic group in Burkina Faso (48.6%) followed by "Peul" ethnic group (7.8%) (Ministere Des Affaires Etrangeres et de la Cooperation Regionale 2011). The population of "Barkoundba" belongs to the "Peul" ethnic group with Islam as main religion and cattle farming as the main livelihood activity, while the inhabitants of "Kologoudiessé" are members of the "Mossi" ethnic group, Christians in general and with crop farming as the main activity.

Several pilot households were identified in the different experimental sites. Total numbers are indicated in Table 17.1 with reference. Each reference gives 3 types of information separated by an underscore 1_2_3: 1 = Rural versus Semi-urban (SemiUrb) site; 2 = first 2 letters of the village name: ko = Kologoudiessé, ba = Barkoundba and ka = Kamboinsin; 3 = first letter of family name and first name of the chief of each pilot household.

The pilot households were chosen by paying particular attention to families who already practice vegetable gardening around their house.

As illustrated in red color in Fig. 17.3, there are some places more or less important around families' houses. These places are used both for rain-fed agriculture (i.e., 3–4 months/year) and small gardening during dry season. However, the limiting factor for gardening activity in dry season is water, since they are not any source of surface or groundwater permanently available.

17.3 Greywater Treatment

The present chapter focuses on results coming from two households: *Rural_ko_IA* and *Rural_ko_IE*. A previous study has shown that greywater is generated from three main sources in this village: laundry, shower, and dishwashing activities with shower greywater being the most produced in a given household (up to 70%) (Maiga et al. 2014). In addition, greywater is poured directly on the ground causing smelly stagnant water (Fig. 17.4). The same study estimated the per capita daily greywater production to 13 L/capita/day. In order to solve this unpleasant situation and allow the collection of all greywater sources produced, a shower room has been designed and constructed. This shower room is designed to allow a separate collection of greywater to avoid mixture of urine. A slanted soil treatment system (SSTS) is designed and constructed to treat the greywater for reuse purposes.

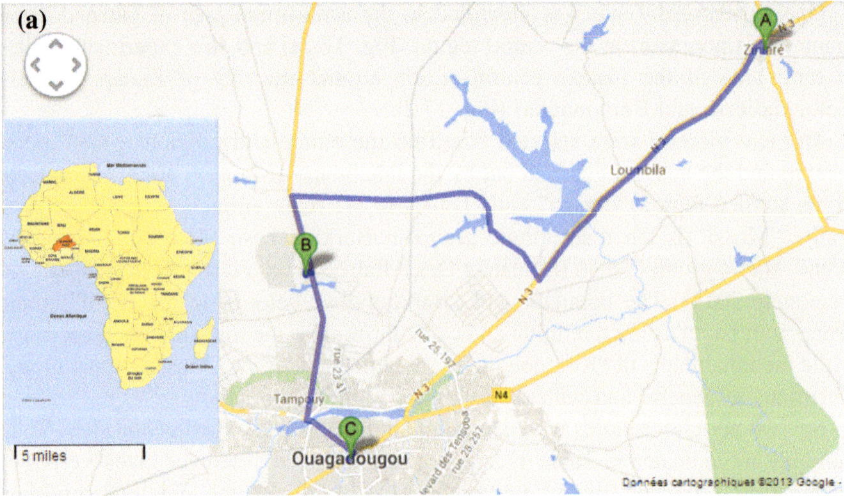

A: city of Ziniaré; B: semi-urban pilot sites in Kamboinsin; C: Ouagadougou, capital of Burkina Faso

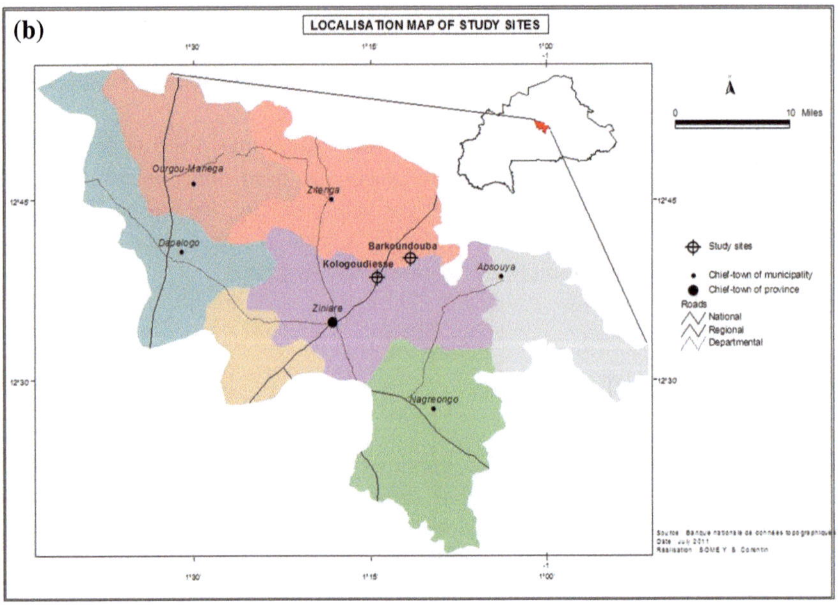

Fig. 17.2 Localization map of the study site. **a** Semi_urban site and the city of Ziniaré; **b** rural sites close to Ziniaré

Table 17.1 Pilot households

	Household reference	Number of person	Sanitation facilities before the project	Additional sanitation facilities
Rural area				
Kologodjesse	Rural_ko_CE	6	1 traditional toilet/1 BSS	1 composting toilet/1 ISR
	Rural_ko_IE	13	O.D/BSS	1 composting toilet/1 ISR
	Rural_ko_IA	4	O.D/BSS	1 ISR
	Rural_ko_IL	7	O.D/BSS	1 composting toilet/1 ISR
Barkoundba	Rural_ba_DA	20	1 traditional toilet/1 BSS	2 composting toilet/1 ISR
	Rural_ba_DR	8	1 traditional toilet (also used as BSS)	1 composting toilet /1 ISR
Semi-urban area				
Kamboinsin	SemiUrb_ka_SB	9	1 traditional toilet/1 ISR	1 composting toilet/1 ISR
	SemiUrb_ka_ZE	7	1 traditional toilet/1 BSS	1 composting toilet/1 ISR

O.D Open Defecation; *BSS* Basic Shower Space; *ISR* Improved Shower Room

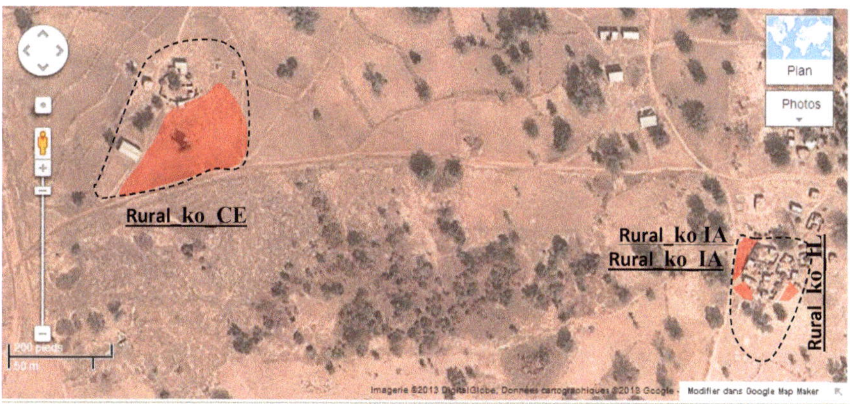

Fig. 17.3 Pilot family spaces in rural area (case of Kologoudiessé)

17.3.1 Developing the Treatment System

Keeping in mind the willingness to collect, treat, and reuse the effluents for gardening, the following points have been considered when developing the treatment system (Maiga et al. 2013):

Fig. 17.4 Current situation of greywater disposal in rural area showing smelly stagnant water (photo by Ynoussa Maïga)

- The treatment system should be able to collect all greywater sources especially shower greywater which contributes up to 70% of the total greywater produced in the household;
- The effluent should be stored in a storage tank for daily watering of the home garden;
- The treated greywater should be safe to be used for gardening;
- The treatment unit should be affordable for local population.

17.3.2 Complex Shower Room—Slanted Soil System

The shower room is composed of two parts: a shower space from which the greywater is collected to the SSTS and a urinary space containing a urinary bowl which is devoted to the collection of urine (Fig. 17.5a). Urine is collected and stored in a tank outside the shower room.

The SSTS (Fig. 17.5b) is composed of three (3) main parts:

- The collection point is composed of two (2) compartments: the first one is a polyethylene tank which is devoted to collect laundry and dishwashing greywater. The second compartment is a conduit that carries the shower greywater from the shower to the filtration area.
- The filtration area is composed of two (2) boxes of 3 m total length: the top box made of plastic has a length of 1 m. The second box built of concrete has a length of 2 m and a width of 20 cm. The boxes are leant of 5% slope to allow greywater drainage. Both boxes are filled with granite as a filter material to a depth of 10 cm;
- The treated greywater is stored in a storage tank of 200 L made of plastic and buried in the soil.

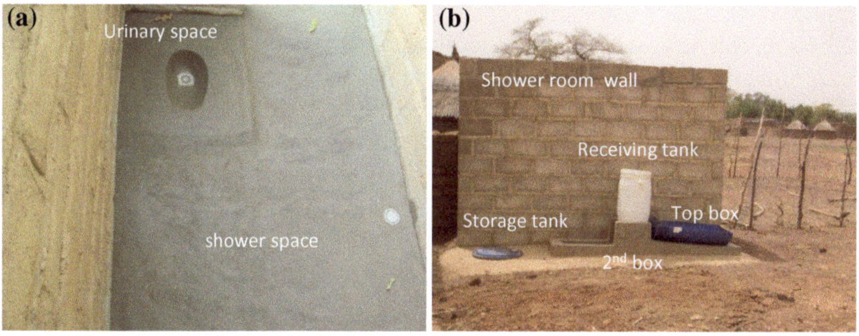

Fig. 17.5 Internal space of the shower room and SSTS connected to the shower room (photo by Ynoussa Maïga)

Hence, the SSTS is designed to collect shower greywater and mixture of laundry and dishwashing greywater. The system receives greywater depending on the activities of the households (morning, noon and night).

17.3.3 Performance

The raw greywater is slightly neutral to alkaline with pH varying from 6.3 to 8.09 for all greywater collected in household ko_IE. In household ko_IA, pH for raw greywater is varying for 7.2–9.38. The treated greywater exhibited pH ranging from 6.3 to 7.9 in household ko_IE and from 7 to 8.22 in household ko_IA.

The raw greywater is highly polluted with bacterial indicators contrary to common consideration. Indeed, the average concentrations of *Escherichia coli* (*E. coli*) in the influent shower greywater are of 1.47×10^7 CFU/100 mL in household ko_IE and. 2.5×10^7 CFU/100 mL in household ko_IA. For mixture of laundry and dishwashing, greywater *E. coli* average concentrations in the influent greywater were 1.62×10^6 and 2.76×10^7 CFU/100 mL in household ko_IE and ko_IA, respectively.

The efficiency of the SSTS in term of organic matter removal and *E. coli* removal is also summarized in Table 17.2. Household ko_IE exhibited the highest Chemical Oxygen Demand (COD) and 5-days Biological Oxygen Demand (BOD_5)

Table 17.2 Performance of the greywater treatment system

	Rural_ko_IE	Rural_ko_IA
Size (persons)	13	4
SS (%)	63	66
COD (%)	54	43
BOD5 (%)	55	34
E. coli (log.u CFU/100 mL)	0.76	1.79

removal compared to household ko_IA. In contrast, household ko_IA exhibited the highest *E. coli* removal. The SSTS designed is adapted to collect greywater produced from the three (3) main sources in rural area but the efficiency of the system should be increased. Further studies should consider grain size distribution and filter media type to increase the efficiency. The grain size should be determined keeping in mind the maintenance issue since lower particle size conducts to easy clogging.

17.3.4 Training of Users

After the construction of the complex shower room SSTS, each household members have been trained on how to use the technology since separate collection of urine and greywater is necessary. In addition, collection and transport of mix laundry–dishwashing greywater to the receiving tank of the SSTS is necessary.

17.4 Urine Treatment

There are at least seven types of urine collection and treatment processes (Pronk and Koné 2009; Maurer et al. 2006) but as indicated by authors, they are limited to one purpose such as phosphorus or nitrogen recovery and major part of these treatment processes are still at laboratory stage. At field stage, two processes can be considered: evaporation and storage. Evaporation process consists of solidifying urine by solar radiation. Urine is put in a still with a cover glass and evaporation is done by combined effect of convection and greenhouse effect. The process mainly focuses on nitrogen and phosphorus recovery and need acid addition to increase recovery rate (Antonini et al. 2012). Storage is actually the most widespread process. It consists of keeping urine in plastic tanks during several months to inactivate pathogen microorganisms. Recognized as efficient to remove major part of pathogens, its major constraint is however the long time and the large storage capacity required. Knowing that plastic tanks remain expensive item for poor rural population even to store drinking water, it is highly unlikely that those people could pay required numbers of tanks to store urine. Current literature recommendation for storage time varies from 1 to 6 months between 4 and 20 °C (Winblad et al. 2004). However, in sub-Saharan countries annual mean temperature is higher than 20 °C meaning that recommended storage time could be reduced in these countries. To allow this reduction, solar disinfection (SODIS) treatment, previously used for drinking water treatment could be adapted to treat urine. SODIS consists of filling water in transparent plastic bottle and exposes it to the sun during 6–48 h (McGuigan et al. 2012). This treatment is known to inactivate several species of bacteria, fungi as well as ova and cysts of helminth and protozoa by combined effect of heating and UV light radiation. Exposition time depends on temperature, turbidity, and radiation intensity for drinking water (McGuigan et al. 2012;

Gomez-Couso et al. 2009) and applying to urine disinfection, pH raising can be also taking into account as improving factor of microorganism inactivation since it already plays important role in "simple storage" process.

17.4.1 Laboratory Experiment

To test SODIS efficiency for urine disinfection, a laboratory and field-scale experiment were performed. The laboratory experiment consists of testing temperature effect on urine disinfection. To do this, sterilized urine was spiked with *E. coli* to obtain concentration of about 6 $\log_{10}$ unit CFU/100 mL. This urine was tested in oven at 45 and 50 °C during 6 h. For each hour, sample was taken to follow *E. coli* decay. Figure 17.6 shows that at 50 °C total disinfection occurs after 6 h whereas 50% disinfection occurs for the same time at 45 °C.

Before testing SODIS effect on urine in field condition, temperature was followed in polyethylene (PET) bottle filled with urine and exposed on roof from 8.00 am to 6.00 pm. This preliminary test shown in Fig. 17.7 indicates that best inactivation time is between 11.00 am to 5.00 pm. During this time (6 h) temperature is between 45 and 50 °C.

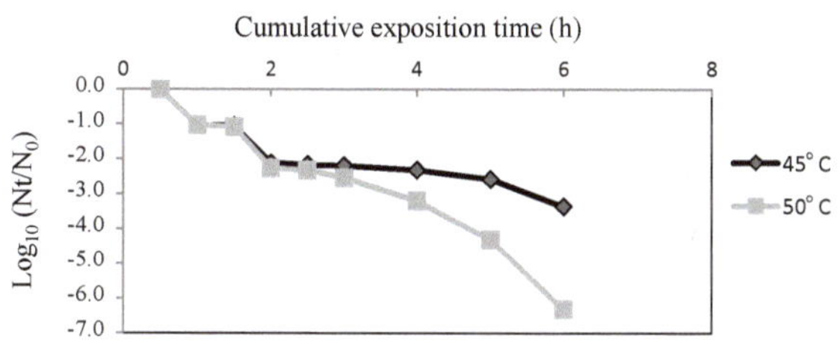

Fig. 17.6 Temperature effect on *E. coli* inactivation in urine

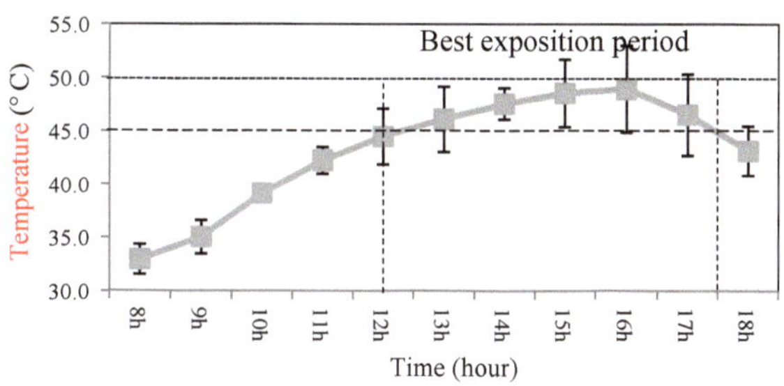

Fig. 17.7 Maximum temperature reached in PET bottle at field condition

17.4.2 Field Tests and Implementation in Pilot Families

Field tests consisted of exposing 26 PET bottles filled with urine during 6 h from 11.00 to 7.00 pm. Results presented in Fig. 17.8 show that 77% of bottles have at least −5 $\log_{10}$ CFU unit reduction after 6 h of exposition. Based on these results, recommendations to household families was exposition of PET urine bottles on roof during three consecutive days.

To implement urine disinfection by SODIS, each pilot household received four small tanks of 5 L, (Fig. 17.9a) 1 bigger tank of 20 L equipped with sluice gate (Fig. 17.9b) and at least 20 PET bottles of 1.5 L (Fig. 17.9c). The small tanks are used to collect urine from urinals installed in shower rooms and from source separated toilet. When small tanks are filled, the urine is diverted in the bigger tank to allow an easy transfer in PET bottles. Pilot household members received instruction during a training to fill bottles with urine and expose them at the toilet roof during three consecutive days. After that bottle must be stocked in dry clean area until the using as fertilizer.

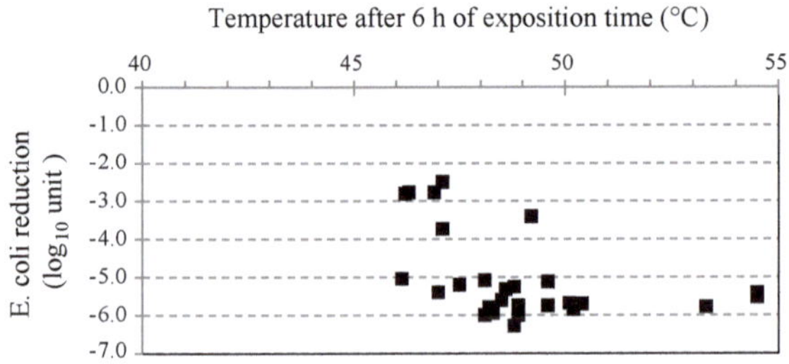

Fig. 17.8 SODIS performance on urine disinfection

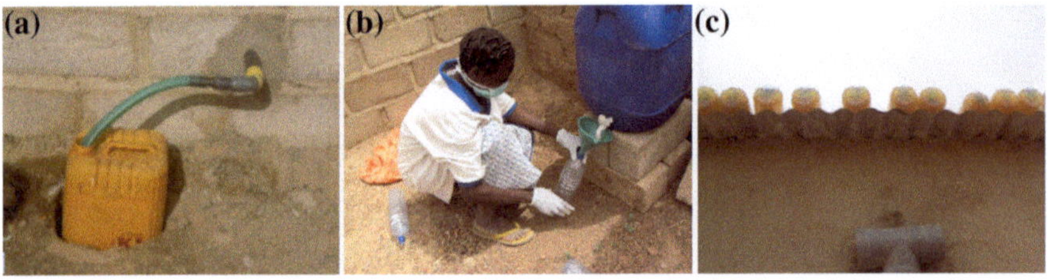

Fig. 17.9 Urine treatment by SODIS method (photo by Mariam Sou/Dakouré)

17.5 Feces Treatment

Composting is an inexpensive sustainable method to convert mainly biodegradable organic matter to carbon dioxide and ammonium under aerobic condition (Lopez Zavala et al. 2004; Hotta and Funamizu 2007), so it represents a cheap and efficient method to treat human waste materials (i.e., feces) through composting use of toilet. Composting process can be performed by using several type of matrixes (as carbon bulky agent). Mixing of such matrix with feces allows bioconversion of human excreta into compost which can be used both as organic fertilizer rich in N, P, and K, and soil conditioner (Kitsui and Terazawa 1999; Del Porto and Steinfeld 2000). The use of composting toilet is becoming increasingly popular in the world because it doesn't require water or drainage compared to a flush toilet. Such advantage is particularly interesting in water shortage countries since it contributes to preserve water resources (Winblad et al. 2004). In Burkina Faso, composting toilet can be introduced with inexpensive investment in rural communities where other toilet systems are problematic (Ushijima et al. 2011).

17.5.1 Construction of Composting Toilet Pilot

A prototype of composting toilet was installed in all pilot households (Fig. 17.10d). The schematic illustration is shown in Fig. 17.10a–c.

The upper part of Fig. 17.10a represents the interface, which is composed of a toilet bowl and a urine diversion system. This interface is illustrated in Fig. 17.10c. The lower part of composting toilet is the reactor (i.e., the place of composting process). The reactor is previously filled with carbon matrix and progressively receives feces during the using of the toilet. A rotation system is also available to allow oxygen provision in the reactor. The rotation is performed manually by users after each use of the toilet. The rotation handle is visible in Fig. 17.10c. A tank (shown in Fig. 17.10a) is used to collect urine. The small quantity of wastewater from body cleaning is discharged out of the toilet room and infiltrate in a hole. Several matrixes have been tested during the composting process and results did not show significant difference on inactivation rate of pathogenic bacteria (Sossou et al. 2014). Therefore, local material, sawdust, shea nut husk, and millet stem husk available in Burkina Faso were used in the composting toilet.

17.5.2 Composting Process Performance

Indicator bacteria (total coliforms, *E. coli*, enterococci, and spores of sulfite-reducing clostridia—SSRC) were present in the compost and were reduced during composting process. Only pathogenic bacteria (*Salmonella* sp.) appeared to

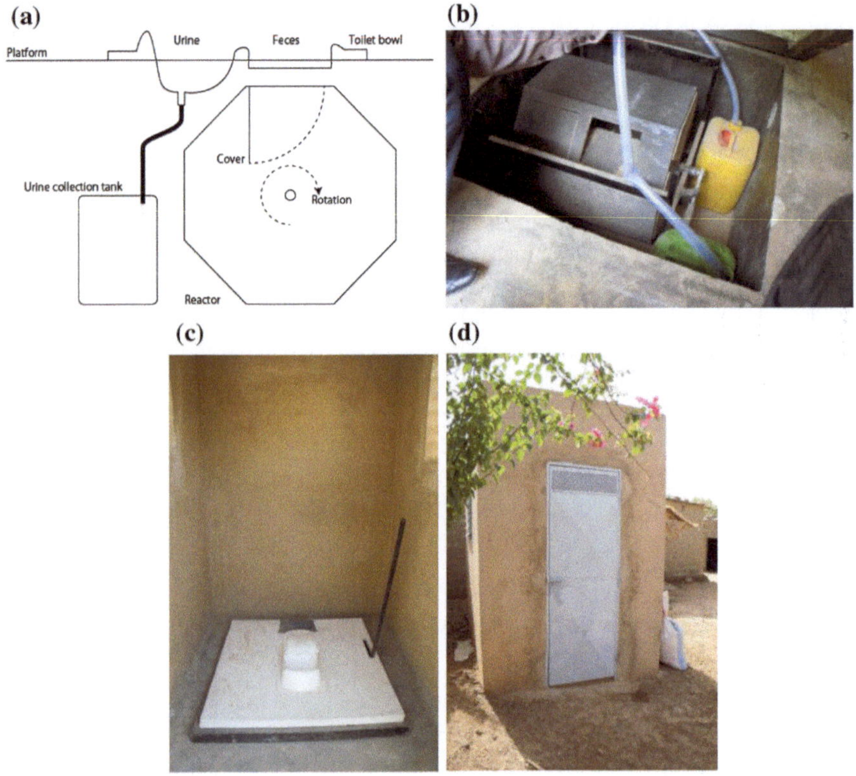

Fig. 17.10 Illustration of squatting type composting toilet reactor (the photographs and illustration are supplied by courtesy of Dr. Ryusei Ito)

be eliminated in compost after 30 days. Enterococcus were still present in compost at temperature monitoring compost (Christensen et al. 2002) and have observed regrowth sometimes in composting toilet (Sossou et al. 2014). Spores of sulfite-reducing clostridia can survive in composting conditions that inactivate other bacteria (Jones and Martin 2003). Such results are explained by the fact that temperature keep at mesophilic stage (i.e., similar to ambient temperature) during composting process, despite alkaline pH condition favorable to promote thermophilic composting. Such temperature condition do not allows adequate pathogens inactivation. In addition, average moisture level of compost (12.5%) was under the ideal moisture level and oxygen utilization rate (OUR) was lower than 350 mg/h with a finished reaction at 60,000 s, meaning an absence of intensive decomposition. According to several authors (Mehl et al. 2011; Redlinger et al. 2001), the majority of composting toilet in developing countries does not reach enough temperatures to complete pathogen destruction, so desiccation at alkaline condition may be the primary mechanism for pathogen inactivation. Others factors, except temperature, also influence microbial counts like microbial competition and the bulking agent used (Kim et al. 2009). However, pathogen destruction at ambient temperatures (tropical conditions 20–35 °C) could be improved if the contact time increases (Strauss and Blumenthal 1990; Cairncross and Feachem 1999).

17.6 Acceptability of the System by Users and Willingness to Reuse By-products

The challenge of Ameli-EAUR project is to treat urine and feces and reuse them as by-products of sanitation in agriculture. Only urine was used in the present experiment because it was done immediately after training on how to treat urine (see Sect. 17.4).

The experiment had two objectives: (i) demonstrate to pilot households that urine addition could significantly increase productivity of crops they use to cultivate without any addition of fertilizer (ii) and observe pilot households behavior and their involvement to introduce urine as fertilizer in their agricultural practices and to manipulate it. The experiment was performed during rainy season firstly because it is a high agriculture activity period in our experimental site and also because project did not start treat greywater and compost reuse at this stage.

17.6.1 Demonstration of Urine Fertilizer Value

Urine fertilizer value was tested on rain-fed cereal (as maize and millet) and vegetable (as okra). Okra is one of the main vegetable locally consumed by people. Instructions to pilot households for cereal cultivation were as follow:

1. Identify two small plots (plot R = reference plot: plot U = urine amended plot) and prepare both plots by themselves as they use to do before planting. It is important to mention that pilot households did not receive any other instructions at this step to prepare the soil. So some of them probably added manure during this preparation, as they use to do.
2. Seeding themselves both plots as they use to so with the same variety of cereal (maize or millet, depending on pilot household choice).
3. Pilot households received instructions concerning urine quantity to be provided in plot U. Quantity of urine was estimated based on the targeted cereal needs in nitrogen and urine mean concentration. Urine was provided three times during the cereal growing period. The first application of urine was done in presence of technicians as training, afterward for others applications, household just receives dose of urine be applied by themselves.

At harvest period, total quantity of cereal in each plots was weighted to show urine effect on cereal yields compared to usual practice. Figure 17.11a (plot U) and b

(a) (b)

Fig. 17.11 Urine effect on maize production (**a**) compared to no fertilizer treatment (**b**) (photo by Mariam Sou/Dakouré)

(plot R) presents total amount of maize harvested in the two different plots of one pilot household (*Rural_ko_CE*).

The demonstration allows families to see real impact of urine addition. Different values of yields obtained with experiments on cereals are summarized in Table 17.3. These values indicated that two pilot households (under total of three households who participated to this experiment) obtained at least 2 times higher quantity of cereals in plot U (amended with urine) compared to reference plots R. These results strongly depend on the rigorous degree of household to implement and respect instructions gave to them, especially for quantity and frequency of urine application. However, this first experiment is encouraging knowing that it was first time for all pilot households to apply urine as biofertilizer during their usual agricultural practice.

Experiment on vegetables' cultivation was slightly different to cereal cultivation. Long-term objective for vegetable cultivation is to have a permanent small garden very close to the households where all sanitary by-products (compost, treated greywater, and treated urine) will be available permanently for agricultural purposes. Figure 17.12 presents an example of small garden in household *Rural_ko_CE*. This garden is behind the house (Fig. 17.12a) and has a mean area of 20 m^2. It is covered with wire mesh to avoid any intrusion of animals (Fig. 17.12b).

Table 17.3 Urine fertilizer value tested on rain-fed cereals

Pilot households	Type of crop	Plots	Treatments	Yield (t ha^{-1})
Rural_ko_CE	Maize	Plot R: 24 m^2	Usual practice	3.7
		Plot U: 24 m^2	Usual practice + urine	7.3
Rural_ba_DR	Maize	Plot R: 35 m^2	Usual practice	13.4
		Plot U: 35 m^2	Usual practice + urine	14.9
Rural_ba_DA	Millet	Plot R: 42 m^2	Usual practice	2.7
		Plot U: 42 m^2	Usual practice + urine	4.4

(a)

(b)

Fig. 17.12 Vegetables' garden space at pilot household sites (site Rural_ko_CE). **a** Imagerie©2013Digital Globe, **b** the photograph is supplied by courtesy of D. Sangaré

Experiment was done on okra (as consensual choice of pilot households) during rainy season also. The experiment was particularly followed in all pilot households of the rural village of Kologoudiessé and in one semi-urban household of Kamboinsin. All others instructions regarding soil preparation and seeding were similar to previous experiments done on cereal except that references plots were note included for vegetables' experiment.

In all pilot household, results of this second experiment show that yields of okra reach minimal level of excepted yields (i.e., yields obtained in optimal conditions of water and nutrients provision). Indeed, Fig. 17.13a shows yields values among 14–17 t ha^{-1} (except in household *Rural_ko_IE*), whereas literature values in similar climate conditions have optimal yields between 14 and 29 t ha^{-1}.

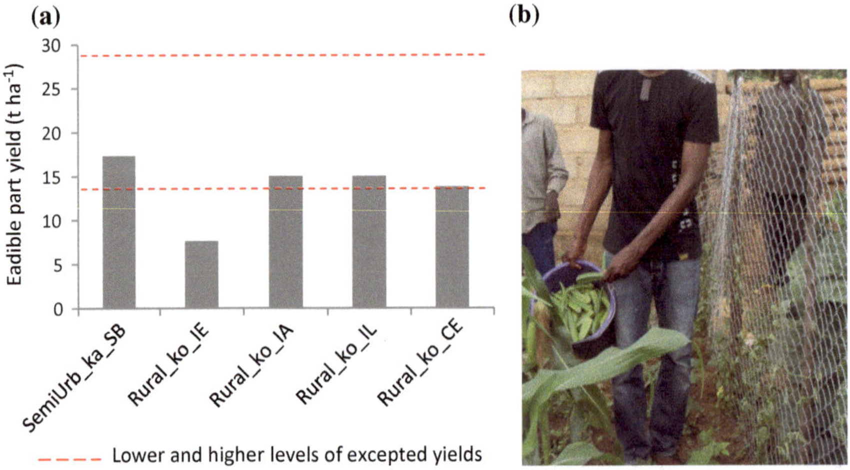

Fig. 17.13 Yields of okra amended with urine in household's gardens (the photograph and illustration are supplied by courtesy of D. Sangaré)

17.6.2 Feedbacks on Acceptability

The bet of reusing sanitary by-products, if not insurmountable, encounter some difficulties arising from the nature of waste targeted. Indeed, in a context where the fear of contamination[1] is still vivid—particularly in rural areas[2]—it is not easy to accept the idea of manipulating the bodily secretions, much less in order to produce food. The village of Kologoudiessé to date is recorded a greater enthusiasm to make use of urine and feces in agriculture. In this village, the population is predominantly Catholic and Mossi. In addition to rain-fed agriculture during the dry season, the people of Kologoudiessé make market gardening activity for which the manure is often sought.

Less than 10 km from this village, at Barkoundba, willingness to use the by-products is not so frankly expressed compared to Kologoudiesse village. The majority population in Barkoundba is Fulani and Muslim. Considering their habits, their beliefs and their representations, it was noticed that they were clearly more reluctant to manipulate and/or use urine and feces as fertilizer. Besides the fear of

[1]"The world is born from the idea of clean, first learning of the child is the cleanliness and the daily gesture refounding of civilization is to wash and store. Being clean is to be clean, be oneself, clearly separated from staining and non-self: get rid of the dirt draws the first existential border." (Kaufmann 1997)

[2]"The spatial organization of domestic life in the universe villagers reflected the desire to reduce the vulnerability to contamination, which comes from a certain type of contacts or meetings with individuals or entities invisible with special status or contact with animals or unclean things. This social construction of occupied space is topocentric design that characterizes the land use in most of rural West Africa." (Bouju and Ouattara 2002)

defilement, all about which reports to the scatological is taboo. This is another obstacle to the acceptance of the principle of manipulation and exploitation of by-products of sanitation. In addition, farmers benefit from this locality to feces of animals to fertilize their fields. These are all elements that make available through the Ameli—EAUR project innovations do not get a very favorable Barkoundba echo.

Despite the difficulties encountered, the idea to clean and reuse by-products of wastewater being in its infancy, it is necessary to continue the experiments that can prove the merits populations.

17.7 Conclusion

The present study is a part of project which global objective is to propose an innovative alternative to conventional sanitation. Indeed, in Sahelian context, implementing an integrated agro-sanitation system which combines hygiene improvement with nutrients and water source recoveries is very advantageous. However, such system much respects some criteria among which: (i) Robustness, simplicity, and efficiency of sanitation facilities; (ii) harmlessness of by-products; (iii) sustainability of the system through users building capacity; and (iv) willingness of users to work with sanitary based fertilizers. At middle stage of the project, results reported in this case study give overview of partial achievement of these criteria. Implementation of greywater facility, composting toilets, and urine treatment are identified and implemented in typical rural households in Sahelian context. Performances of these facilities are assessed and weaknesses identified to be improved. Users' capacity and willingness to manage such agro-sanitation system are partially tested on urine collection, treatment, and reuse. This assessment shows that users' willingness to appropriate the system cannot be dissociated to their social cultural and religious factors. So further investigations and scaling up must integrate these dimensions in the proposal of the system to other rural communities.

References

Antonini S, Nguyen PT, Arnold U, Eichert T, Clemens J (2012) Solar thermal evaporation of human urine for nitrogen and phosphorus recovery in Vietnam. Sci Total Environ 414(59):2–9. https://doi.org/10.1016/j.scitotenv.2011.11.055

Bouju J, Ouattara F (2002) «Une anthropologie politique de la fange. Conceptions culturelles, pratiques sociales et enjeux institutionnels de la propreté urbaine à Ouagadougou et Bobo-Dioulasso (Burkina Faso)». In Programme de recherche Gestion durable des déchets et de l'assainissement urbain, Action de recherche n° 4, SHADYC (Marseille)-GRIL (Ouagadougou)

Cairncross S, Feachem RG (1993, reprinted 1999) Environmental health engineering in the tropics: an introductory text, 2nd edn. Wiley, Chichester

Christensen KK, Carlbaek M, Kron E (2002) Strategies for evaluating the sanitary quality of composting. J Appl Microbiol 92:1143–1158. https://doi.org/10.1046/j.1365-2672.2002.01648.x

Del Porto D, Steinfeld C (2000) The composting toilet system book. Centre for Ecological Pollution Prevention, USA

ENA (2010) Enquête nationale sur l'accès des ménages aux ouvrages d'assainissement familial. Ministère de l'Agriculture et de l'Hydraulique – Burkina Faso. 4 pp [in french]

Galan DI, Kim S, Graham JP (2013) Exploring changes in open defecation prevalence in sub-Saharan Africa based on national level indices. BMC Public Health 13:527. https://doi.org/10.1186/1471-2458-13-527

Gomez-Couso H, Fontan-Sainz M, McGuigan KG, Ares-Mazas E (2009) Effect of the radiation intensity, water turbidity and exposure time on the survival of cryptosporidium during simulated solar disinfection of drinking water. Acta Trop 112:43–48. https://doi.org/10.1016/j.actatropica.2009.06.004

Hotta S, Funamizu N (2007) Biodegradability of fecal nitrogen in composting process. Biores Technol 98(17):3412–3414. https://doi.org/10.1016/j.biortech.2006.10.045

Jones P, Martin M (2003) A review of the literature on the occurrence and survival of pathogens of animals and humans in green compost. Ed. The waste and resources action programme. The Old Academy, Banbury

Kaufmann JC (1997) Le cœur à l'ouvrage: théorie de l'action ménagère. Nathan, Paris

Kim J, Shepherd M, Jiang X (2009) Evaluating the effect of environmental factors on pathogen regrowth in compost extract. Microb Ecol 58(3):498–508. https://doi.org/10.1007/s00248-009-9524-x

Kitsui T, Terazawa M (1999) Environmentally-friendly toilets for the 21st century, Biotoilet. In: Proceedings of the 10th ISWPC, Yokohama, vol 3, pp 120–121

Lopez Zavala MA, Funamizu N, Takakuwa T (2004) Temperature effect on aerobic biodegradation of feces using sawdust as a matrix. Water Res 38(9):2406–2416. https://doi.org/10.1016/j.watres.2004.02.026

Maiga Y, Moyenga D, Nikiema B, Ushijima K, Sou M, Maiga AH (2013) Pilot study of greywater treatment for small communities in rural area: the case of Kologoudiessé. In: 3rd IWA development congress and exhibition. Nairobi, Kenya, 14–17 Oct 2013

Maiga Y, Moyenga D, Ushijima K, Sou M, Maiga AH (2014) Greywater characteristics in rural areas of the Sahelian region for reuse purposes: the case of Burkina Faso. Revue des Sciences de l'Eau [J Water Sci] (in press)

Maurer M, Pronk W, Larsen TA (2006) Treatment processes for source-separated urine. Water Res 40(3):151–166. https://doi.org/10.1016/j.watres.2006.07.012

McGuigan KG, Conroy RM, Mosler H-J, du Preez M, Ubomba-Jaswa E, Fernandez-Ibanez P (2012) Solar water disinfection (SODIS): a review from bench-top to roof-top. J Hazard Mater 235–236:29–46. https://doi.org/10.1016/j.jhazmat.2012.07.053

Mehl J, Kaiser J, Hurtado D, Gibson D, Izurieta R, Mihelcic J (2011) Pathogen destruction and solids decomposition in composting latrines: study of fundamental mechanisms and user operation in rural Panama. J Water Health 9(1):187–199. https://doi.org/10.2166/wh.2010.138

Ministere des affaires Etrangeres et de la Cooperation Regionale (2011) Population et groupe ethnique du Burkina Faso. http://www.mae.gov.bf/population%20ethnie.html

Pronk W, Koné D (2009) Options for urine treatment in developing countries. Desalination 248:360–368. https://doi.org/10.1016/j.desal.2008.05.076

Redlinger T, Graham J, Corella-Barud V, Avitia R (2001) Survival of fecal coliforms in dry-composting toilets. Appl Environ Microbiol 67:4036–4040. https://doi.org/10.1128/aem.67.9.4036-4040.2001

Sossou SK, Hijikata N, Sou M, Tezuka R, Maiga AH, Funamizu N (2014) Inactivation mechanisms of pathogenic bacteria in several matrixes during the composting process in a composting toilet. Environ Technol 35(6):674–680. https://doi.org/10.1080/09593330.2013.841268

Strauss M, Blumenthal UJ (1990) Use of human wastes in agriculture and aquaculture; utilization practices and health perspectives. IRCWD report, 8

Ushijima K, Yabui K, Hijikata N, Ito R, Funamizu N (2011) Development self-buildable simple composting toilet. In: Proceedings of IWA aspire (USB-memory) conference and exhibition, 2–6 Oct, Tokyo, Japan

Winblad U, Hebert MS, Calvert P, Morgan P, Rosemarin A, Sawyer R, Xiao J (2004) Ecological sanitation, revised and enlarged edn. Stockholm Environment Institute, Stockholm

Chapter 18
Experiment of a Resource-Oriented Agro-Sanitation System for Urban Slum Area: Case of Indonesia

Neni Sintawardani, Ken Ushijima, Umi Hamidah, Yosuke Deguchi, Jovita Triastuti, Naoyuki Funamizu, Mitsuteru Irie and Tadaharu Ishikawa

Abstract This chapter describes case study of applicability assessment for resource-oriented agro-sanitation business for urban slum in Indonesia, based on interdisciplinary field survey, (1) economic feasibility assessment, (2) demand potential assessment, and (3) social acceptability. The value flow analysis evaluated garbage disposal cost and excreta disposal cost at 0.7 and 1.1%, respectively, of household income, which represents a reasonable cost to disposers. However, under the current disposal scheme, negative value due to excreta disposal to river appears to be passed on to downstream populations. The structure of this value flow indicates that little attention is paid to the negative externalities, indicating that a

N. Sintawardani · U. Hamidah · J. Triastuti
Research Unit for Clean Technology, Indonesian Institute of Sciences, Bandung, Indonesia

K. Ushijima (✉)
Northern Regional Building Research Institute, Hokkaido Research Organization (HRO), 3-1-20, Midorigaoka Higashi 1, Asahikawa 078-8801, Japan
e-mail: ushijima-ken@hro.or.jp

Y. Deguchi
Nippon Koei Co., Ltd, 5-4 Kojimachi, Chiyoda-ku, Tokyo 102-8539, Japan

N. Funamizu
Graduate School of Global Food Resources, Hokkaido University, Kita-9, Nishi-9, Kita-ku, Sapporo 060-8589, Japan

N. Funamizu
Research Institute of Humanity and Nature, Motoyama Kamigamo, Kita-ku, Kyoto 603-8047, Japan

M. Irie
Alliance for Research on North Africa, University of Tsukuba, Tsukuba, Ibaraki 305-8572, Japan

T. Ishikawa
Department of Environment and Technology, Tokyo Institute of Technology, Yokohama 226-8502, Japan

N. Funamizu (ed.), *Resource-Oriented Agro-sanitation Systems*,
https://doi.org/10.1007/978-4-431-56835-3_18

solution in this case may be difficult to obtain. The affordable initial cost of a new resource recycling system was evaluated and compared with the costs of a currently available compost toilet by evaluating equipment lifetimes and potential income streams available from the sale of human excrement fertilizers. To estimate a value for the human excrement fertilizer, real-world case study in Indonesia, in which money was paid in exchange for human urine, was referred to. The evaluated affordable initial cost derived was sufficiently comparable to the price of a cheaper composting toilet currently available in the market. This implies that such a resource recycling system is feasible even in an urban slum. Results of demand and supply assessment showed that if resource-oriented sanitation system was diffused, supply would excess demands in the circle of 32 km radius, therefore cost discussion should assume more than 32 km transportation. According to the results of case estimation of tea plantation, total cost for fertilizer provided by resource-oriented sanitation system is possibly competitive to the normal price of synthetic fertilizer. However, if compared to subsidized price for low-income farmers, it seems difficult to compete. Regarding social aspect, the context of Islamic law is important. Islamic authorities regarded the resource-oriented sanitation system as no problematic concept and acceptable in the context of Islamic law.

Keywords Material flow · Value flow · Sustainable design · Initial cost Urban slum

18.1 Introduction

Indonesia is one of the countries who experiencing rapid development. Generally, rapid development causes rapid urbanization with population explosion. As a result, urban slum has high population density and poor infrastructure. Bad sanitary condition is one of serious issues. From material flow viewpoint, highly populated urban areas require large material inflows of food or other basic needs for people. Although this incoming materials is consumed, it does not disappear, because waste materials are generated, which need to be disposed of in some manner, mainly as solid waste and as wastewater, including toilet wastewater. If the urban system does not introduce systems to manage these material outflows, unfavorable wastes accumulate in the urban area or flow to unfavorable locations, such as into surface and groundwater, both of which are usually important water sources. Such problems can be observed in many developing countries (Metcalf and Eddy 1998; Tanaka 1992; Tsuzuki et al. 2007). However, investment to outflow systems tends to be given lower priority because outflow system does not give clear benefits to individual, like supplying system.

On the other hand, sustainability has recently been recognized as important factor for societies' system. Several types of resource recycling toilets and wastewater treatment systems for reuse have been proposed as sustainable sanitation technologies (Winblad and Simpson-Hebert 2004; Lopez et al. 2002; Ushijima et al. 2013a). Although these technologies have the advantage to create value from unwanted waste, they need to be embedded in the system design and suitably to manage equipment outflow. It is therefore important to understand current and future material flows in and around targeted urban areas. Material flow analysis (MFA) and substance flow analysis (SFA) are suitable methodologies for this purpose. The application of the MFA to sanitation systems has been attempted in several studies, some of which considered sanitation systems as a part of material flows (Harada et al. 2010; Aramaki and Thuy 2010), and some of which focused mainly on sanitation system (Zhou et al. 2010; Li et al. 2010). However, these studies describe current material flows and only providing the conceptual description of possible improvements.

In an actual application, the optimization of the material flow design is not sufficient to drive the system. Successful material flow programs need to be designed to give benefit for all stakeholders, and value chains for this material flow need to be designed also. Therefore, it is believed that both material flow and value flow should be considered simultaneously. This chapter describes application assessment of resource-oriented agro-sanitation business for urban slum in and around Kiaracondong district, Bandung city, Indonesia, based on interdisciplinary field survey, (1) economic feasibility assessment, (2) demand potential assessment, and (3) social acceptability.

18.2 Framework of Agro-Sanitation Model for Urban Slum in Indonesia

18.2.1 Overview of Case Field

Bandung city is the fourth largest city in Indonesia, with a population of 2.5 million (BPS 2005). It is in the inland part of west Java and is located upstream to the Citarum River, a major source of water for the capital, Jakarta (Fig. 18.1a). Only 16% of the domestic wastewater generated in Bandung city is treated, as the city has only one treatment plant, located in the suburbs (World Bank 2003). The Kiaracondong sub-district is located near the center of Bandung city and is known as the city's largest slum. According to 2006 local statistics, Kiaracondong has a population of 116,000 living within an area of 6.1 km^2.

Two types of field surveys were conducted. One was conducted in the Jondol Canal (Ushijima et al. 2008a), which flows through central Kiaracondong into the Cicadas River, which is a tributary of the Citarum River. In the area surveyed, there were no houses connected to the sewerage system feeding the treatment plant. Figure 18.1c shows the measurement locations and catchment of the studied section of the Jondol Canal, with a distance of 960 m between the two observation sites. This area of the catchment was measured to be 0.41 km^2 and, as determined by

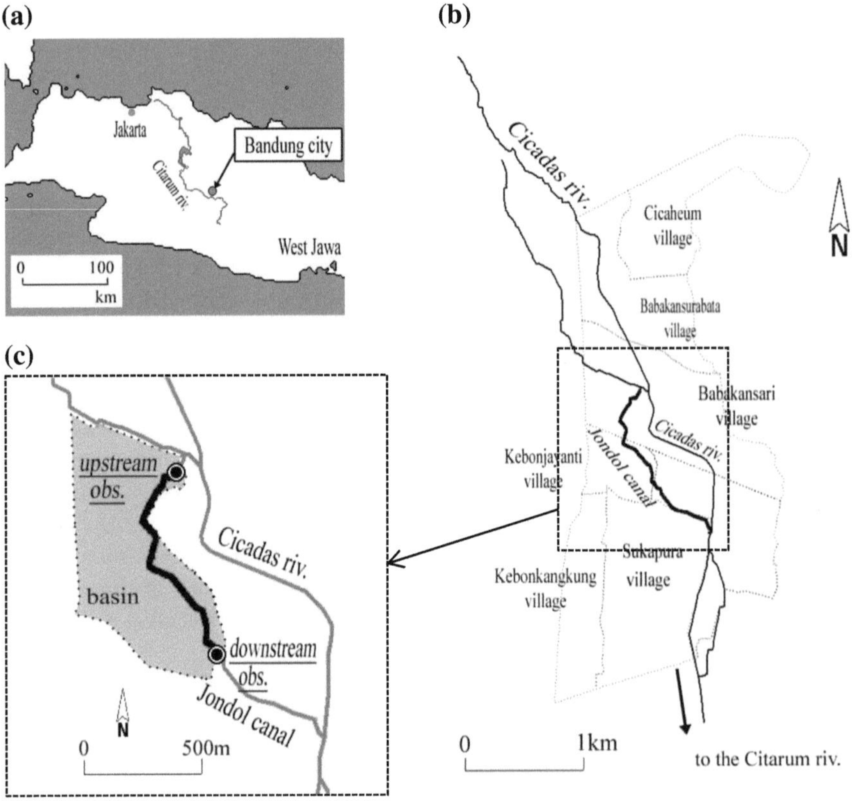

Fig. 18.1 Location of studied area: **a** location of Bandung city; **b** Kiaracondong sub-district; **c** observatories and catchment. Reprinted from Journal of Hydroscience and Hydraulic Engineering 26(1):129–140 (2008)

previous investigations and local statistics, has a population of 16,240 people and 3585 households. The other survey was conducted in the Jondol Canal catchment area. In this survey, 62 families living near the canal were interviewed about their water use, their wastewater disposal, and other related matters (Ushijima 2007).

Figure 18.2 shows income distribution based on questionnaire survey for 62 households (Ushijima et al. 2008b). Averaged per capita income was approximately 300,000 Rupiah/month, and mode was 100,000–199,000 Rupiah/month. Additional detailed interview was conducted to 5 of 62 households, regarding to water fee, electricity fee, and garbage collection fee. Garbage collection fee means the money each households pay to the community for garbage collection, as mentioned below (Table 18.1). Percentage of water fee, electricity fee, and garbage collection fee in total income was 1–6, 4–7, and 0.1–0.7%, respectively. Water fee

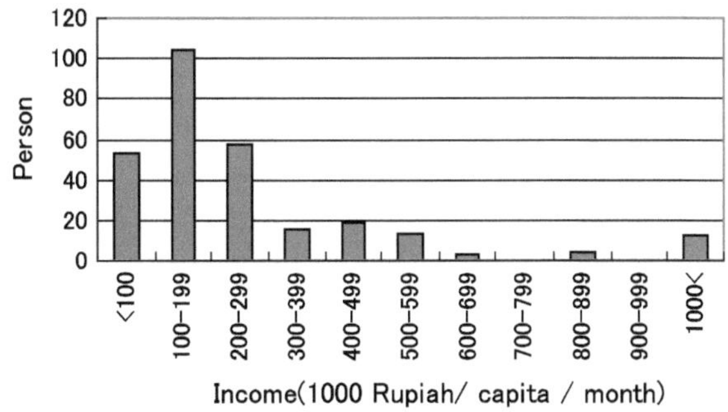

Fig. 18.2 Income range distribution of study area ($N = 62$). Reprinted from Proceedings of the 6th International Symposium on Sustainable Sanitation, Changchun. Changchun: Northeast Normal University (NENU) 27–33 (2008)

showed wide variety. One of probable reasons is unit price of water (0–30,000 Rupiah/m^3). It has wide variety according to source, since water source for people in this area came from communal open well, private tap, shared tap, private well, etc., that implied to the price.

18.2.2 Agro-Sanitation Model

It is clear that urban area, especially overpopulated slum area does not have sufficient agricultural land to consume fertilizer from human excreta generated in that area. Agricultural production locates mostly in rural or suburb area. Therefore, agro-sanitation system for urban slum has to include some kind of transportation of sanitation materials from slum to rural or suburban area. In this study, the model of compost collection system should be designed. Basic framework of collection and transportation system was designed using the framework of present garbage collection system; compost and urine are collected by manpower and transferred to truck at middle station, then transported to farmland by truck.

18.2.2.1 Compost Collection—Current Garbage Collection System

In Indonesia, city or province has a responsibility of solid waste management from the study by law; however, from the study it was showed that primary collection is carried out by community. In many case, community employs garbage collector by the money collected from each house (Unisuga and Watanabe 2004). Garbage collector collected garbage from each house with handcart (Fig. 18.3) and bring them to middle station as temporary site for garbage collection (Tempat Penampungan Sampah = TPS) prepared by city or province. Garbage was uploaded to truck (Fig. 18.4) and transported to dump site.

In this study, detailed survey was carried out focusing on the community managing part in 2006 and 2012. Interview survey and measurement are combined to understand actual situation. The observed community, RW02 Sukapura village,

Table 18.1 Expenditure for water, electricity, and garbage

	Family size	Household income (10^3 Rupiah/month)	Per capita income (10^3 Rupiah/month/capita)	Water (10^3 Rupiah/month)		Electricity (10^3 Rupiah/month)		Garbage collection (10^3 Rupiah/month)	
				Per household	Percentage in income (%)	Per household	Percentage in income (%)	Per household	Percentage in income (%)
(1)	5	2500	500	150.0	6.0	150	6.0	2.5	0.1
(2)	4	1700	425	17.5	1.0	70	4.1	2.5	0.1
(3)	6	1700	283	90.0	5.3	75	4.4	2.0	0.1
(4)	4	300	75	18.0	6.0	20	6.7	2.0	0.7
(5)	6	1600	267	50.0	3.1	80	5.0	2.5	0.2

Reproduced from Proceedings of the 6th International Symposium on Sustainable Sanitation, Changchun. Changchun: Northeast Normal University (NENU) 27–33 (2008)

Fig. 18.3 Handcart for garbage collection. Reprinted from Proceedings of the 6th International Symposium on Sustainable Sanitation, Changchun. Changchun: Northeast Normal University (NENU) 27–33 (2008)

Fig. 18.4 Garbage transportation by truck. Reprinted from Dissertation for the Doctoral Degree. Tokyo: Tokyo Institute of Technology (2007)

employs one worker for garbage collection. People directly handed the garbage to collector or put it in front of their house. Collector's handcart has a capacity of 1.3 m^3 (700 × 1150 × 1550 mm). Collection was carried out three times in a week, and collected amount per collection day is always corresponds to two handcarts. Working time is always morning, sometimes it continues to afternoon.

The garbage collector of RW02 Sukapura village was paid 400,000 Rupiah/month. Money for garbage collector, including maintenance cost for handcart, is collected by community from each household. Basic price is 2000 Rupiah/month/household, and some extra-charge according to income. However, the extra-charging rule was not clear and applied not so strictly.

Garbage amount from 62 households were measured and showed the target of questionnaire survey. The households should provide garbage generated in one day to be measured their total weight and components. Summary of the result is that total amount was 184 g/day/capita and 75% was organic waste in average. Table 18.2 showed summary of the result related to designing collection system. Statistical data of Bandung city and other reported values were also shown as reference. Total amount per capita was estimated by dividing the collecting amount in volume. Smaller amount of the result than reference appears to be appropriate, because reference value is average of Bandung city but living level is rather low in slum area. Percentage of organic waste is almost the same as reference data. Collection capacity is smaller than reference. Because a collector in the area works only 3 days in a week and he works only morning, it could be assumed that in present situation he had enough margin. Collection fee is smaller than reference data reported on Asia and Latin America, and percentage of collection fee in income is much smaller than reference data reported as general affordability to pay (ATP) (JICA 2002). Of course ATP can change by country, community, income level, and so on, however at least, present fee seems not to be too expensive for people.

Table 18.2 Summary of field survey

	Results	Reference data	Note
Total amount	0.9 L/day/capita (0.184 kg/day/capita)	1.8 L/day/capita (0.36 kg/day/capita)	Estimated from statistics of Bandung city (BPS-Kota Bandung 2004) Weight was estimated by JICA data (JICA 2005) reporting the density of 0.2 kg/L
Organic waste	75%	73.98%	Statistics of Bandung city (BPS-Kota Bandung 2004)
Collection capacity of collector	1.1 m^3/day	3 m^3/day	Yamamura (1987) reported on the case of Jakarta
Collection fee	556 Rupiah/month/capita	800–2500 Rupiah/month/capita	Estimated cost on Asia and Latin America (Bartone 2000)
Percentage of collection fee in income	0.3–0.6%*	2%	Estimated value of ATP (JICA 2002)

Reproduced from Proceedings of the 6th International Symposium on Sustainable Sanitation, Changchun. Changchun: Northeast Normal University (NENU) 27–33 (2008)
*Estimated for mode value in the study site

18.2.2.2 Urine Collection—Current Water Tank Carrier

Regarding urine collection, 20-L plastic tank is assumed to be used. Many of households in Kiaracondong do not have private water tap, and they use public tap called CORSEN to meet their daily water demand. To take and bring water from CORSEN to their houses, people use 20-L plastic tanks as necessary equipment and they use handcart which is specially designed to carry 20-L tanks (Fig. 18.5). The handcart can bring 8–12 tanks. In fact, there is water vendor who obtains income by carrying and selling tank of water; however, this study did not refer their income as assumed collector. We just applied idea of water tank carrier and we set unit worker cost for urine collection same as garbage collector.

18.2.2.3 Collection Capacity

We assumed the frequency and volume of sawdust replacement as twice in a year and 0.025 m^3 at one time, following to the specification of commercial composting toilet. It was assumed averaged family size as four persons/family that based on actual average of family size of 3.6 persons/family in local statistics of the RW. In this case, replacing amount of sawdust will be 0.1 m^3/household/time. If all 410 households in targeted area use composting toilet and replace sawdust twice in a year, then total amount is 82 m^3/year. Thus, the required collection capacity was 0.22 m^3/day in average. If it was assumed that collection frequency three times/ week, the required collection capacity was 0.52 m^3 per collection day. Required time for sawdust replacement greatly depends on the design of the machine. Unfortunately, present commercial model does not have useful design in this aspect. By one of the author's experience, the screw in the tank becomes big obstacle, and replacement takes always more than 30 min. However, if the screw

Fig. 18.5 Handcart specially designed for carrying water tanks through narrow streets in slum area. Reprinted from Dissertation for the Doctoral Degree. Tokyo: Tokyo Institute of Technology (2007)

can be removable, replacement work does not take so much. This point is one of issue for mechanical design.

Regarding urine collection, if five persons/family, one 20-L tank becomes full by about 4 days. Therefore, 410 households in target area produce 102 tanks/day in total. If we assume that urine collector can collect three carts/day, and that one cart carries ten tanks, 3–4 workers are required.

Regarding to the cost, there is no assurance that community can employ compost collector by same salary as for garbage collector. However, according to interview results to three garbage collectors, they given their consent to join the work as compost collector in same salary like garbage collector, since they need more job in same salary because they need more job, and one of them mentioned that compost collection seems to have lower risk of injury, such as injury by glass or needles in garbage.

18.3 Economic Feasibility Assessment of Agro-Sanitation Business Model

18.3.1 Materials and Methods

18.3.1.1 Data Source

The results of field surveys on river water (Ushijima et al. 2008a), domestic wastewater (Ushijima et al. 2008a; Ushijima 2007), and garbage disposal and collection (Ushijima et al. 2008b), in addition to official statistics for Indonesia (Gunawan 1997), were used as the main data.

18.3.1.2 Methods

MFA and SFA were used to analyze current and future sanitation systems from a comprehensive and practical viewpoint. The subjects covered in the MFA included excrement as well as all related materials, namely water and food. The SFA was conducted along a similar framework, focusing on nitrogen and phosphorus which, as major components of fertilizers, are important factors in resource recycling systems. When data for the slum area were unavailable, data from the statistics and literature on Indonesia or other Southeast Asian countries were used for extrapolation. The project boundary was set as the catchment area described in Fig. 18.1c. All the data were expressed in terms of per capita years, because both material and value flows will be discussed together in this study and the use of such unit is meaningful to the study of value flow.

MFA and SFA were also used for estimating value flow. In contrast to material and substance flows, value flow can be measured in a number of ways. Some can be definitely measured in quantitative factors such as prices or monetary value, whereas some can be only evaluated in qualitative factors, such as status or comfort derived. In this study, the quantitative values were evaluated based on the available data, whereas qualitative values were simply determined to be positive or negative by assessing the directions of material and value flows.

Material, substance, and value flow analysis were conducted for two scenarios: one representing current conditions, and the other assuming the implementation of a resource recycling sanitation system involving the installation of a urine diversion aerobic composting toilet (Ushijima et al. 2011) in each household. The framework proposed and discussed quantitatively by Ushijima (2007) for the collection of urine and compost for integration with existing garbage collection systems, wherein organic solid waste is processed in a composting toilet along with feces, was applied.

Successful diffusion of this resource recycling system primarily depends on each household's decision of whether to install the composting toilet. This decision is based on the following criterion:

$$\text{(Initial cost/expected lifetime)} < \text{(WTP for excreta disposal)} + \text{(income from selling excreta)}, \tag{18.1}$$

where WTP is "willingness to pay."

Here, WTP for excreta disposal is estimated by

$$\begin{aligned}\text{(WTP for excreta disposal)} = {} & \text{(Current excreta disposal cost)} \\ & + \text{(Initial cost for water flush toilet/expected lifetime)}\end{aligned} \tag{18.2}$$

where the current excreta disposal cost is equal to the sum $O_c11 + O_c12$, given in Table 18.3.

18.3.2 Results

18.3.2.1 Current Material Flow

Figure 18.6 shows current material flows. Total mass flow, phosphorus mass flow, nitrogen mass flow, and total value flows are summarized in Table 18.3. Note that several value flows, namely I_c21, I_c22, I_c23, I_c24, and I_c25 were not quantified in this study.

Table 18.3 Mass, phosphorus, nitrogen, and value of each arrow in current material flow

ID	Supplied item	Mass/	P/	N/	Value/	Data source
		(kg (year capita)$^{-1}$)			(Rp (year capita)$^{-1}$)/(% against income)	
I_c1	Water	32,000	0	0	+110,000 (2.8%)	Mass, value: average of three households in Ushijima (2007). P, N: regarded as ignorable
I_c2	Food	170	1.3	6.0	+1,400,000 (35.9%)	Mass, value: Gunawan (1997), P, N: blackwater + solid waste
I_c3	Electricity	0	0	0	+170,000 (4.4%)	Value: average of three households in Ushijima (2007), mass, P, N: meaningless
O_c1	Blackwater	14,000	1.2	5.9	±0 (0%)	Mass: $I_c1 \times 0.41$ (Ushijima 2007) + O_c11 + O_c12, P, N: O_c11 + O_c12
O_c11	Feces	33	0.3	0.9	−13,000 (0.3%)	Mass, P, N: Ushijima (2007), value: water consumption for defecation (water cost × percentage of water use for defecation)
O_c12	Urine	370	0.9	5.0	−32,000 (0.8%)	Mass, P, N: Ushijima(2007), value: water consumption for urination (water cost × percentage of water use for urination)
O_c2	Gray water	19,000	1.0	0.0	±0 (0%)	Mass: I_c1-O_c1, P, N: total wastewater [maximum in Ushijima (2007)]—O_c1
O_c3	Solid waste	67	0.1	0.2	−28,000 (0.7%)	Mass: Ushijima (2007), P, N (unit): Aramaki and Thuy (2010), value: collection fee (Ushijima 2007)
O_c31	Junk material	(no data)	0	0	+	P, N: Non-organic is regarded as 0
O_c32	Solid waste	(no data)	0.1	0.2	–	D3–D31
O_c33	Junk material	(no data)	0	0	+	P, N: Non-organic is regarded as 0
O_c34	Accumulation	(no data)	0.1	0.2	–	D32–D33
O_c4	Work force	0	0	0	+3,900,000 (100%)	Mass, P, N: meaningless value: average of three households in Ushijima (2007)
R_c1	River inflow	52,000	0.4	1.3	±0 (0%)	Ushijima (2007), N: NH_4–N + NO_2–N + NO_3–N P: PO_4–P
R_c2	River outflow	110,000	1.2	3.5	±0 (0%)	Ushijima (2007), N: NH_4–N + NO_2–N + NO_3–N P: PO_4–P
R_c3	Mosque	8300	0	0	±0 (0%)	Ushijima (2007)

Reproduced from Frontiers of Environmental Science and Engineering 7(1):120–126 (2013)

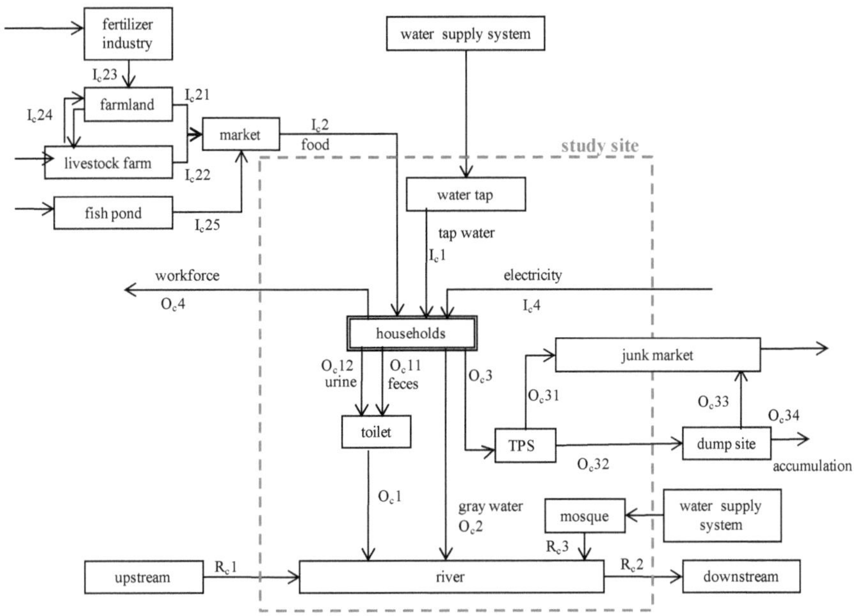

Fig. 18.6 Current material flow. Reprinted by permission from Springer Nature: Springer Nature, Ushijima et al. (2013b)

18.3.2.2 Material Flow of Resource-Oriented Sanitation System

Figure 18.7 shows material flows under a future scenario in which a resource recycling sanitation system is applied. Total mass flow, phosphorus mass flow, nitrogen mass flow, and total value flows are summarized in Table 18.4. Note that several mass flows, namely I_r21, I_r22, I_r23, I_r24, and I_r25 were not quantified in this study.

18.3.3 Discussion

18.3.3.1 Current Material Flow

Using material flow and value flow analysis, the value flow in and around an urban slum family was analyzed, not only in terms of easy-to-price factors or quantitative flows but also in terms of difficult-to-price factors or qualitative flows, such as garbage disposal and excreta disposal. Based on the result, "costs for disposal" for garbage and wastewater was determined. These values, listed in Table 18.3, are approximately 0.7 and 1.1% of per capita income for garbage disposal and for urine and excreta disposal, respectively. The World Bank has roughly evaluated

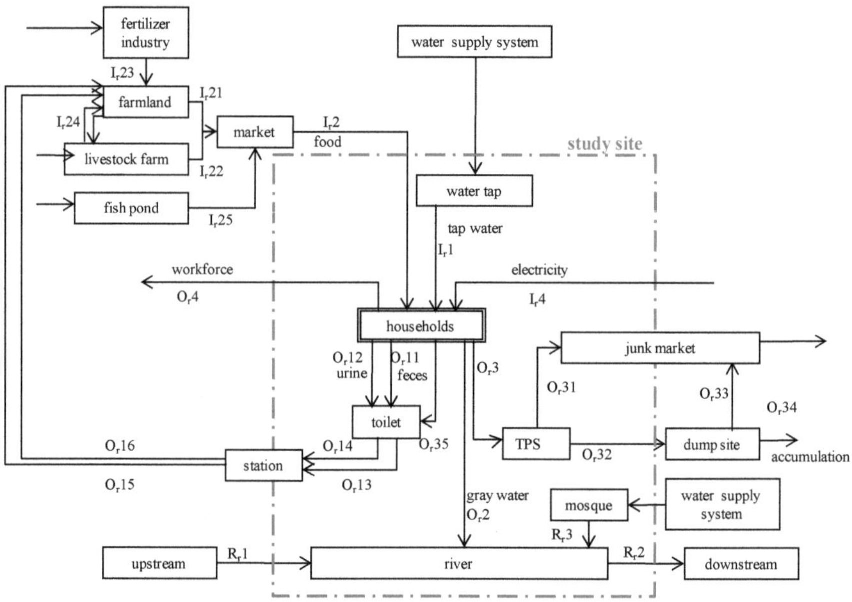

Fig. 18.7 Material flow of resource recycling scenario. Reprinted by permission from Springer Nature: Springer Nature, Ushijima et al. (2013b)

benchmarks for affordability to pay (ATP) as 2 and 1% of consumable per capita income for garbage collection service and sewerage service, respectively (JICA 2002), similar to the current costs derived here. Thus, current value flows for garbage and excreta disposal seem reasonable for people in light of such benchmarks. However, the material flow of excreta into the river is clearly causing deterioration of the water environment. Because this value loss has a negligible direct effect, and because the disadvantages apply to populations downstream instead of to disposers, it would seem that the structure of the current value flow would impede efforts to improve the water environment.

18.3.3.2 Resource-Oriented Sanitation System and Mass Balance

A proposed resource recycling sanitation system would not discharge excreta to the river; rather such waste would be sent to farms, where it would have positive value for a fertilizer. From the catchment area studied, the system would remove almost all the nitrogen and half of the phosphorus discharged, and therefore, the net flux of nitrogen and phosphorus through the river under a recycling scenario (see Fig. 18.7: R_r2–R_r1) would be less than under current conditions (Fig. 18.6: R_c2–R_c1). The positive impacts on river water quality resulting from the introduction of resource recycling sanitation would be potentially large.

Table 18.4 Mass, phosphorus, nitrogen, and value of each arrow in material flow of resource recycling sanitation system

ID	Supplied item	Mass/	P/	N/	Value/	Data source
		(kg (year capita)$^{-1}$)			(Rp (year capita)$^{-1}$)/ (% against income)	
I_r1	Water	32,000	0	0	+110,000 (2.8%)	Mass, value: average of three households in Ushijima (2007) P, N: regarded as ignorable
I_r2	Food	170	1.3	6.0	+1,400,000 (35.9%)	Mass, value: Gunawan (1997), P, N: blackwater + solid waste
I_r3	Electricity	0	0	0	+170,000 (4.4%)	Value: average of three households in Ushijima (2007), mass, P, N: meaningless
O_r11	Feces	33	0.3	0.9	Variable	Mass, P, N: Ushijima (2007), value: water consumption for defecation (water cost × percentage of water use for defecation)
O_r12	Urine	370	0.9	5.0	Variable	Mass, P, N: Ushijima (2007), value: water consumption for urination (water cost × percentage of water use for urination)
O_r13	Processed feces	Variable	0.3	<0.9	Variable	P, N: = O_r11
O_r14	Processed urine	Variable	0.9	<5.0	Variable	P, N: = O_r12
O_r15	Collected feces	Variable	0.3	<0.9	Variable	P, N: = O_r11
O_r16	Collected urine	Variable	0.9	<0.9	Variable	P, N: = O_r12
O_r2	Gray water	19,000	1.0	0.0	±0 (0%)	Mass: I_c1-O_c1, P, N: total wastewater [maximum in Ushijima (2007)]—O_c1
O_r3	Solid waste	17	0.0	0.0	−28,000 (0.7%)	Mass: Ushijima(2007), P, N: Non-organic is regarded as 0, value: collection fee (Ushijima 2007)

(continued)

Table 18.4 (continued)

ID	Supplied item	Mass/ (kg (year capita)$^{-1}$)	P/	N/	Value/ (Rp (year capita)$^{-1}$)/ (% against income)	Data source
O_r31	Junk material	(no data)	0	0	+	P, N: Non-organic is regarded as 0
O_r32	Solid waste	(no data)	0.0	0.0	–	D3–D31
O_r33	Junk material	(no data)	0	0	+	P, N: Non-organic is regarded as 0
O_r34	Accumulation	(no data)	0.0	0.0	–	D32–D33
O_r35	Organic waste	50	0.1	0.1	+	Mass: Ushijima (2007), P, N (unit): Aramaki and Thuy (2010)
O_r4	Work force	0	0	0	+3,900,000 (100%)	Mass, P, N: meaningless, value: average of three households in Ushijima (2007)
R_r1	River inflow	52,000	0.4	1.3	±0 (0%)	Ushijima (2007), N: NH_4–N + NO_3–N P: PO_4–P
R_r2	River outflow	97,000	0.8	3.5	±0 (0%)	Ushijima (2007), N: NH_4–N + NO_3–N P: PO_4–P, reaching ratio of N, P: 0.36
R_r3	Mosque	8300	0	0	±0 (0%)	Ushijima (2007)

Reproduced from Frontiers of Environmental Science and Engineering 7(1):120–126 (2013)

As it is a source of nutrients, excreta has good potential for the generation of value through use as fertilizer. However, the value of compost made from human excreta may also be discounted relative to chemical fertilizers because of psychological resistance. Further, the quality or handling of the processed excreta may also affect the value of the resulting compost. However, the price of synthetic fertilizers has been rising the last decade, and if this rise continues, the incentive of saving costs through substitution may overcome the psychological barriers.

In any event, the exact value of human excreta is currently difficult to evaluate. A case study of one company selling liquid fertilizer made from human urine seems suitable as a reference (Ushijima 2011). The company pays 1000 Rupiah L^{-1} to people as compensation for urine offer. Assuming that one liter of urine is produced per capita per day, this price corresponds to 365,000 Rupiah (capita year)$^{-1}$, equivalent to 9% of per capita income, which is listed in Table 18.3 as O_c4.

In the area studied, the initial cost of a water flush toilet is reported to be approximately 2,000,000 Rupiah (Ushijima 2007). Assuming a family size of four people and a toilet lifetime of 15 years (which is standard in Japan), the WTP is thus estimated at 79,000 Rupiah (capita year)$^{-1}$. If we apply the excreta price mentioned above as an optimum estimation, affordable initial cost for resource recycling sanitation system is estimated as

$$(\text{Initial cost}/\text{lifetime}) < 4,400,000\,\text{Rupiah}(\text{capita year})^{-1}. \tag{18.3}$$

For a 15-year lifetime composting toilet, the estimated affordable initial cost for a four person family would be 27,000,000 Rupiah, equivalent to approximately 3200 USD according to the June 2011 exchange rate. This is similar in price to a cheaper composting toilet (approximately 2000 USD) currently available in the market. For people who live in slum area, this price consumes about 56% from their annual income per capita.

18.4 Demand Potential Assessment

18.4.1 Assessment Scheme

Balance of fertilizer demand and supply is fundamental assessment to evaluate resource-oriented agro-sanitation system's feasibility. As a simple assumption, this study regarded total phosphorus contained in annual human excreta production in each area as a potential supply. It was estimated by multiplying population and unit value of phosphorus excretion from human body (Ushijima et al. 2013b). This study also regarded annual total consumption of phosphorus in fertilizer as potential demand. It was estimated by multiplying standard fertilizer use (FAO 2005) and cultivation area (data sources are listed Table 18.5) for each products. Those are analyzed in district level, which is called as "kecamatan" in Indonesian language.

As a practical assessment, this study also evaluated the price of fertilizer made from excreta. In this concept, cost factor of fertilizer is collection and transportation fee. This evaluation assumed to use those excreta fertilizers for tea plantation which locates surrounding area of Bandung city.

18.4.2 Demand and Supply Balance

Figure 18.8 shows potential balance of demand and supply of excreta fertilizer in each district. Inside of Bandung city and surrounding some districts show supply excess, on the other hand other districts shoed demand excess. It is clear that human

Table 18.5 List of agricultural data source

BPS-Bandung Barat (2010)	Bandung Barat Dalam Angka
BPS-Bandung (2009)	Bandung Dalam Angka
BPS-Cianjur (2010)	Cianjur Dalam Angka
BPS-Galu (2010)	Galu Dalam Angka
BPS-Subang (2010)	Subang Dalam Angka
BPS-Purwakarta (2010)	Purwakarta Dalam Angka
BPS-Cimahi (2010)	Cimahi Dalam Angka
BPS-Bandung(2010)	Panyileukan Dalam Angka
BPS-Bandung (2010)	Panyileukan Dalam Angka
BPS-Bandung (2010)	Arcamanik Dalam Angka
BPS-Bandung (2010)	Cidapap Dalam Angka
BPS-Bandung (2010)	Coblong Dalam Angka
BPS-Bandung (2010)	Antapani Dalam Angka
BPS-Bandung (2010)	Cinambo Dalam Angka
BPS-Bandung(2010)	Gedebage Dalam Angka
BPS-Bandung (2010)	Regol Dalam Angka
BPS-Bandung (2010)	Bandung Wetan Dalam Angka
BPS-Bandung (2010)	Lengkong Dalam Angka
BPS-Bandung (2010)	Astanaanyar Dalam Angka
BPS-Bandung (2010)	Cibeunying Dalam Angka
BPS-Bandung (2010)	Sumur Bandung Dalam Angka
BPS-Bandung (2010)	Kiaracondong Dalam Angka
BPS-Bandung (2010)	Cibiru Dalam Angka
BPS-Bandung (2010)	Andir Dalam Angka
BPS-Bandung (2010)	Bojongloa Kaler Dalam Angka
BPS-Bandung (2010)	Babakan Ciparay Dalam Angka
BPS-Bandung (2010)	Bandung Kulon Dalam Angka
BPS-Bandung (2010)	Sukasari Dalam Angka
BPS-Bandung (2010)	Rancasari Dalam Angka
BPS-Bandung (2010)	Cibeunying Kaler Dalam Angka
BPS-Bandung (2010)	Sukajadi Dalam Angka
BPS-Bandung (2010)	Bandung Kidul Dalam Angka
BPS-Bandung (2010)	Mandalajati Dalam Angka
BPS-Bandung (2010)	Batununggal Dalam Angka
BPS-Bandung (2010)	Bojongloa Kidul Dalam Angka
BPS-Bandung (2010)	Buahbatu Dalam Angka

excreta generated in Bandung city cannot be consumed all in Bandung city, and transportation is necessary. Figure 18.9 shows accumulated demand and supply by distance from Kiaracondong district. According to this graph, 32-km circle is one of indicative area scale, which balances potential demand and potential supply.

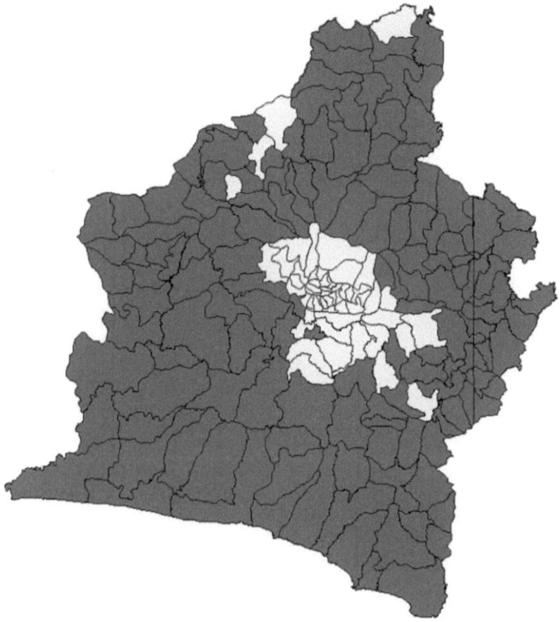

Fig. 18.8 Potential balance between nitrogen supply and demand in each district

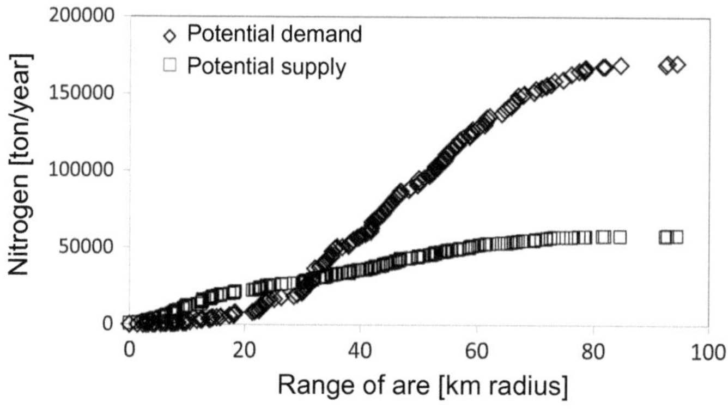

Fig. 18.9 Nitrogen supply potential and demand potential in circle area by its diameter

18.4.3 Fertilizer Price Evaluation

Price of human excreta fertilizer, liquid fertilizer from urine, and compost from feces was evaluated. Simple cost factors were set: collection cost and transportation cost. Additional to these factors, we have to assume a cost for supplemental

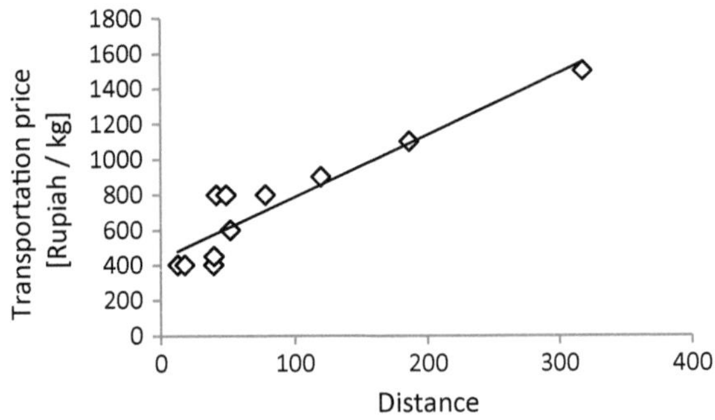

Fig. 18.10 Price-distance curve of truck transportation

fertilizer, which is necessary for adjusting nitrogen phosphorus ratio for plants requirment. Because transportation cost varies by distance, here one existing tea plantation site which is 47 km far from Kiaracondong was set as case site.

Unit collection cost for urine and feces was estimated by following the scheme described in 18.2.2. In current garbage collection in RW02, community collects total 820,000 Rupiah/month from households for one collection worker. That cost includes maintenance cost for handcart. On the other hand, assumed resource-oriented sanitation system seems to require eight collectors for urine collection and one collector for compost collection. Based on those information and assumption, unit cost for urine and compost collection was estimated as 130 and 308 Rupiah/kg.

Unit transportation costs were estimated by using Fig. 18.10, which obtained by interview to truck drivers. According to this price-distance curve, 47-km transportation costs 154 Rupiah/kg.

Using information of required nitrogen and phosphorus for tea plantation, which obtained by interview to tea plantation company, the required amount of urine and compost is estimated as 7236 and 380.7 kg/ha. Those are the estimation based on phosphorus demand, and additional 271 kg/ha of urea fertilizer is required to adjust the N/P ratio appropriately.

Therefore, total cost of fertilizer provided by resource-oriented sanitation system was estimated as 3,318,000 Rupiah/ha. This is almost same as price of synthetic fertilizer, 3,193,000 Rupiah/ha (Fig. 18.11); therefore, fertilizer produced by resource-oriented sanitation is possibly feasible. However, currently low-income Indonesian farmer can buy synthetic fertilizer in subsidized price. If we take it into account, price of synthetic fertilizer becomes less than half, and it is impossible to compete for the fertilizer from resource-oriented sanitation system. In order to overcome this situation, possible idea is reducing collection and transportation cost. According to Fig. 18.11, apparently urine collection and urine transportation cost much. If we could concentrate urine on-site, for example, by evapotranspiration, cost can be decreased.

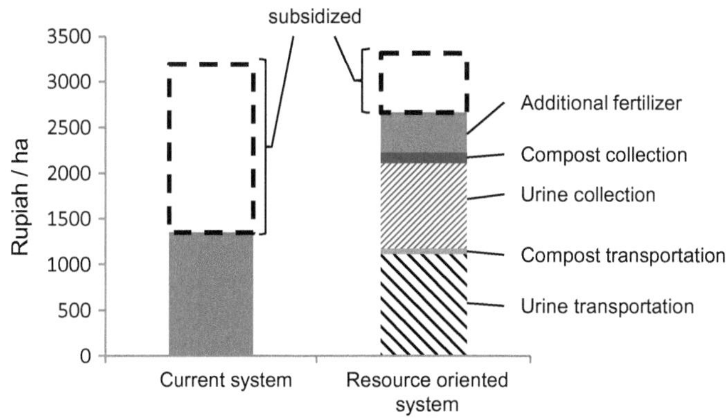

Fig. 18.11 Cost evaluation in required amount for 1 ha of tea plantation

18.5 Discussion on Social Aspect

Cultural concept affects people's sanitation habits and sense of hygiene. A study to measure people's acceptance on a composting dry toilet showed that such model could be accepted easily by the community, if it is appropriate and to their local habits, culture, and religion. In the Islamic rule, human excreta are clearly defined as dirty material called "Najis," and "Najis" have to be completely removed from their bodies before pray. This concept seems to make some misunderstandings that Muslims hard to join to resource recycling sanitation system which uses human excreta as resource.

In Islamic world, rightness of people's habit can be logically discussed and judged by Islamic scholars or authorities. New sanitation system also should be evaluated by this process. Of course, there may be gap between judgment by Islamic scholars or authorities and peoples' sense; however, authorized judgment can affect people's habit and sense. If the sanitation system to be applied could be authorized, it makes application and diffusing process much easier.

This study evaluated concept of resource recycling sanitation system in light of Islamic context in Indonesia, which has largest number of Muslims in the world. On the basis of this concept, interviews were carried out to four religious authorities in Indonesia.

Majelis Ulama Indonesia (MUI) was established to organize the committee covering all major Islamic groups or key persons in Indonesia (Hosen 2004). This committee discusses and judges rightness of the habit which is difficult to judge only by Qur'an and Sunna.

We interviewed two MUI members and two leaders of big Islamic groups. After explanation of the system concept, following questions were asked.

(1) If some process let user or worker touch feces and urine, is it a problem?
(2) Is it a problem to store urine and compost in a house?
(3) Is it a problem to give compensations of urine and feces to users?
(4) Is it a problem to eat food grown by human excreta?
(5) When and how the feces change into non-Najis?

As a result, all Islamic authorities judged whole process of this concept as no problem in light of Islamic context. The answers have minor difference but common points were (1) if feces and urine touched, washing process is enough to remove this Najis, (2) storage is no problem, (3) trade of Najis is prohibited but it is no problem if it was traded as fertilizer, (4) eating food grown by human excreta is no problem. Regarding question (5), no common answer was obtained, but by natural transformation, dilution, and transportation, some "Najis" stuffs could be changed to non-Najis products.

18.6 Conclusions

In this study, for an urban slum, the material and value flows of water, sanitation-related materials, nitrogen, and phosphorus were calculated and plotted. The value flow analysis evaluated garbage disposal cost and excreta disposal cost at 0.7 and 1.1%, respectively, of household income, which represents a reasonable cost to disposers. However, under the current disposal scheme, negative value due to excreta disposal appears to be passed on to downstream populations. The structure of this value flow suggests that little attention is paid to the negative externalities, indicating that a solution in this case may be difficult to obtain.

The affordable initial cost of a new resource recycling system was evaluated and compared with the costs of a currently used commercial compost toilet by evaluating equipment lifetimes and potential income streams available from the sale of human excrement fertilizers. To estimate a value for the human excrement fertilizer, real-world case study in Indonesia, in which money was paid in exchange for human urine, was referred to. The evaluated affordable initial cost derived was sufficiently comparable to the price of a cheaper composting toilet currently available in the market. This implies that such a resource recycling system is feasible even in an urban slum.

Results of demand and supply assessment showed that if resource-oriented sanitation system was diffused, supply would excess demands in the circle of 32 km radius, and therefore, cost discussion should assume more than 32 km transportation. According to the results of case estimation of tea plantation, total cost for fertilizer provided by resource-oriented sanitation system is possibly competitive to the normal price of synthetic fertilizer. However, if compared to subsidized price for low-income farmers, it seems difficult to compete.

Regarding social aspect, the context of Islamic law is important. Islamic authorities regarded the resource-oriented sanitation system as no problematic concept and acceptable in the context of Islamic law.

Acknowledgements This research was financially supported by the Core Research for Evolutional Science and Technology of the Japan Science and Technology Agency (JST-CREST), the Science and Technology Research Partnership for Sustainable Development of the Japan Science and Technology Agency and Japan International Cooperation Agency (JST-JICA-SATREPS), and JSPS KAKENHI Grants-in-Aid for Scientific Research (S).

References

Aramaki T, Thuy NTT (2010) Material flow analysis of nitrogen and phosphorus for regional nutrient management: case study in Haiphong, Vietnam. Sustain Food Water Alliance for Glob Sustain Book Series 18(6):391–399. https://doi.org/10.1007/978-90-481-9914-3_38

Bartone CR (2000) Financial management of urban solid waste services: lessons from a decade of World Bank Lending, World Bank Partnership in Municipal Solid Waste Management workshop

BPS (2005) Seri Kependudukan Jawa Barat Tahun 1920–2005. BPS, Jakarta, Indonesia

BPS-Kota Bandung (2004) Bandung Dalam Angka 2003 116–119

FAO (2005) Fertilizer use in Indonesia, FAO

Hosen N (2004) Behind the science: Fatwa of Majelis Ulama Indonesia (1975–1998). J Islamic Stud 15(2):147–179. https://doi.org/10.1093/jis/15.2.147

Gunawan M (1997) Market prospects for upland crops in Indonesia, the CGPRT Centre Working Paper Series 25. Regional co-ordination centre for research and development of coarse Grains, Pulses, Roots and Tuber Crops in the Humid Tropics of Asia and the Pacific (CGPRT), Bogor

Harada H, Adachi T, Fujii S, Lien NPH, Hai HT (2010) Phosphorus flow analysis in Hanoi focusing on wastewater, agriculture and stockbreeding. Environ Eng Res JSCE 47:465–474 (in Japanese)

Japan International Cooperation Agency (JICA) (2002) Study on Method of Economic Evaluation in Development Studies. JICA, Tokyo (in Japanese)

Japan International Cooperation Agency (JICA) (2005) For international aid to capacity development of solid waste management in developing countries

Li S, Yuan Z, Bi J, Wu H (2010) Anthropogenic phosphorus flow analysis of Hefei City, China. Sci Total Environ 408(23):5715–5722. https://doi.org/10.1016/j.scitotenv.2010.08.052 PMID: 20863550

Lopez Zavala MA, Funamizu N, Takakuwa T (2002) Onsite wastewater differentiable treatment system: modeling approach. Water Sci Technol 46(6–7):317–324

Metcalf & Eddy International Inc. in association with Brockman Tym International and TCGI Engineers (1998) Pasig river development plan, Final Report to ADB, vol 1. Metcalf & Eddy International Inc. in association with Brockman Tym International and TCGI Engineers, Quezon City

Tanaka N (1992) Environment of developing countries in Asia-Pacific, Thai (1). J Res Environ 28(11):1065–1076 (in Japanese)

Tsuzuki Y, Koottatep T, Rahman M D M, Ahmed F (2007) Preliminary study on pollutant discharge from domestic wastewater and pollutant load in the river: a comparative study between Japan, Thailand and Bangladesh. In: Proceedings of the water environment partnership in Asia (WEPA) Forum on "Water Environmental Governance in Asia", Bangkok. Water Environment Partnership in Asia (WEPA), Kanagawa

Unisuga S, Watanabe T (2004) Present condition of environment control in Indonesia (2). Life Environ 49:57–62

Ushijima K (2007) Study on water environment rehabilitation by introducing new toilet system in urban area of Southeast Asia. Dissertation for the Doctoral Degree. Tokyo Institute of Technology,Tokyo (in Japanese)

Ushijima K, Irie M, Sintawardani N, Triastuti J, Ishikawa T (2008a) Field research on domestic wastewater from a slum in Bandung City. J Hydrosci Hydraul Eng 26(1):129–140

Ushijima K, Irie M, Sintawardani N, Triastuti J, Ishikawa T (2008b) Practical model design of sustainable sanitation system in urban slum. In: Proceedings of the 6th International Symposium on Sustainable Sanitation, Changchun. Northeast Normal University (NENU), Changchun, pp 27–33

Ushijima K, Irie M, Sintawardani N, Triastuti J, Ishikawa T, Funamizu N (2011) Sanitation model for urban slum in Southeast Asia. In: Proceedings of IWA 2nd development congress, Kuala Lumpur, Malaysia. IWA, London

Ushijima K, Ito K, Ito R, Funamizu N (2013a) Greywater treatment by slanted soil system. Ecol Eng 50:62–68. https://doi.org/10.1016/j.ecoleng.2012.04.028

Ushijima K, Irie M, Sintawardani N, Triastuti J, Hamidah U, Ishikawa T, Funamizu N (2013b) Sustainable design of sanitation system based on material and value flow analysis for urban slum in Indonesia. Frontiers Environ Sci Eng 7(1):120–126. https://doi.org/10.1007/s11783-012-0460-5

Winblad U, Simpson-Hebert M (2004) Ecological sanitation—revised and enlarged edition. SEI, Stockholm

World Bank (2003) Indonesia environment monitor 2003. World Bank Indonesia Office, Jakarta

Yamamura T (1987) Comparison analysis on solid waste data of Indonesia and Japan. J Solid Liquid Waste 17(7):15–23

Zhou C, Liu J, Wang R, Yang W, Jin J (2010) Ecological-economic assessment of ecological sanitation development in the cities of Chinese Loess Plateau. Ecol Complexity 7(2):162–169. https://doi.org/10.1016/j.ecocom.2009

MIX
Papier aus verantwortungsvollen Quellen
Paper from responsible sources
FSC® C105338

If you have any concerns about our products,
you can contact us on
ProductSafety@springernature.com

In case Publisher is established outside the EU,
the EU authorized representative is:
Springer Nature Customer Service Center GmbH
Europaplatz 3, 69115 Heidelberg, Germany

Printed by Libri Plureos GmbH
in Hamburg, Germany